Power System State Estimation

POWER ENGINEERING

1. Power Distribution Planning Reference Book, *H. Lee Willis*
2. Transmission Network Protection: Theory and Practice, *Y. G. Paithankar*
3. Electrical Insulation in Power Systems, *N. H. Malik, A. A. Al-Arainy, and M. I. Qureshi*
4. Electrical Power Equipment Maintenance and Testing, *Paul Gill*
5. Protective Relaying: Principles and Applications, Second Edition, *J. Lewis Blackburn*
6. Understanding Electric Utilities and De-Regulation, *Lorrin Philipson and H. Lee Willis*
7. Electrical Power Cable Engineering, *William A. Thue*
8. Electric Systems, Dynamics, and Stability with Artificial Intelligence Applications, *James A. Momoh and Mohamed E. El-Hawary*
9. Insulation Coordination for Power Systems, *Andrew R. Hileman*
10. Distributed Power Generation: Planning and Evaluation, *H. Lee Willis and Walter G. Scott*
11. Electric Power System Applications of Optimization, James A. Momoh
12. Aging Power Delivery Infrastructures, *H. Lee Willis, Gregory V. Welch, and Randall R. Schrieber*
13. Restructured Electrical Power Systems: Operation, Trading, and Volatility, *Mohammad Shahidehpour and Muwaffaq Alomoush*
14. Electric Power Distribution Reliability, *Richard E. Brown*
15. Computer-Aided Power System Analysis, *Ramasamy Natarajan*
16. Power System Analysis: Short-Circuit Load Flow and Harmonics, *J. C. Das*
17. Power Transformers: Principles and Applications, *John J. Winders, Jr.*
18. Spatial Electric Load Forecasting: Second Edition, Revised and Expanded, *H. Lee Willis*
19. Dielectrics in Electric Fields, *Gorur G. Raju*
20. Protection Devices and Systems for High-Voltage Applications, *Vladimir Gurevich*
21. Electrical Power Cable Engineering: Second Edition, Revised and Expanded, *William A. Thue*
22. Vehicular Electric Power Systems: Land, Sea, Air, and Space Vehicles, *Ali Emadi, Mehrdad Ehsani, and John M. Miller*

23. Power Distribution Planning Reference Book: Second Edition, Revised and Expanded, *H. Lee Willis*
24. Power System State Estimation: Theory and Implementation, *Ali Abur and Antonio Gómez Expósito*

ADDITIONAL VOLUMES IN PREPARATION

Power System State Estimation

Theory and Implementation

Ali Abur
Texas A&M University
College Station, Texas, U.S.A.

Antonio Gómez Expósito
University of Seville
Seville, Spain

MARCEL DEKKER, INC. NEW YORK • BASEL

Library of Congress Cataloging-in-Publication Data
A catalog record for this book is available from the Library of Congress.

ISBN: 0-8247-5570-7

This book is printed on acid-free paper.

Headquarters
Marcel Dekker, Inc., 270 Madison Avenue, New York, NY 10016, U.S.A.
tel: 212-696-9000; fax: 212-685-4540

Distribution and Customer Service
Marcel Dekker, Inc., Cimarron Road, Monticello, New York 12701, U.S.A.
tel: 800-228-1160; fax: 845-796-1772

Eastern Hemisphere Distribution
Marcel Dekker AG, Hutgasse 4, Postfach 812, CH-4001 Basel, Switzerland
tel: 41-61-260-6300; fax: 41-61-260-6333

World Wide Web
http://www.dekker.com

The publisher offers discounts on this book when ordered in bulk quantities. For more information, write to Special Sales/Professional Marketing at the headquarters address above.

Current printing (last digit):

10 9 8 7 6 5 4 3 2 1

PRINTED IN THE UNITED STATES OF AMERICA

To Our Parents

Foreword

One of the major causes of the New York power outage of 1987 was ultimately traced to incorrect information about the status of a circuit in the system. The operation of a major new market, such as the PJM market, would be nearly impossible without the capabilities afforded by state estimation. It is not yet known to what extent the blackout of 2003 may have been in part caused by missing information. Undoubtedly, thus, the theme of this book is an important one. From its origins as a mathematical curiosity in the 1970's to its limited use during the 1980's to its expanded but not yet central role in the operation of the system in 1990's, nowadays state estimation has become nothing less than the cornerstone upon which a modern control center for a power system is built. Furthermore, to the extent that markets must be integrated with reliable system operation, state estimation has acquired a whole new role: it is the foundation for the creation and operation of real time markets in power systems, and thus the foundation for all markets, real time or not, since ultimately all markets must derive their valuations from real time information. Among the most important properties of a properly operated market is something that I shall call "auditability," that is, the ability to go back and verify why certain things were done the way they were. Without an accurate and ongoing knowledge of the status of every flow and every voltage in the system at all times, it would be impossible to "go back" and explain why, for example, prices were what they were at a particular time.

This book, written by two of the most prominent researchers in the field, brings a fresh perspective to the problem of state estimation. The book offers a blend of theory and mathematical rigor that is unique and very exciting. In addition to the more traditional topics associated with weighted least squares estimation (including such *de rigueur* topics as bad data detection and topology estimation), this book also brings forth several new aspects of the problem of state estimation that have not been presented in a systematic manner prior to this effort. Most notable among these are the chapters on robust estimation and the work on ampere measurements, to name just two. In this sense the book distinguishes itself from the other state estimation book known to this writer, the book by the late great Alcir Monticelli. In such way this book is a great complement to the efforts of Monticelli.

The readers of the book will also find it quite pleasing to have a nice review of a number of topics relating to efficient computation. The book provides excellent material for those wishing to review the topic of efficient computation and sparsity in general. Proper attention is paid throughout the book to computational efficiency issues. Given that computational efficiency is the key to making state estimation work in the first place, the importance of this topic cannot be understressed.

Although the bibliography associated with every chapter and with the appendix is short, it is all quite pertinent and very much to the point. In this sense, the readers can get focused and rapid access to additional original material should they wish to investigate a topic further.

I am particularly pleased to have had the opportunity to comment on both the theme of the book and the book itself, since the authors of this book are unquestionably respected leaders in the field and are themselves the originators of many of the ideas that are in present use throughout the field of state estimation and beyond. I am sure readers will share with me these sentiments after reading this book.

Fernando L. Alvarado

Preface

Power system state estimation is an area that matured in the past three decades. Today, state estimators can be found in almost every power system control center. While there have been numerous papers written on many different aspects of state estimation, ranging from its mathematical formulation to the implementation and start-up issues at the control centers, relatively few books have been published on this subject.

This book is the product of a long-term collaboration between the authors, starting from the summer of 1992 when they worked at the University of Seville on a joint project that was sponsored by the Ministry of Science and Education of the Spanish Government. Since then, they have spent two summers working together on different projects related to state estimation and continued their collaboration. They each taught regular and short courses on this topic and developed class notes, which make up most of the material presented in this book.

The chapters of the book are written in such a way that it can be used as a textbook for a graduate-level course on the subject. However, it may also be used as a supplement in an undergraduate-level course in power system analysis. Professionals working in the field of power systems may also find the chapters of the book useful as self-contained references on specific issues of interest.

The book is organized into nine chapters and two appendices. The introductory chapter provides a broad overview of power system operation and the role of state estimators in the overall energy management system configurations. The second chapter describes the modeling of electric networks during steady state operation and formulates one of the most commonly used state estimation methods in power systems, namely the weighted least squares (WLS) method. Application of the WLS method to power system state estimation presents several challenges ranging from numerical instabilities to the handling of measurements with special constraints. Chapter 3 presents various techniques for addressing these problems. Network observability is analyzed in Chapter 4, where a brief review of networks and graphs is followed by the description of alternative methods for network observability determination. Chapter 5 is concerned with detecting and identifying incorrect measurements. In this chapter, it is assumed that the WLS method is used for state estimation and bad data processing takes place after the convergence of the WLS state estimator. In Chapter 6, the topic of robust estimation is introduced and some robust estimation methods which have already been investigated for power system applications are presented. Chapter 7 is about different methods of estimating transmission line parameters and transformer taps. These network parameters are typically assumed to be perfectly known, despite the fact that errors in them significantly affect the state estimates. The problem of topology error identification is the topic of Chapter 8. Topology errors cause state estimators to diverge or converge to incorrect solutions. The challenges in detecting and identifying such errors and methods of overcoming them are presented in this chapter. Finally, Chapter 9 discusses the use of ampere measurements and various issues associated with their presence in the measurement set. The book also has two appendices, one on basic statistics and the other on sparse linear equations.

All chapters, except for the first one, end with some practice problems. These may be useful if the book is adopted for teaching a course at either the graduate or undergraduate level. The first five chapters are recommended to be read in the given order since each one builds on the previously covered material. However, the last four chapters can be covered in any arbitrary order.

Parts of the work presented in this book have been funded by the United States National Science Foundation projects ECS-9500118 and ECS-8909752 and by the Spanish Government, Directory of Scientific and Technical Investigations (DGICYT) Summer Research Grants No. SAB 95-0354 and SAB 92-0306, and Research Project No. PB94-1430.

It has been a pleasure to work with our many graduate students who have contributed to the development and implementation of some of the ideas in this book. Specifically, we are happy to acknowledge the contributions made by Esther Romero, Francisco González, Antonio de la Villa, Mehmet Kemal Çelik, Hongrae Kim, Fernando Hugo Magnago and Bei Gou in their respective research projects.

Finally, we are also grateful for the constant encouragement and support that we have received from our spouses, Ayşen and Cati, during the preparation of this book.

Ali Abur
Antonio Gómez Expósito

Contents

Chapter 1

Introduction

Power systems are composed of transmission, sub-transmission, distribution and generation systems. Transmission systems may contain large numbers of substations which are interconnected by transmission lines, transformers, and other devices for system control and protection. Power may be injected into the system by the generators or absorbed from the system by the loads at these substations. The output voltages of generators typically do not exceed 30-kV. Hence, transformers are used to increase the voltage levels to levels ranging from 69-kV all the way up to 765-kV at the generator terminals for efficient power transmission. High voltage is preferred at the transmission system for different reasons one of which is to minimize the copper losses that are proportional to the ampere flows along lines. At the receiving end, the transmission systems are connected to the sub-transmission or distribution systems which are operated at lower voltage levels ranging from 115-KV to 4.16-KV. Distribution systems are typically configured to operate in a radial configuration, where feeders stretch from distribution substations and form a tree structure with their roots at the substation and branches spreading over the distribution area.

1.1 Operating States of a Power System

The operating conditions of a power system at a given point in time can be determined if the network model and complex phasor voltages at every system bus are known. Since the set of complex phasor voltages fully specifies the system, it is referred to as the static state of the system. According to [1], the system may move into one of three possible states, namely normal, emergency and restorative, as the operating conditions change.

A power system is said to operate in a normal state if all the loads in the system can be supplied power by the existing generators without violating any operational constraints. Operational constraints include the limits on the transmission line flows, as well as the upper and lower limits on bus voltage magnitudes. A normal state is said to be *secure* if the system can remain in a normal state following the occurrence of each contingency from a list of critical contingencies. Common contingencies of interest are transmission line or generator outages due to unexpected failures of equipment or natural causes such as storms. Otherwise, the normal state is classified as *insecure* where the power balance at each bus and all operating inequality constraints are still satisfied, yet the system remains vulnerable with respect to some of the considered contingencies. If the system is found to be in a normal but *insecure* operating state then, preventive actions must be taken to avoid its move into an emergency state. Such preventive controls can be determined typically by the help of a security constrained optimal power flow program which accounts for a list of critical contingencies.

Operating conditions may change significantly due to an unexpected event which may cause the violation of some of the operating constraints, while the power system continues to supply power to all the loads in the system. In such a situation the system is said to be operating in an emergency state. Emergency state requires immediate corrective action to be taken by the operator so as to bring the system back to a normal state.

While the system is in the emergency state, corrective control measures may be able to avoid system collapse at the expense of disconnecting various loads, lines, transformers or other equipment. As a result, the operating limit violations may be eliminated and the system may recover stability with reduced load and reconfigured topology. Then, the load versus generation balance may have to be restored in order to start supplying power to all the loads. Such an operating state is called the restorative state, and the actions to be taken in order to transform it into a normal state are referred to as restorative controls. The state diagram in Figure 1.1 illustrates the possible transitions between the different operating states defined above.

1.2 Power System Security Analysis

Power systems are operated by system operators from the area control centers. The main goal of the system operator is to maintain the system in the normal secure state as the operating conditions vary during the daily operation. Accomplishing this goal requires continuous monitoring of the system conditions, identification of the operating state and determination of the necessary preventive actions in case the system state is found to be *insecure*. This sequence of actions is referred to as the security analysis of the system.

The first step of security analysis is to monitor the current state of the system. This involves acquisition of measurements from all parts of the

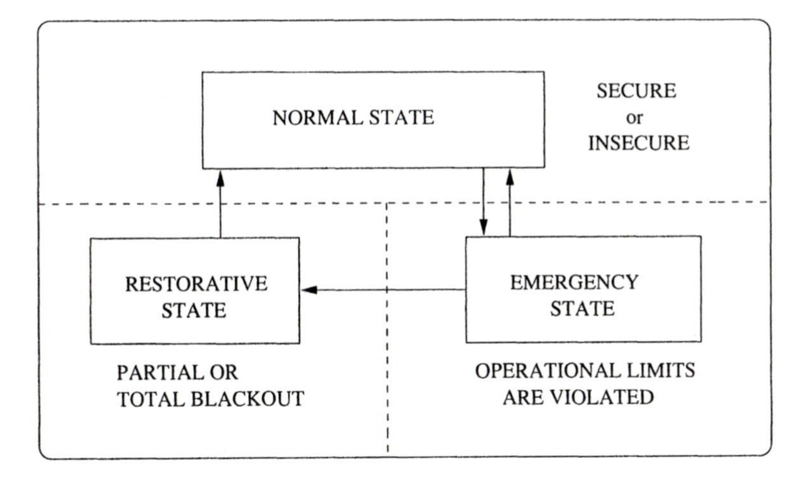

Figure 1.1. State Diagram for Power System Operation

system and then processing them in order to determine the system state. The measurements may be both of analog and digital (on/off status of devices) type. Substations are equipped with devices called remote terminal units (RTU) which collect various types of measurements from the field and are responsible for transmitting them to the control center. More recently, the so-called intelligent electronic devices (IED) are replacing or complementing the existing RTUs. It is possible to have a mixture of these devices connected to a local area network (LAN) along with a SCADA front end computer, which supports the communication of the collected measurements to the host computer at the control center. The SCADA host computer at the control center receives measurements from all the monitored substations' SCADA systems via one of many possible types of communication links such as fiber optics, satellite, microwave, etc. Figure 1.2 shows the configuration of the EMS/SCADA system for a typical power system.

Measurements received at the control center will include line power flows, bus voltage and line current magnitudes, generator outputs, loads, circuit breaker and switch status information, transformer tap positions, and switchable capacitor bank values. These raw data and measurements are processed by the state estimator in order to filter the measurement noise and detect gross errors. State estimator solution will provide an optimal estimate of the system state based on the available measurements and on the assumed system model. This will then be passed on to all the energy management system (EMS) application functions such as the contingency analysis, automatic generation control, load forecasting and optimal power flow, etc. The same information will also be available via a LAN connection

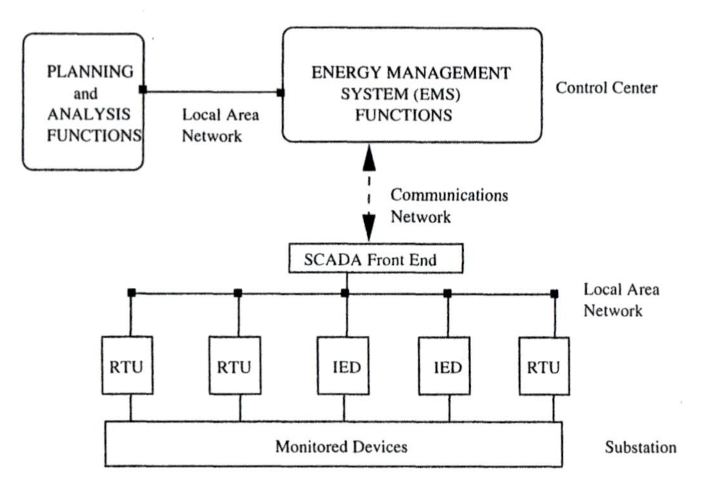

Figure 1.2. EMS/SCADA system configuration.

to the corporate offices where other planning and analysis functions can be executed off-line.

Initially, power systems were monitored only by supervisory control systems. These are control systems which essentially monitor and control the status of circuit breakers at the substations. Generator outputs and the system frequency were also monitored for purposes of Automatic Generation Control (AGC) and Economic Dispatch (ED). These supervisory control systems were later augmented by real-time system-wide data acquisition capabilities, allowing the control centers to gather all sorts of analog measurements and circuit breaker status data from the power system. This led to the establishment of the first Supervisory Control and Data Acquisition (SCADA) Systems. The main motivation behind this development was the facilitation of security analysis. Various application functions such as contingency analysis, corrective real and reactive power dispatch could not be executed without knowing the real-time operating conditions of the system. However, the information provided by the SCADA system may not always be reliable due to the errors in the measurements, telemetry failures, communication noise, etc. Furthermore, the collected set of measurements may not allow direct extraction of the corresponding A.C. operating state of the system. For instance, bus voltage phase angles are not typically measured, and not all the transmission line flows are available. Besides, it may not be economically feasible to telemeter all possible measurements even if they are available from the transducers at the substations.

1.3 State Estimation

The foregoing concerns were first recognized and subsequently addressed by Fred Schweppe, who proposed the idea of state estimation in power systems [2, 3, 4]. Introduction of the state estimation function broadened the capabilities of the SCADA system computers, leading to the establishment of the Energy Management Systems (EMS), which would now be equipped with, among other application functions, an on-line State Estimator (SE).

In order to identify the current operating state of the system, state estimators facilitate accurate and efficient monitoring of operational constraints on quantities such as the transmission line loadings or bus voltage magnitudes. They provide a reliable real-time data base of the system, including the existing state based on which, security assessment functions can be reliably deployed in order to analyze contingencies, and to determine any required corrective actions.

The state estimators typically include the following functions:

- Topology processor: Gathers status data about the circuit breakers and switches, and configures the one-line diagram of the system.

- Observability analysis: Determines if a state estimation solution for the entire system can be obtained using the available set of measurements. Identifies the unobservable branches, and the observable islands in the system if any exist.

- State estimation solution: Determines the optimal estimate for the system state, which is composed of complex bus voltages in the entire power system, based on the network model and the gathered measurements from the system. Also provides the best estimates for all the line flows, loads, transformer taps, and generator outputs.

- Bad data processing: Detects the existence of gross errors in the measurement set. Identifies and eliminates bad measurements provided that there is enough redundancy in the measurement configuration.

- Parameter and structural error processing: Estimates various network parameters, such as transmission line model parameters, tap changing transformer parameters, shunt capacitor or reactor parameters. Detects structural errors in the network configuration and identifies the erroneous breaker status provided that there is enough measurement redundancy.

Thus, power system state estimator constitutes the core of the on-line security analysis function. It acts like a filter between the raw measurements received from the system and all the application functions that require the most reliable data base for the current state of the system. Figure 1.3

describes the data and functional interfaces between the various application functions involved in the on-line static security assessment procedure. Raw measurements which include the switch and circuit breaker positions in the substations, are processed by the topology processor, which in turn generates a bus/branch model of the power system. This model not only includes all buses within the area of the control center EMS, but also selected buses from the neighboring systems. The information and measurements obtained from the neighboring systems are used to build and update the external system model. Furthermore, there may be unobservable pockets within one's own area due to temporary loss of telemetry, rejected bad data or other unexpected failures. Such areas whether physically located within the control area or part of the external system, will be estimated via the use of pseudo measurements. Pseudo measurements can be generated based on short term load forecasts, generation dispatch, historical records or other similar approximation methods. Naturally, they are assigned high variances (low weights) or they can be forced to be critical measurements by design. Definition and properties of a critical measurement will be discussed in detail in chapter 5. In addition, there may be passive buses with no generation or load, having net zero real and reactive power injection. Such bus injections, even though not measured, can be used as error free measurements in the state estimation formulation and referred to as "virtual" measurements. The results obtained by the state estimator will be checked in order to classify the system state into one of the three categories shown in Figure 1.1. If it is found to be in the *normal* state, then contingency analysis will be carried out to determine the system security against a set of predetermined contingencies. In case of insecurity, preventive control actions have to be calculated via the use of a software tool such as a security constrained optimal power flow. Implementing these preventive measures will move the system into the desired *normal and secure* state. Figure 1.3 also indicates the emergency and restorative control actions which will be deployed under *abnormal* operating conditions, however these topics are beyond the scope of this book and will not be discussed any further.

1.4 Summary

Power systems are continuously monitored in order to maintain the operating conditions in a normal and secure state. State estimation function is used for this purpose. It processes redundant measurements in order to provide an optimal estimate of the current operating state. State estimation problem has been investigated by several researchers since its introduction in the late 1960s. Being an on-line function, computational issues related to speed, storage and numerical robustness of the solution algorithms have been carefully studied. Measurement configuration and its effect on

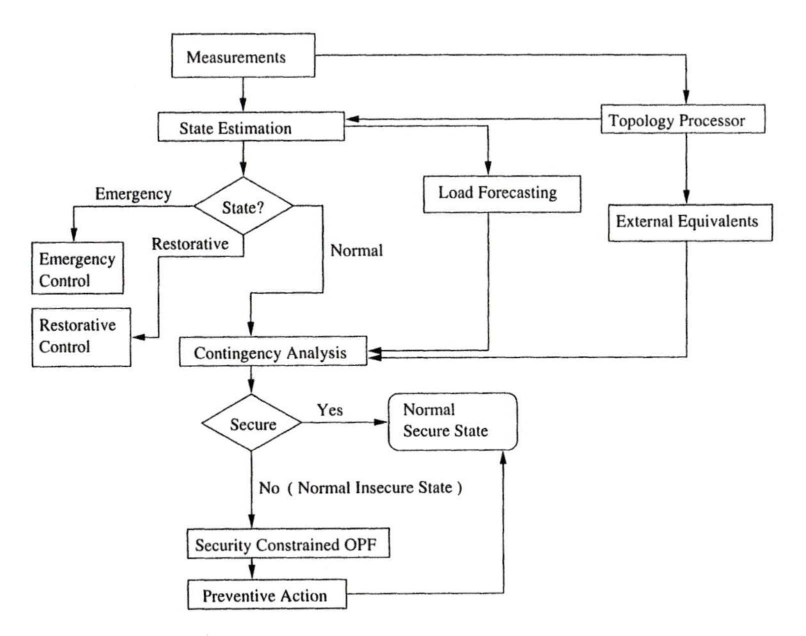

Figure 1.3. On-line Static Security Assessment: Functional Diagram

state estimation have been addressed by the developed observability analysis methods. State estimators also function as filters against incorrect measurements, data and other information received through the SCADA system. Hence, the subject of bad data processing has been investigated and detection/identification algorithms for errors in analog measurements have been developed. Special methods also exist for the identification of those errors related to the topology information and/or network parameters. On the other hand, the use of ampere measurements present some problems which do not exist in their absence from the measurement set. In the following chapters, these issues will be presented in more detail and methods which are developed to address them will be described.

References

[1] Dy Liacco T.E., "Real-Time Computer Control of Power Systems", Proceedings of the IEEE, Vol. 62, No.7, July 1974, pp.884-891.

[2] Schweppe F.C. and Wildes J., "Power System Static-State Estimation, Part I: Exact Model", IEEE Transactions on Power Apparatus and Systems, Vol.PAS-89, January 1970, pp.120-125.

[3] Schweppe F.C. and Rom D.B., "Power System Static-State Estimation, Part II: Approximate Model", IEEE Transactions on Power Apparatus and Systems, Vol.PAS-89, January 1970, pp.125-130.

[4] Schweppe F.C., "Power System Static-State Estimation, Part III: Implementation", IEEE Transactions on Power Apparatus and Systems, Vol.PAS-89, January 1970, pp.130-135.

[5] Fink L.H. and Carlsen K.,"Operating under Stress and Strain", IEEE Spectrum, March 1978.

[6] N. Balu et al. "On-line Power System Security Analysis", Proc. of the IEEE, vol. 80(2), pp. 262-280.

Chapter 2

Weighted Least Squares State Estimation

2.1 Introduction

Static state estimation refers to the procedure of obtaining the voltage phasors at all of the system buses at a given point in time. This can be achieved by direct means which involve very accurate synchronized phasor measurements of all bus voltages in the system. However, such an approach would be very vulnerable to measurement errors or telemetery failures. Instead, state estimation procedure makes use of a set of redundant measurements in order to filter out such errors and find an optimal estimate. The measurements may include not only the conventional power and voltage measurements, but also those others such as the current magnitude or synchronized voltage phasor measurements as well. Simultaneous measurement of quantities at different parts of the system is practically impossible, hence a certain amount of time skew between measurements is commonly tolerated. This tolerance is justified due to the slowly varying operating conditions of the power systems under normal operating conditions.

The definition of the system state usually includes the steady state bus voltage phasors only. This implies that the network topology and parameters are perfectly known. However, errors in the network parameters or topology do exist occasionally, due to various reasons such as unreported outages, transmission line sags on hot days, etc. Detection and correction of such errors will be separately discussed later on in chapters 7 and 8.

2.2 Component Modeling and Assumptions

Power system is assumed to operate in the steady state under balanced conditions. This implies that all bus loads and branch power flows will be three phase and balanced, all transmission lines are fully transposed, and all other series or shunt devices are symmetrical in the three phases. These assumptions allow the use of single phase positive sequence equivalent circuit for modeling the entire power system. The solution that will be obtained by using such a network model, will also be the positive sequence component of the system state during balanced steady state operation. As in the case of the power flow, all network data as well as the network variables, are expressed in the per unit system. The following component models will thus be used in representing the entire network.

2.2.1 Transmission Lines

Transmission lines are represented by a two-port π-model whose parameters correspond to the positive sequence equivalent circuit of transmission lines. A transmission line with a positive sequence series impedance of $R+jX$ and total line charging susceptance of $j2B$, will be modelled by the equivalent circuit shown in Figure 2.1.

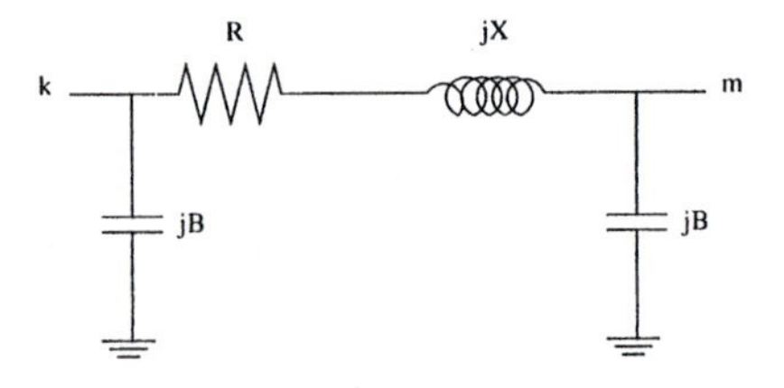

Figure 2.1. Equivalent circuit for a transmission line

2.2.2 Shunt Capacitors or Reactors

Shunt capacitors or reactors which may be used for voltage and/or reactive power control, are represented by their per phase susceptance at the corresponding bus. The sign of the susceptance value will determine the type of the shunt element. It will be positive or negative corresponding to a shunt capacitor or reactor respectively.

2.2.3 Tap Changing and Phase Shifting Transformers

Transformers with off-nominal but in-phase taps, can be modeled as series impedances in series with ideal transformers as shown in Figure 2.2. The two transformer terminal buses m and k are commonly designated as the impedance side and the tap side bus respectively.

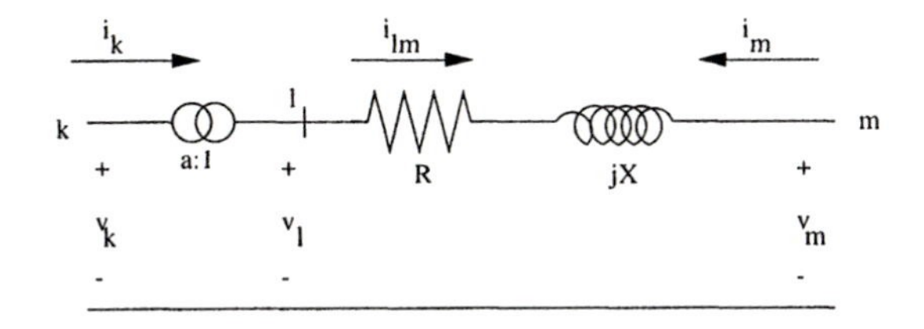

Figure 2.2. Equivalent circuit for an off-nominal tap transformer

The nodal equations of the two port circuit of Figure 2.2 can be derived by first expressing the current flows $i_{\ell m}$ and i_m at each end of the series branch $R + jX$. Denoting the admittance of this branch $\ell - m$ by y, the terminal current injections will be given by:

$$\begin{bmatrix} i_{\ell m} \\ i_m \end{bmatrix} = \begin{bmatrix} y & -y \\ -y & y \end{bmatrix} \begin{bmatrix} v_\ell \\ v_m \end{bmatrix} \tag{2.1}$$

Substituting for $i_{\ell m}$ and v_ℓ:

$$\begin{aligned} i_{\ell m} &= a \cdot i_k \\ v_\ell &= v_k / a \end{aligned}$$

the final form will be obtained as follows:

$$\begin{bmatrix} i_k \\ i_m \end{bmatrix} = \begin{bmatrix} y/a^2 & -y/a \\ -y/a & y \end{bmatrix} \begin{bmatrix} v_k \\ v_m \end{bmatrix} \tag{2.2}$$

where a is the in phase tap ratio. Figure 2.3 shows the corresponding two port equivalent circuit for the above set of nodal equations.

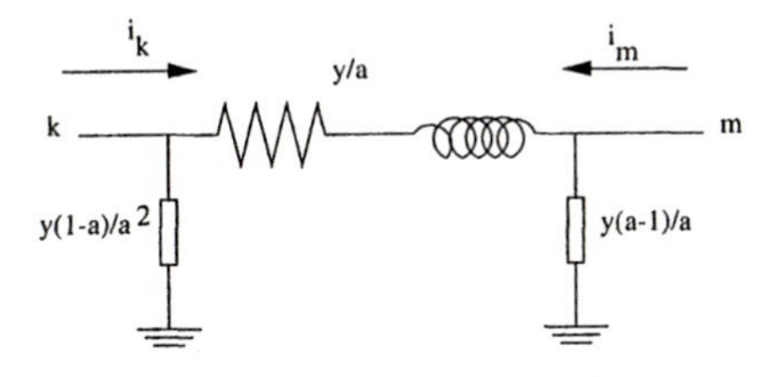

Figure 2.3. Equivalent circuit of an in-phase tap changer

For a phase shifting transformer where the off-nominal tap value a, is complex, the equations will slightly change as:

$$\begin{aligned} a^* i_k &= i_{\ell m} \\ a v_\ell &= v_k \end{aligned}$$

yielding the following set of nodal equations:

$$\begin{bmatrix} i_k \\ i_m \end{bmatrix} = \begin{bmatrix} y/\mid a\mid^2 & -y/a^* \\ -y/a & y \end{bmatrix} \begin{bmatrix} v_k \\ v_m \end{bmatrix} \tag{2.3}$$

Note the loss of reciprocity as the admittance matrix is no longer symmetrical. Therefore, a passive equivalent circuit such as the one shown in Figure 2.3 for the in-phase tap changer, can no longer be realized for the phase shifting transformer. However, the circuit equations can still be solved as before by only modifying the admittance matrix which is no longer symmetrical.

2.2.4 Loads and Generators

Loads and generators are modeled as equivalent complex power injections and therefore have no effect on the network model. Exceptions are constant impedance type loads which are included as shunt admittances at the corresponding buses.

2.3 Building the Network Model

The above-described component models can be used to build the network model for the entire power system. This is accomplished by writing a set of nodal equations which are derived by applying Kirchhoff's current law at each bus. Denoting the vector of net current injections by I, and the vector of bus voltage phasors by V, these equations will take the following form:

$$I = \begin{bmatrix} i_1 \\ i_2 \\ \vdots \\ i_N \end{bmatrix} = \begin{bmatrix} Y_{11} & Y_{12} & \cdots & Y_{1N} \\ Y_{21} & Y_{22} & \cdots & Y_{2N} \\ \vdots & \vdots & \vdots & \vdots \\ Y_{N1} & Y_{N2} & \cdots & Y_{NN} \end{bmatrix} \begin{bmatrix} v_1 \\ v_2 \\ \vdots \\ v_N \end{bmatrix} = Y \cdot V \tag{2.4}$$

where
i_k is the net current injection phasor at bus k.
v_k is the voltage phasor at bus k.
Y_{km} is the (k,m)th element of Y.

Note that, as a convention, currents (or power) entering a bus will be assumed to be positive injections throughout the rest of the book. Matrix Y is referred to as the bus admittance matrix, and has the following properties:

1. It is in general complex, and can be written as $G + jB$.
2. It is structurally symmetric. It may also be numerically symmetric depending upon the absence of certain network components such as phase shifters, with non-symmetrical nodal equations.

3. It is very sparse.
4. It is non-singular provided that each island in the network has at least one shunt connection to ground.

Equation (2.4) is valid for any N-port passive circuit with external current injections defined by the vector I. This representation of the network by nodal equations, facilitates the modification of equations in case of topology changes. Adding or removing a k-port sub-circuit can be easily done by adding or subtracting the corresponding entries of the admittance matrix.

As an example, consider a two-port model of a transformer connected between bus k and m, having a series admittance of y_t and a tap ratio of a, represented by the following nodal equations:

$$\begin{bmatrix} i_k \\ i_m \end{bmatrix} = \begin{bmatrix} y_t/\mid a \mid^2 & -y_t/a^* \\ -y_t/a & y_t \end{bmatrix} \begin{bmatrix} v_k \\ v_m \end{bmatrix} \tag{2.5}$$

Given the bus admittance matrix Y for the entire system, the transformer model can be introduced by modifying the following 4 entries in Y:

$$\begin{aligned} Y_{kk}^{new} &= Y_{kk} + y_t/\mid a \mid^2 \\ Y_{km}^{new} &= Y_{km} - y_t/a^* \\ Y_{mk}^{new} &= Y_{mk} - y_t/a \\ Y_{mm}^{new} &= Y_{mm} + y_t \end{aligned}$$

Hence, the bus admittance matrix Y of a large power system can be built from scratch by introducing one subsystem at a time and modifying the corresponding entries of Y until all branches are processed. One of the simplest subsystems is a two-port network such as the model of a transformer or a transmission line as shown in Figures 2.1 and 2.3.

Example 2.1:

Consider the 4-bus power system whose one-line diagram is given in Figure 2.4. Network data and the steady state bus voltages are listed below. The susceptance of the shunt capacitor at bus 3 is given as 0.5 per unit.

From Bus	To Bus	R pu	X pu	Total Line Charging Susceptance	Tap a	Tap Side Bus
1	2	0.02	0.06	0.20	–	–
1	3	0.02	0.06	0.25	–	–
2	3	0.05	0.10	0.00	–	–
2	4	0.00	0.08	0.00	0.98	2

Bus No.	Voltage Mag. pu	Phase Angle degrees
1	1.0000	0.00
2	0.9629	-2.76
3	0.9597	-3.58
4	0.9742	-3.96

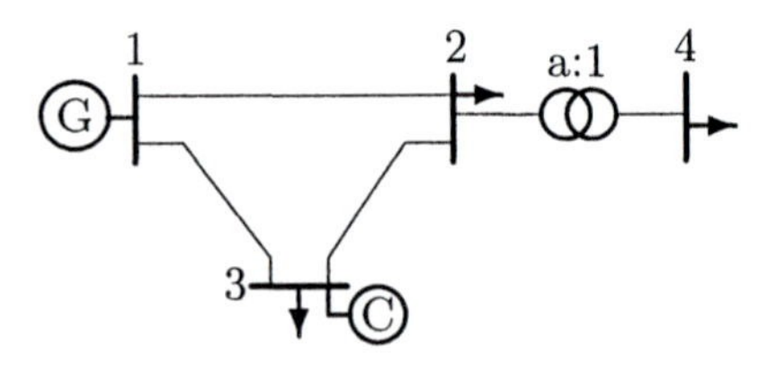

Figure 2.4. One-line diagram of a 4-bus power system

- Write the nodal equations for the 2-port π-model of the transformer connected between bus 2 and 4.
- Form the bus admittance matrix, Y for the entire system.
- Calculate the net complex power injections at each bus.

Solution:

The nodal equations for the transformer branch will be obtained by substituting for y and a in Equation (2.2):

$$\begin{bmatrix} i_2 \\ i_4 \end{bmatrix} = \begin{bmatrix} -j13.02 & j12.75 \\ j12.75 & -j12.50 \end{bmatrix} \begin{bmatrix} v_2 \\ v_4 \end{bmatrix}$$

Bus admittance matrix for the entire system can be obtained by including one branch at a time and expanding the above admittance matrix to a 4x4 matrix:

$$\begin{bmatrix} i_1 \\ i_2 \\ i_3 \\ i_4 \end{bmatrix} = \begin{bmatrix} 10.00 - j29.77 & -5.00 + j15.00 & -5.00 + j15.00 & 0 \\ -5.00j15.00 & 9.00 - j35.91 & -4.00 + j8.00 & j12.75 \\ -5.00j15.00 & -4.00 + j8.00 & 9.00 - j22.37 & 0 \\ 0 & j12.75 & 0 & -j12.50 \end{bmatrix} \begin{bmatrix} v_1 \\ v_2 \\ v_3 \\ v_4 \end{bmatrix}$$

Complex power injection at bus k will be given by:

$$S_k = v_k \cdot i_k^*$$

Substituting for i_k from the above nodal equation:

$$S_k = v_k \cdot \sum_{j=1}^{4} Y_{kj}^* v_j^*$$

Evaluating them for $k = 1, \ldots, 4$ yields:

$$\begin{aligned} S_1 &= 2.00 + j0.45 \\ S_2 &= -0.50 - j0.30 \\ S_3 &= -1.20 - j0.80 \\ S_4 &= -0.25 - j0.10 \end{aligned}$$

2.4 Maximum Likelihood Estimation

The objective of state estimation is to determine the most likely state of the system based on the quantities that are measured. One way to accomplish this is by maximum likelihood estimation (MLE), a method widely used in statistics. The measurement errors are assumed to have a known probability distribution with unknown parameters. The joint probability density function for all the measurements can then be written in terms of these unknown parameters. This function is referred to as the likelihood function and will attain its peak value when the unknown parameters are chosen to be closest to their actual values. Hence, an optimization problem can be set up in order to maximize the likelihood function as a function of these unknown parameters. The solution will give the maximum likelihood estimates for the parameters of interest.

The measurement errors are commonly assumed to have a Gaussian (Normal) distribution and the parameters for such a distribution are its mean, μ and its variance, σ^2. The problem of maximum likelihood estimation is then solved for these two parameters. The Gaussian probability density function (p.d.f.) and the corresponding probability distribution function (d.f.) will be reviewed below briefly before describing the maximum likelihood estimation method.

2.4.1 Gaussian (Normal) probability density function

The Normal probability density function for a random variable z is defined as:

$$f(z) = \frac{1}{\sqrt{2\pi}\sigma} e^{-\frac{1}{2}\left\{\frac{z-\mu}{\sigma}\right\}^2}$$

where z : random variable
μ : mean (or expected value) of $z = \mathrm{E}(z)$
σ : standard deviation of z

The function $f(z)$ will change its shape depending on the parameters μ and σ. However, its shape can be standardized by using the following change of variables:

$$u = \frac{z - \mu}{\sigma}$$

which yields:

$$\begin{aligned} E(u) &= \frac{1}{\sigma}(E(z) - \mu) = 0 \\ Var(u) &= \frac{1}{\sigma^2}Var(z - \mu) = \frac{\sigma^2}{\sigma^2} = 1.0 \end{aligned}$$

Hence, the new function becomes:

$$\Phi(u) = \frac{1}{\sqrt{2\pi}}e^{-\frac{u^2}{2}}$$

A plot of $\Phi(u)$, which is referred to as the Standard Normal (Gaussian) Probability Density Function, is shown in Figure 2.5.

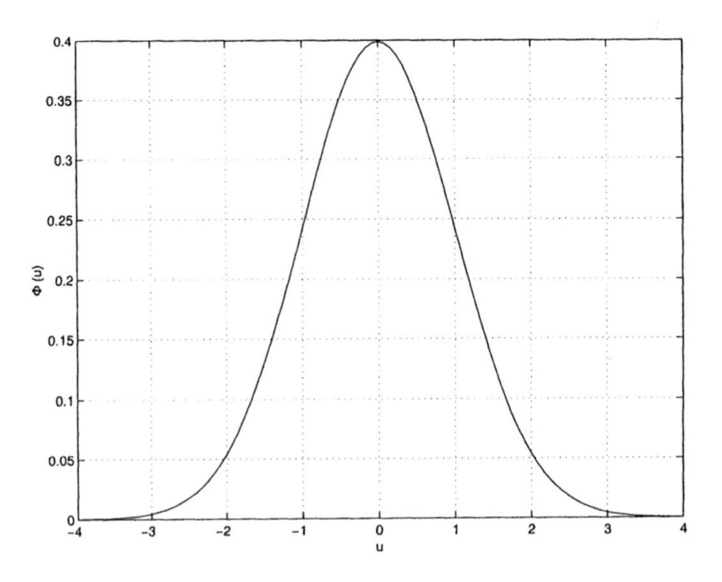

Figure 2.5. Standard Gaussian (Normal) Probability Density Function, $\Phi(u)$

2.4.2 The likelihood function

Consider the joint probability density function which represents the probability of measuring m independent measurements, each having the same Gaussian p.d.f. The joint p.d.f can simply be expressed as the product of individual p.d.f's if each measurement is assumed to be independent of the rest:

$$f_m(z) = f(z_1)f(z_2)\cdots f(z_m)$$

$$\text{where} \quad z_i : i\text{th measurement}$$
$$z^T : [z_1, z_2, \cdots, z_m]$$

The function $f_m(z)$ is called the likelihood function for z. Essentially it is a measure of the probability of observing the particular set of measurements in the vector z.

The objective of maximum likelihood estimation is to maximize this likelihood function by varying the assumed parameters of the density function, namely its mean μ and its standard deviation σ. In determining the optimum parameter values, the function is commonly replaced by its logarithm, in order to simplify the optimization procedure. The modified function is called the Log-Likelihood Function, $\mathcal{L}$ and is given by:

$$\begin{aligned} \mathcal{L} = \log f_m(z) &= \sum_{i=1}^{m} \log f(z_i) \\ &= -\frac{1}{2}\sum_{i=1}^{m}(\frac{z_i - \mu_i}{\sigma_i})^2 - \frac{m}{2}\log 2\pi - \sum_{i=1}^{m}\log \sigma_i \end{aligned}$$

MLE will maximize the likelihood (or log-likelihood) function for a given set of observations $z_1, z_2, \ldots, z_m$. Hence, it can be obtained by solving the following problem:

$$\begin{aligned} &\text{maximize} \quad \log f_m(z) \\ &\qquad\text{OR} \\ &\text{minimize} \quad \sum_{i=1}^{m}(\frac{z_i - \mu_i}{\sigma_i})^2 \end{aligned} \tag{2.6}$$

This minimization problem can be re-written in terms of the *residual* r_i of measurement i, which is defined as:

$$r_i = z_i - \mu_i = z_i - E(z_i)$$

where the mean μ_i, or the expected value $E(z_i)$ of the measurement z_i can be expressed as $h_i(x)$, a nonlinear function relating the system state vector x to the ith measurement. Square of each residual r_i^2 is weighted by W_{ii}

$= \sigma_i^{-2}$, which is inversely related to the assumed error variance for that measurement. Hence, the minimization problem of Equation (2.6) will be equivalent to minimizing the weighted sum of squares of the residuals or solving the following optimization problem for the state vector x:

$$\text{minimize} \quad \sum_{i=1}^{m} W_{ii} r_i^2 \tag{2.7}$$

$$\text{subject to} \quad z_i = h_i(x) + r_i, \quad i = 1, \ldots, m. \tag{2.8}$$

The solution of the above optimization problem is called the *weighted least squares* (WLS) estimator for x. A review of the measurement model and the associated assumptions will be given next, before discussing the numerical solution methods.

2.5 Measurement Model and Assumptions

Consider the set of measurements given by the vector z:

$$z = \begin{bmatrix} z_1 \\ z_2 \\ \vdots \\ z_m \end{bmatrix} = \begin{bmatrix} h_1(x_1, x_2, \ldots, x_n) \\ h_2(x_1, x_2, \ldots, x_n) \\ \vdots \\ h_m(x_1, x_2, \ldots, x_n) \end{bmatrix} + \begin{bmatrix} e_1 \\ e_2 \\ \vdots \\ e_m \end{bmatrix} = h(x) + e \tag{2.9}$$

where:
$h^T = [h_1(x), h_2(x), \ldots, h_m(x)]$
$h_i(x)$ is the nonlinear function relating measurement i to the state vector x
$x^T = [x_1, x_2, \ldots, x_n]$ is the system state vector
$e^T = [e_1, e_2, \ldots, e_m]$ is the vector of measurement errors.

The following assumptions are commonly made, regarding the statistical properties of the measurement errors:

- $\mathrm{E}(e_i) = 0, \quad i = 1, \ldots, m.$
- Measurement errors are independent, i.e. $E[e_i e_j] = 0$.
  Hence, $Cov(e) = E[e \cdot e^T] = R = \text{diag}\ \{ \sigma_1^2, \sigma_2^2, \cdots, \sigma_m^2 \}$.

The standard deviation σ_i of each measurement i is calculated to reflect the expected accuracy of the corresponding meter used.

The WLS estimator will minimize the following objective function:

$$\begin{aligned} J(x) &= \sum_{i=1}^{m} (z_i - h_i(x))^2 / R_{ii} \\ &= [z - h(x)]^T R^{-1} [z - h(x)] \end{aligned} \tag{2.10}$$

At the minimum, the first-order optimality conditions will have to be satisfied. These can be expressed in compact form as follows:

$$g(x) = \frac{\partial J(x)}{\partial x} = -H^T(x)R^{-1}[z - h(x)] = 0 \tag{2.11}$$

$$\text{where} \quad H(x) = \left[\frac{\partial h(x)}{\partial x}\right]$$

Expanding the non-linear function $g(x)$ into its Taylor series around the state vector x^k yields:

$$g(x) = g(x^k) + G(x^k)(x - x^k) + \cdots = 0$$

Neglecting the higher order terms leads to an iterative solution scheme known as the Gauss-Newton method as shown below:

$$x^{k+1} = x^k - [G(x^k)]^{-1} \cdot g(x^k)$$

where k is the iteration index,

x^k is the solution vector at iteration k,

$$G(x^k) = \frac{\partial g(x^k)}{\partial x} = H^T(x^k) \cdot R^{-1} \cdot H(x^k)$$

$$g(x^k) = -H^T(x^k) \cdot R^{-1} \cdot (z - h(x^k)).$$

$G(x)$ is called the *gain matrix*. It is sparse, positive definite and symmetric provided that the system is fully observable. The issue of observability will be discussed in detail in Chapter 4. The matrix $G(x)$ is typically not inverted (the inverse will in general be a full matrix, whereas $G(x)$ itself is quite sparse), but instead it is decomposed into its triangular factors and the following sparse linear set of equations are solved using forward/back substitutions at each iteration k:

$$[G(x^k)]\Delta x^{k+1} = H^T(x^k)R^{-1}[z - h(x^k)] \tag{2.12}$$

where $\Delta x^{k+1} = x^{k+1} - x^k$. The set of equations given by Equation (2.12) is also referred to as the Normal Equations.

Example 2.2:

Consider the 3-bus power system shown in Figure 2.6. The network data are presented in the table below:

Line		Resistance	Reactance	Total Susceptance
From Bus	To Bus	R (pu)	X (pu)	$2b_s$ (pu)
1	2	0.01	0.03	0.0
1	3	0.02	0.05	0.0
2	3	0.03	0.08	0.0

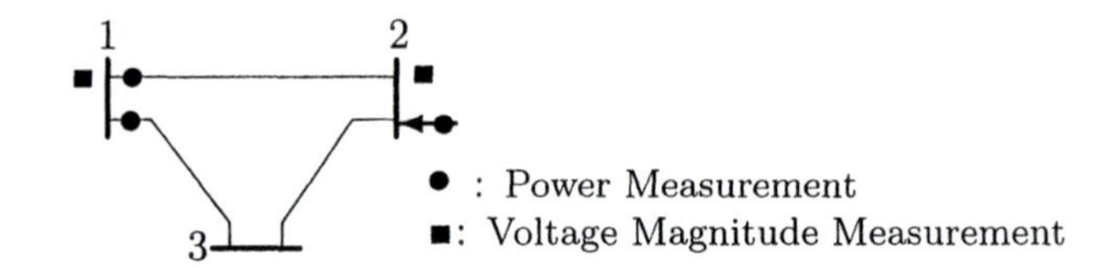

Figure 2.6. One-line diagram and measurement configuration of a 3-bus power system

The system is monitored by 8 measurements, hence $m = 8$ in Equation (2.9). Measurement values and their associated error standard deviations $\sqrt{R_{ii}} = \sigma_i$, are given as:

Measurement, i	Type	Value (pu)	$\sqrt{R_{ii}}$ (pu)
1	p_{12}	0.888	0.008
2	p_{13}	1.173	0.008
3	p_2	-0.501	0.010
4	q_{12}	0.568	0.008
5	q_{13}	0.663	0.008
6	q_2	-0.286	0.010
7	V_1	1.006	0.004
8	V_2	0.968	0.004

The state vector x will have 5 elements in this case ($n = 5$),

$$x^T = [\theta_2, \theta_3, V_1, V_2, V_3]$$

$\theta_1 = 0$ is chosen as the arbitrary reference angle.

2.6 WLS State Estimation Algorithm

WLS State Estimation involves the iterative solution of the Normal equations given by Equation (2.12). An initial guess has to be made for the state vector x^0. As in the case of the power flow solution, this guess typically corresponds to the flat voltage profile, where all bus voltages are assumed to be 1.0 per unit and in phase with each other.

The iterative solution algorithm for WLS state estimation problem can be outlined as follows:

1. Start iterations, set the iteration index $k = 0$.
2. Initialize the state vector x^k, typically as a flat start.

3. Calculate the gain matrix, $G(x^k)$.
4. Calculate the right hand side $t^k = H(x^k)^T R^{-1}(z - h(x^k))$
5. Decompose $G(x^k)$ and solve for Δx^k.
6. Test for convergence, $\max \mid \Delta x^k \mid \leq \epsilon$?
7. If no, update $x^{k+1} = x^k + \Delta x^k$, $k = k + 1$, and go to step 3. Else, stop.

The above algorithm essentially involves the following computations in each iteration, k:

1. Calculation of the right hand side of Equation (2.12).
   (a) Calculating the measurement function, $h(x^k)$.
   (b) Building the measurement Jacobian, $H(x^k)$.
2. Calculation of $G(x^k)$ and solution of Equation (2.12).
   (a) Building the gain matrix, $G(x^k)$.
   (b) Decomposing $G(x^k)$ into its Cholesky factors.
   (c) Performing the forward/back substitutions to solve for Δx^{k+1}.

2.6.1 The Measurement Function, $h(x^k)$

Measurements can be of a variety of types. Most commonly used measurements are the line power flows, bus power injections, bus voltage magnitudes and line current flow magnitudes. These measurements can be expressed in terms of the state variables either using the rectangular or the polar coordinates. When using the polar coordinates for a system containing N buses, the state vector will have $(2N - 1)$ elements, N bus voltage magnitudes and $(N - 1)$ phase angles, where the phase angle of one reference bus is set equal to an arbitrary value, such as 0. The state vector x will have the following form assuming bus 1 is chosen as the reference:

$$x^T = [\theta_2 \theta_3 \dots \theta_N V_1 V_2 \dots V_N]$$

The expressions for each of the above types of measurements are given below, assuming the general two-port π-model for the network branches, shown in Figure 2.7:

- Real and reactive power injection at bus i:

$$\begin{aligned} P_i &= V_i \sum_{j \in \aleph_i} V_j (G_{ij} \cos\theta_{ij} + B_{ij} \sin\theta_{ij}) \\ Q_i &= V_i \sum_{j \in \aleph_i} V_j (G_{ij} \sin\theta_{ij} - B_{ij} \cos\theta_{ij}) \end{aligned}$$

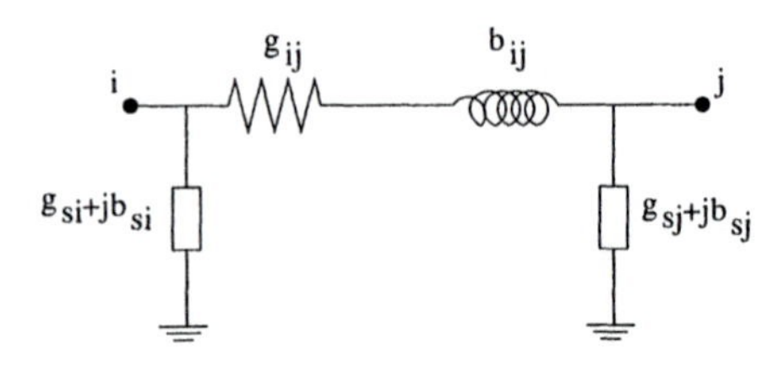

Figure 2.7. Two-port π-model of a network branch

- Real and reactive power flow from bus i to bus j:

$$
\begin{aligned}
P_{ij} &= V_i^2(g_{si} + g_{ij}) - V_iV_j(g_{ij}\cos\theta_{ij} + b_{ij}\sin\theta_{ij}) \\
Q_{ij} &= -V_i^2(b_{si} + b_{ij}) - V_iV_j(g_{ij}\sin\theta_{ij} - b_{ij}\cos\theta_{ij})
\end{aligned}
$$

- Line current flow magnitude from bus i to bus j:

$$I_{ij} = \frac{\sqrt{P_{ij}^2 + Q_{ij}^2}}{V_i}$$

or ignoring the shunt admittance $(g_{si} + jb_{si})$:

$$I_{ij} = \sqrt{(g_{ij}^2 + b_{ij}^2)(V_i^2 + V_j^2 - 2V_iV_j\cos\theta_{ij})}$$

where
V_i, θ_i is the voltage magnitude and phase angle at bus i.
$\theta_{ij} = \theta_i - \theta_j$.
$G_{ij} + jB_{ij}$ is the ijth element of the complex bus admittance matrix.
$g_{ij} + jb_{ij}$ is the admittance of the series branch connecting buses i and j.
$g_{si} + jb_{si}$ is the admittance of the shunt branch connected at bus i as shown in Figure 2.7.
$\aleph_i$ is the set of bus numbers that are directly connected to bus i.

2.6.2 The Measurement Jacobian, H

The structure of the measurement Jacobian H will be as follows:

$$H = \begin{bmatrix} \frac{\partial P_{inj}}{\partial \theta} & \frac{\partial P_{inj}}{\partial V} \\ \frac{\partial P_{flow}}{\partial \theta} & \frac{\partial P_{flow}}{\partial V} \\ \frac{\partial Q_{inj}}{\partial \theta} & \frac{\partial Q_{inj}}{\partial V} \\ \frac{\partial Q_{flow}}{\partial \theta} & \frac{\partial Q_{flow}}{\partial V} \\ \frac{\partial I_{mag}}{\partial \theta} & \frac{\partial I_{mag}}{\partial V} \\ 0 & \frac{\partial V_{mag}}{\partial V} \end{bmatrix}$$

The expressions for each partition are given below:

- Elements corresponding to real power injection measurements:

$$\begin{aligned} \frac{\partial P_i}{\partial \theta_i} &= \sum_{j=1}^{N} V_i V_j (-G_{ij} \sin \theta_{ij} + B_{ij} \cos \theta_{ij}) - V_i^2 B_{ii} \\ \frac{\partial P_i}{\partial \theta_j} &= V_i V_j (G_{ij} \sin \theta_{ij} - B_{ij} \cos \theta_{ij}) \\ \frac{\partial P_i}{\partial V_i} &= \sum_{j=1}^{N} V_j (G_{ij} \cos \theta_{ij} + B_{ij} \sin \theta_{ij}) + V_i G_{ii} \\ \frac{\partial P_i}{\partial V_j} &= V_i (G_{ij} \cos \theta_{ij} + B_{ij} \sin \theta_{ij}) \end{aligned}$$

- Elements corresponding to reactive power injection measurements:

$$\begin{aligned} \frac{\partial Q_i}{\partial \theta_i} &= \sum_{j=1}^{N} V_i V_j (G_{ij} \cos \theta_{ij} + B_{ij} \sin \theta_{ij}) - V_i^2 G_{ii} \\ \frac{\partial Q_i}{\partial \theta_j} &= V_i V_j (-G_{ij} \cos \theta_{ij} - B_{ij} \sin \theta_{ij}) \\ \frac{\partial Q_i}{\partial V_i} &= \sum_{j=1}^{N} V_j (G_{ij} \sin \theta_{ij} - B_{ij} \cos \theta_{ij}) - V_i B_{ii} \\ \frac{\partial Q_i}{\partial V_j} &= V_i (G_{ij} \sin \theta_{ij} - B_{ij} \cos \theta_{ij}) \end{aligned}$$

- Elements corresponding to real power flow measurements:

$$\begin{aligned}
\frac{\partial P_{ij}}{\partial \theta_i} &= V_i V_j (g_{ij} \sin\theta_{ij} - b_{ij} \cos\theta_{ij}) \\
\frac{\partial P_{ij}}{\partial \theta_j} &= -V_i V_j (g_{ij} \sin\theta_{ij} - b_{ij} \cos\theta_{ij}) \\
\frac{\partial P_{ij}}{\partial V_i} &= -V_j (g_{ij} \cos\theta_{ij} + b_{ij} \sin\theta_{ij}) + 2\,(g_{ij} + g_{si}) V_i \\
\frac{\partial P_{ij}}{\partial V_j} &= -V_i (g_{ij} \cos\theta_{ij} + b_{ij} \sin\theta_{ij})
\end{aligned}$$

- Elements corresponding to reactive power flow measurements:

$$\begin{aligned}
\frac{\partial Q_{ij}}{\partial \theta_i} &= -V_i V_j (g_{ij} \cos\theta_{ij} + b_{ij} \sin\theta_{ij}) \\
\frac{\partial Q_{ij}}{\partial \theta_j} &= V_i V_j (g_{ij} \cos\theta_{ij} + b_{ij} \sin\theta_{ij}) \\
\frac{\partial Q_{ij}}{\partial V_i} &= -V_j (g_{ij} \sin\theta_{ij} - b_{ij} \cos\theta_{ij}) - 2\,V_i (b_{ij} + b_{si}) \\
\frac{\partial Q_{ij}}{\partial V_j} &= -V_i (g_{ij} \sin\theta_{ij} - b_{ij} \cos\theta_{ij})
\end{aligned}$$

- Elements corresponding to voltage magnitude measurements:

$$\frac{\partial V_i}{\partial V_i} = 1, \frac{\partial V_i}{\partial V_j} = 0, \frac{\partial V_i}{\partial \theta_i} = 0, \frac{\partial V_i}{\partial \theta_j} = 0$$

- Elements corresponding to current magnitude measurements (ignoring the shunt admittance of the branch):

$$\begin{aligned}
\frac{\partial I_{ij}}{\partial \theta_i} &= \frac{g_{ij}^2 + b_{ij}^2}{I_{ij}} V_i V_j \sin\theta_{ij} \\
\frac{\partial I_{ij}}{\partial \theta_j} &= -\frac{g_{ij}^2 + b_{ij}^2}{I_{ij}} V_i V_j \sin\theta_{ij} \\
\frac{\partial I_{ij}}{\partial V_i} &= \frac{g_{ij}^2 + b_{ij}^2}{I_{ij}} (V_i - V_j \cos\theta_{ij}) \\
\frac{\partial I_{ij}}{\partial V_j} &= \frac{g_{ij}^2 + b_{ij}^2}{I_{ij}} (V_j - V_i \cos\theta_{ij})
\end{aligned}$$

Example 2.3:

Consider the same system and measurement configuration shown in example 2.2. Assume flat start conditions, where the state vector is equal to:

$$x^0 = \begin{matrix} \theta_2 \\ \theta_3 \\ V_1 \\ V_2 \\ V_3 \end{matrix} \quad \begin{bmatrix} 0 \\ 0 \\ 1.0 \\ 1.0 \\ 1.0 \end{bmatrix}$$

Then, the measurement Jacobian can be evaluated as follows, using the expressions given above:

$$H(x^0) = \begin{matrix} \\ \partial p_{12} \\ \partial p_{13} \\ \partial p_2 \\ \partial q_{12} \\ \partial q_{13} \\ \partial q_2 \\ \partial V_1 \\ \partial V_2 \end{matrix} \begin{matrix} \begin{matrix} \partial\theta_2 & \partial\theta_3 & \partial V_1 & \partial V_2 & \partial V_3 \end{matrix} \\ \left[\begin{array}{cc|ccc} -30.0 & & 10.0 & -10.0 & \\ & -17.2 & 6.9 & & -6.9 \\ 40.9 & -10.9 & -10.0 & 14.1 & -4.1 \\ \hline 10.0 & & 30.0 & -30.0 & \\ & 6.9 & 17.2 & & -17.2 \\ -14.1 & 4.1 & -30.0 & 40.9 & -10.9 \\ & & 1.0 & & \\ & & & 1.0 & \end{array}\right] \end{matrix}$$

Note that the dimension of H is $m \times n = 8 \times 5$, and it is a sparse matrix. Its sparsity becomes more pronounced for large scale systems, where the number of nonzeros per row stays fairly constant, irrespective of the system size.

2.6.3 The Gain Matrix, G

Gain matrix is formed using the measurement Jacobian H and the measurement error covariance matrix, R. The covariance matrix is assumed to be diagonal having measurement variances as its diagonal entries. Since G is formed as:

$$G(x^k) = H^T R^{-1} H$$

it has the following properties:

1. It is structurally and numerically symmetric.
2. It is sparse, yet less sparse compared to H.
3. In general it is a non-negative definite matrix, i.e. all of its eigenvalues are non-negative. It is positive definite for fully observable networks.

G is built and stored as a sparse matrix for computational efficiency and memory considerations. It is built by processing one measurement at a time. Consider the measurement jacobian H and the covariance matrix for a set of m measurements, each one corresponding to one row, as shown below:

$$H = \begin{bmatrix} H_1 \\ H_2 \\ \vdots \\ H_m \end{bmatrix}, R = \begin{bmatrix} R_{11} & 0 & \cdots & 0 \\ 0 & R_{22} & 0 & 0 \\ 0 & 0 & \ddots & 0 \\ 0 & 0 & \cdots & R_{mm} \end{bmatrix}$$

Then, the gain matrix can be re-written as follows:

$$G = \sum_{i=1}^{m} H_i^T R_{ii}^{-1} H_i$$

Since H_i arrays are very sparse row vectors, their product will also yield a sparse matrix. Nonzero terms in G can thus be calculated and stored in sparse form.

Example 2.4:

Using the measurement jacobian $H(x^0)$ evaluated in example 2.3, the gain matrix $G(x^0)$ will be obtained as follows:

$$G(x^0) = 10^7 \left[\begin{array}{rr|rrr} 3.4392 & -0.5068 & 0.0137 & & -0.0137 \\ -0.5068 & 0.6758 & -0.0137 & 0.0137 & 0.0000 \\ \hline 0.0137 & -0.0137 & 3.1075 & -2.9324 & -0.1689 \\ & 0.0137 & -2.9324 & 3.4455 & -0.5068 \\ -0.0137 & 0.0000 & -0.1689 & -0.5068 & 0.6758 \end{array} \right]$$

Gain matrix is 5×5, symmetric and less sparse than the corresponding measurement jacobian $H(x^0)$. Its eigenvalues can be computed as:

$$Eigen(G) = 10^7 \begin{bmatrix} 3.5293 \\ 6.2254 \\ 0.5857 \\ 0.9992 \\ 0.0042 \end{bmatrix}$$

confirming that it is positive definite.

2.6.4 Cholesky Decomposition of G

The gain matrix G can be written as a product of a lower triangular sparse matrix and its transpose. This is called the Cholesky decomposition of G, details of which are given in Appendix B. Decomposed form of G will be:

$$G = L \cdot L^T$$

Note that this decomposition may not exist for systems which are not fully observable. As a result, a state estimation solution can not be obtained for such unobservable systems. Chapter 4 describes methods for testing observability and related topics.

Example 2.5:

Cholesky decomposition of the gain matrix $G(x^0)$ in example 2.4, is given as follows:

$$G(x^0) = LL^T$$

where :

$$L = 10^3 \begin{bmatrix} 5.8645 & 0 & 0 & 0 & 0 \\ -0.8643 & 2.4517 & 0 & 0 & 0 \\ 0.0234 & -0.0476 & 5.5743 & 0 & 0 \\ -0.0000 & 0.0559 & -5.2600 & 2.6045 & 0 \\ -0.0234 & -0.0082 & -0.3030 & -2.5579 & 0.3503 \end{bmatrix}$$

Triangular factors of G are not unique and their sparsity depends heavily on the way the decomposition is carried out. There are several ordering methods for optimizing the sparsity of the resulting L factors. These are discussed in detail in Appendix B.

2.6.5 Performing the Forward/Back Substitutions

Assuming that the gain matrix is properly decomposed into its Cholesky factors L and L^T, the next step is to solve the Normal equation for Δx^k:

$$LL^T \Delta x^k = t^k$$

where, t^k denotes the right hand side of Equation (2.12). This solution is obtained in two steps:

1. Forward substitution: Let $L^T \Delta x^k =$ u, and obtain the elements of u starting from u_1 by using substitutions in the transformed equation

$Lu = t^k$. The top row will yield the solution for u_1 as t_1/L_{11}. Substituting for u_1 in the rest of the rows will reduce the set of equations by one. Repeating the same procedure for u_2 and others sequentially, will yield the entire solution for u.

2. Back substitution: Now that u is available, use $L^T \Delta x^k = \text{u}$, to back substitute and solve for the entries of Δx^k. This time, the substitutions should start at the bottom row, where the last element of the solution vector is obtained as $\Delta x^k(n) = u_n / L_{nn}$. Substituting for it in the remaining rows, the back substitution process can continue until all entries are calculated.

Note that both the forward and back substitution steps proceed very efficiently due to the sparse structure of the triangular factor L.

Example 2.6:

Consider the iterative solution of the WLS state estimation problem for the system of example 2.2. Applying the algorithm described in section 2.6, the state vector can be solved iteratively. The convergence criteria will be chosen as 10^{-4} for the state variable updates. Starting from the flat start, and using the jacobian and gain matrices already initialized in examples 2.3 and 2.4, the solution is obtained in 3 iterations. The convergence summary for the objective function $J(x^k)$ and the state updates Δx^k are given in the below table.

Iterations, k	1	2	3
$\Delta\theta_2^k$	$-2.10e-2$	$-6.00e-4$	$-0.02e-5$
$\Delta\theta_3^k$	$-4.52e-2$	$-2.70e-3$	$2.81e-6$
ΔV_1^k	$3.00e-4$	$-1.09e-4$	$-1.65e-6$
ΔV_2^k	$-2.57e-2$	$-1.06e-4$	$-1.63e-6$
ΔV_3^k	$-5.72e-2$	$1.15e-3$	$1.87e-6$
Objective Function, $J(x^k)$	49,123	59.6	8.6

The algorithm converges to the following state estimation solution:

Bus i	$\hat{V}_i$ (pu)	$\hat{\theta}_i$ (degrees)
1	0.9996	0.0
2	0.9742	−1.2475
3	0.9439	−2.7457

Finally, we can also compute the estimated measurements and their residual vector given by:

$$r = z - h(\hat{x})$$

These values are shown below:

Measurement No. i	Type	Measured Value (pu)	Estimated Value (pu)	Residual (pu)
1	p_{12}	0.888	0.8930	−0.0050
2	p_{13}	1.173	1.1711	0.0019
3	p_2	−0.501	−0.4959	−0.0051
4	q_{12}	0.568	0.5588	0.0092
5	q_{13}	0.663	0.6677	−0.0047
6	q_2	−0.286	−0.2977	0.0117
7	V_1	1.006	0.9996	−0.0064
8	V_2	0.968	0.9742	−0.0062

The gain matrix G evaluated in the 3rd iteration is given by:

$$G(x^3) = 10^7 \left[\begin{array}{cc|ccc} 3.2086 & -0.4472 & -0.0698 & -0.0314 & 0.0038 \\ -0.4472 & 0.5955 & -0.0451 & 0.0045 & 0.0000 \\ \hline -0.0698 & -0.0451 & 3.2011 & -2.8862 & -0.2160 \\ -0.0314 & 0.0045 & -2.8862 & 3.3105 & -0.4760 \\ 0.0038 & 0.0000 & -0.2160 & -0.4760 & 0.6684 \end{array}\right]$$

Note that this matrix is not very different from the initial $G(x^0)$ matrix evaluated at flat start in example 2.3. While there are exceptions, in general the gain matrix elements do not change significantly during the iterative solution procedure.

2.7 Decoupled Formulation of the WLS State Estimation

The main computational burden associated with the WLS state estimation solution algorithm presented in section 2.6 is the calculation and triangular decomposition of the gain matrix. One way to reduce this burden is to maintain a constant but approximate gain matrix. This approximation is in line with the observation in exercise 2.6, that the elements of the gain matrix do not significantly change between flat start initialization and the converged solution. Furthermore, as observed earlier for the power flow problem [1], sensitivity of the real (reactive) power equations to changes in the magnitude (phase angle) of bus voltages is very low, especially for high voltage transmission systems. These two observations lead to the fast decoupled formulation of the state estimation problem [2, 3]. In this formulation, the measurement equations are partitioned into two parts:

- Real power measurements, including the real power bus injections and real power flows in branches. These measurements will be denoted by the subscript A, meaning the active measurements.

- Reactive power measurements, including the reactive power bus injections, reactive power flows in branches and bus voltage magnitude measurements. These measurements will be denoted by the subscript R, meaning the reactive measurements.

Note that branch current magnitude measurements are not included in either of these groups of measurements. This is intentional due to the fact that such measurements do not lend themselves as readily as the others to the decoupled formulation. This is one of the shortcomings of the decoupled formulation and the problems associated with the use of current magnitude measurements will be discussed further in Chapter 9.

Thus, the measurement and their related arrays can be partitioned based on the above designation:

$$\begin{aligned} z^T &= [z_A^T \ z_R^T] \\ H &= \begin{bmatrix} H_{AA} & H_{AR} \\ H_{RA} & H_{RR} \end{bmatrix} \\ R &= \begin{bmatrix} R_A & 0 \\ 0 & R_R \end{bmatrix} \end{aligned}$$

The following assumptions are used to obtain the fast decoupled state estimation algorithm:

1. Assume flat start operating conditions, i.e. all bus voltages being at nominal magnitude of 1.0 pu and in phase with each other.

2. Ignore the off diagonal blocks H_{AR} and H_{RA} in the measurement jacobian H, and compute the gain matrix using this approximation. This will also eliminate the off diagonal blocks in the gain matrix, yielding a constant and decoupled gain matrix evaluated at flat start:

$$\begin{aligned} G &= \begin{bmatrix} G_{AA} & 0 \\ 0 & G_{RR} \end{bmatrix} \\ G_{AA} &= H_{AA}^T R_A^{-1} H_{AA} \\ G_{RR} &= H_{RR}^T R_R^{-1} H_{RR} \end{aligned}$$

3. Repeat the same approximation for the jacobian entries when calcu-

lating the right hand side vector:

$$T = \begin{bmatrix} H_{AA}^T R_A^{-1} \Delta z'_A \\ H_{RR}^T R_R^{-1} \Delta z'_R \end{bmatrix} = \begin{bmatrix} T_A \\ T_R \end{bmatrix}$$

$$\begin{aligned} \text{where}: \quad \Delta z'_A &= \Delta z_A / V \\ \Delta z'_R &= \Delta z_R / V \\ \Delta z_A &= z_A - h_A(\hat{x}) \\ \Delta z_R &= z_R - h_R(\hat{x}) \end{aligned}$$

There are two variations to the basic assumptions listed above. These variations essentially pertain to the submatrices H_{AA} and H_{RR}. Ignoring the branch series resistances in forming H_{AA} or H_{RR} will lead to the so-called XB or BX formulation of the fast decoupled state estimation respectively. Details of the justification of these formulations can be found in [4, 5].

The above assumptions lead to a decoupled solution algorithm using the polar coordinates in the calculations. Hence, the solution for the phase angle $\Delta\theta$ and magnitude ΔV updates are obtained alternatingly and convergence is tested based on the max. changes in both of these arrays. The steps of the solution algorithm are given below:

1. Initialize all bus voltages at flat start, $V_i = 1$ pu, $\theta_i = 0$ for all $i = 1, \ldots, N$.

2. Build and perform triangular decomposition of G_{AA} and G_{RR}.

3. Calculate T_A.

4. Solve $G_{AA}\, \Delta\theta = T_A$.

5. Check if both $\Delta\theta$ and ΔV are less than the convergence tolerance. If yes, stop. Else, continue.

6. Update $\theta^{k+1} = \theta^k + \Delta\theta$.

7. Calculate T_R.

8. Solve $G_{RR}\, \Delta V = T_R$.

9. Check if both $\Delta\theta$ and ΔV are less than the convergence tolerance. If yes, stop. Else, continue.

10. Update $V^{k+1} = V^k + \Delta V$.

11. Go to step 3.

Note that the gain sub-matrices G_{AA} and G_{RR} are computed and decomposed into their triangular factors only once at the beginning of the iterative solution. Solutions for $\Delta\theta$ and ΔV are carried out very efficiently using the forward and back substitutions, since the triangular factors need not be updated during the iterations. Furthermore, the dimension of the two gain sub-matrices are half the size of the fully coupled gain matrix, further reducing the computational effort.

Example 2.7:

The constant gain matrices used in the fast decoupled state estimation algorithm are given below for the same network of example 2.2:

$$G_{AA} = 10^7 \begin{bmatrix} 3.837 & -0.5729 \\ -0.5729 & 0.7812 \end{bmatrix}$$

$$G_{RR} = 10^7 \begin{bmatrix} 2.777 & -2.635 & -0.1357 \\ -2.635 & 3.090 & -0.4489 \\ -0.1357 & -0.4489 & 0.5846 \end{bmatrix}$$

In this example, the fast decoupled state estimation algorithm converges in 3.5 iterations, i.e. 3 real power and 4 reactive power iterations are needed. Estimated state is given by:

Bus i	$\hat{V}_i$ (pu)	$\hat{\theta}_i$ (degrees)
1	1.000	0.0
2	0.97438	−1.24
3	0.94401	−2.71

Fast decoupled state estimation has found wide acceptance in the industry and various versions have been implemented in control centers all over the world. When compared with the full (coupled) WLS solution algorithm, the decoupled version has the following advantages:

- It requires less memory.
- It is computationally faster, since the gain sub-matrices are smaller and constant requiring the triangular decomposition to be carried out only once at the first iteration.

On the other hand, it has the below given limitations which should be carefully considered before using it for a particular system and measurement set:

- There may be cases when the network parameters or operating conditions violate the stated decoupling assumptions. Such cases may not converge or converge to significantly inaccurate solutions. However, such cases are rather rare in practice.
- Branch current magnitude (ampere) measurements do not have the same type of decoupling properties as the rest of the measurement types. Hence, the decoupled formulation can not be reliably used in the presence of branch current magnitude measurements.

2.8 DC State Estimation Model

It is often helpful to work with a simplified DC approximation model for the measurement equations in analyzing the inherent limitations of various methods related solely to the measurement configuration. DC approximation is obtained by assuming that the bus voltage magnitudes are already known and are all equal to 1.0 per unit. Neglecting all shunt elements and branch resistances, the real power flow measured from bus k to m can be approximated by the first order Taylor expansion around $\theta = 0$, given by:

$$P_{km} = \frac{\theta_k - \theta_m}{x_{km}} + e \tag{2.13}$$

where x_{km} is the reactance of branch $k - m$, θ_k is the phase angle at bus k and e is the measurement error. Similarly, a power injection measurement at a given bus i can be expressed as a sum of flows along incident branches to that bus:

$$P_i = \sum_{j \in \aleph_j} P_{ij} + e \tag{2.14}$$

where $\aleph_j$ is the set of buses connected to bus j.

Hence, the DC model for the real power measurements can be expressed in matrix form as follows:

$$z_A = H_{AA}\theta + e_A \tag{2.15}$$

where z_A includes flow and injection measurements, H_{AA} is a function of the branch reactances only, and e_A is the vector of random errors. Note that the reference bus phase angle is typically excluded from θ and the corresponding column will be missing in H_{AA}. We will refer to this DC model frequently in the subsequent chapters.

2.9 Problems

1. The measurement model for a certain application is given as follows:

$$y_i = ax_i^2 + b + e_i$$

where, y_i is the ith measurement, x_i is a known parameter for measurement i, e_i is the error in the ith measurement, a and b are unknown parameters of the measurement model.

Assume that e_i's are independent random variables distributed according to a Normal distribution with $N(0, 0.4)$, and calculate the MLE of a and b for the measurements given in the below table:

i	y_i	x_i
1	5.2	1
2	15.3	2
3	33.0	3
4	58.0	4

2. Determine the minimum value of n for which

$$Pr(\mid \bar{X}_n - \mu \mid > 1.2) \leq 0.01$$

if the random sample is taken from a Normal distribution whose mean μ is unknown and variance is 0.16.

3. Suppose that the voltage at a certain substation has a normal distribution with mean 345,000 V and variance 225,000,000. If five independent measurements of the voltage are made, what is the probability that all five measurements will lie between 340 kV and 360 kV?

4. Suppose that a random sample of size N is to be taken from a normal distribution with mean μ and standard deviation 0.04. Determine the smallest value of N such that:

$$Pr(\mid \overline{X_N} - \mu \mid \leq 0.04) \geq 0.85 \tag{2.16}$$

5. Suppose that $X_1, \cdots, X_N$ form a random sample from a distribution for which the probability density function (p.d.f) is given as follows:

$$f(x \mid \theta) = \begin{cases} \theta \cdot x^{\theta-1} & \text{for } 0 < x < 1 \\ 0 & \text{otherwise} \end{cases}$$

Also, suppose that the value of θ is unknown ($\theta > 0$) Find the MLE of θ.

6. Suppose that a wattmeter has errors distributed according to a normal distribution with mean 0.5 W and variance 0.25 W. If 100 independent power measurements are made using this instrument, what is the probability that all 100 measurements will have errors less than 0.6W? Assume that wattmeters can measure negative watts.

7. One hundred independent voltage measurements will be made at a power system bus.

   (a) Assuming that the bus voltage is distributed according to a Normal distribution with a mean of 230-kV and a variance of 50 million square volts, find the probability that all one hundred measurements will lie between 220-kV and 235-kV.

   (b) Repeat part (a) this time assuming a variance of 0.5 million square volts.

8. Suppose that

$$\{0.2, 0.45, 0.86, 0.55, 0.01, 0.94, 0.33, 0.66, 0.32, 0.75, 0.24\}$$

form a random sample from a distribution for which the probability density function (p.d.f) is given as follows:

$$f(y \mid \theta) = \begin{cases} \sqrt{\theta} \cdot y^{2\theta - 1} & \text{for } 0 < y < 1 \\ 0 & \text{otherwise} \end{cases}$$

Also, suppose that the value of the parameter θ is unknown ($\theta > 0$). Find the MLE of θ.

9. A small power system is monitored through 7 measurements whose errors are independently distributed according to a Normal distribution with zero mean and a variance as given below for each measurement:

Measurement, i	σ_i^2
1	6.4010^{-5}
2	9.0010^{-6}
3	1.6010^{-5}
4	4.9010^{-6}
5	1.0010^{-6}
6	1.4410^{-6}
7	2.5010^{-5}

What is the probability that the absolute value of the error of at least one of the measurements will be greater than 0.01?

10. A 2-bus power system is shown in Figure 2.8. Assume that the following measurement set is available for estimation:

$$[z]^T = [P_2 Q_2 V_1] = [-0.30, -0.15, 1.0]$$

Assume that the measurements are equally accurate.

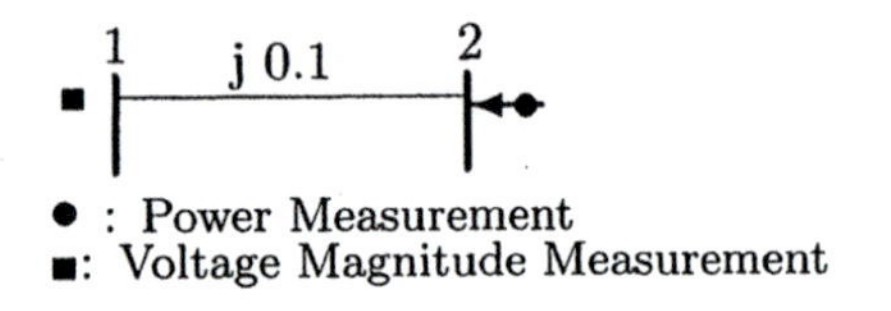

Figure 2.8. 2-bus system diagram for Problem 10

(a) Find the WLS estimator for V_2 and θ_2.

(b) What is the value of the objective function $J(x)$ at the optimal solution?

(c) Does $J(x)$ have a unique minimum? If not, find all other possible solutions.

References

[1] B. Stott and O. Alsaç, "Fast Decoupled Load Flow", IEEE Transactions on Power Apparatus and Systems, Vol. PAS-93, No.3, pp.859-867, May/June 1974.

[2] A. Garcia, A. Monticelli and P. Abreu, "Fast Decoupled State Estimation and Bad Data Processing", IEEE Transactions on Power Apparatus and Systems, Vol. PAS-98, No.5, pp.1645-1652, September/October 1979.

[3] J.J. Allemong, L. Radu and A.M. Sasson, "A Fast and Reliable State Estimation Algorithm for AEP's New Control Center", IEEE Transactions on Power Apparatus and Systems, Vol. PAS-101, No.3, pp.933-944, April 1982.

[4] R.A.M. Van Amerongen, "A General Purpose Version of the Fast Decoupled Load Flow", IEEE Transactions on Power Apparatus and Systems, Vol.4, pp.760-770, May 1989.

[5] A. Monticelli and A. Garcia, "Fast Decoupled State Estimators", IEEE Transactions on Power Systems, Vol.5, No.2, pp.556-564, May 1990.

Chapter 3

Alternative Formulations of the WLS State Estimation

Solution of the WLS State Estimation problem via the use of the Normal Equations (NE), as explained in the former chapter, can almost always be successfully carried out, especially on modern extended-wordlength computers. However, it is well known that under certain circumstances that are likely to occur in actual systems, the NE will become prone to numerical instabilities. Such situations will prohibit the solution algorithm from reaching an acceptable solution or even will cause divergence.

In this chapter, the limitations of NE will be first discussed and illustrated. Then, several, numerically more robust, alternative techniques, will be presented.

3.1 Weaknesses of the Normal Equations Formulation

Let us first recall, from the previous chapter, that the WLS State Estimator leads to the iterative solution of the so-called NE:

$$G(x^k)\Delta x^k = H^T(x^k)W\Delta z^k \tag{3.1}$$

where

k denotes the iteration index
$\Delta z = z - h(x)$ is the residual vector
H is the Jacobian of $h(x)$

$W = R^{-1} = \text{diag}^{-1}(\sigma_i^2)$ is the weighting matrix
$G = H^T W H$ is the gain matrix

Equation (3.1) is solved by Cholesky factorization of G followed by forward/backward substitutions on the right hand side vector. Since G is positive definite for observable systems, there is no need to worry about pivoting to preserve numerical stability.

However, prior to the decomposition of G, its rows/columns must be symmetrically permuted according to the minimum degree criterion so as to maintain its sparse structure as much as possible. The sparsity pattern of G can be directly deduced from that of H which, in turn, is determined by the network topology and measurement configuration. Every injection measurement brings in second-neighbor adjacency for the corresponding bus, as shown by the examples of Chapter 2. This implies that, G will in general be less sparse than the bus admittance matrix. Consequently, solving the NE will involve significantly more computations than those required by the power flow solution for the same network.

Another and perhaps a more important difference between the state estimation and the load flow problems, for the matters discussed in this chapter, is the numerical conditioning of the solution equations. A linear equation system is said to be *ill-conditioned* if small errors in the entries of the coefficient matrix and/or the right hand side vector translate into significant errors in the solution vector. The more singular a matrix is, the more ill-conditioned its associated system will be. The degree to which a system is ill-conditioned, can be quantified by a measure called the condition number, which is defined as:

$$\kappa(A) = ||A|| \cdot ||A^{-1}||$$

where $||\cdot||$ represents a given matrix norm. This value is equal to unity for identity matrices and tends to infinity for matrices approaching singularity. Condition numbers are typically approximately computed, due to the high computing cost of κ as evident from its definition above. One such approximation which yields a good estimate of the condition number is the ratio $\lambda_{max}/\lambda_{min}$, where λ_{max}, λ_{min} are the largest and smallest absolute eigenvalues respectively of a normalized matrix.

It can be shown that

$$\kappa(A^T A) = [\kappa(A)]^2 \tag{3.2}$$

which means that the NE are intrinsically ill-conditioned. Consider as an example the following matrix:

$$H = \begin{bmatrix} 1 & 1 & 1 & 1 & 1 \\ \varepsilon & & & & \\ & \varepsilon & & & \\ & & \varepsilon & & \\ & & & \varepsilon & \\ & & & & \varepsilon \end{bmatrix}$$

If the floating-point accuracy is $1e-10$ and $\varepsilon = 0.5e-5$, then $\text{rank}(H) = 5$. However, for such an accuracy, the corresponding gain matrix reduces to

$$H^T H = \begin{bmatrix} 1+\varepsilon^2 & 1 & 1 & 1 & 1 \\ 1 & 1+\varepsilon^2 & 1 & 1 & 1 \\ 1 & 1 & 1+\varepsilon^2 & 1 & 1 \\ 1 & 1 & 1 & 1+\varepsilon^2 & 1 \\ 1 & 1 & 1 & 1 & 1+\varepsilon^2 \end{bmatrix} \approx \begin{bmatrix} 1 & 1 & 1 & 1 & 1 \\ 1 & 1 & 1 & 1 & 1 \\ 1 & 1 & 1 & 1 & 1 \\ 1 & 1 & 1 & 1 & 1 \\ 1 & 1 & 1 & 1 & 1 \end{bmatrix}$$

which is a rank-one matrix.

Although such extreme cases are never found in practice, a combination of a too low termination threshold, poor word-length and severe ill-conditioning may cause convergence problems or even divergence.

Furthermore, for the particular case of the WLS state estimation, the following specific sources of ill-conditioning have been described in the literature:

- Very large weighting factors used to enforce virtual measurements.
- Short and long lines simultaneously present at the same bus.
- A large proportion of injection measurements.

The following three examples will illustrate these ill-conditioning mechanisms, for very simple cases.

Example 3.1:

Consider the three-bus system shown in figure 3.1, where we are interested in the linear DC state estimation problem.

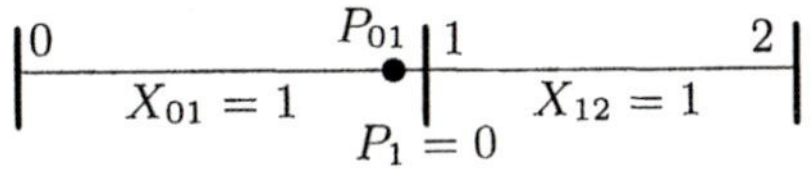

Figure 3.1. 3-bus system for example 3.1

For simplicity, only two critical measurements are considered to estimate θ_1 and θ_2, namely the regular power flow measurement P_{01} and the exact null injection P_1. If H and C denote the Jacobian of regular and virtual measurements respectively, we have in this case:

$$\begin{bmatrix} H \\ C \end{bmatrix} = \begin{bmatrix} -1 & 0 \\ 2 & -1 \end{bmatrix}$$

$$G = \begin{bmatrix} H \\ C \end{bmatrix}^T \begin{bmatrix} W & 0 \\ 0 & V \end{bmatrix} \begin{bmatrix} H \\ C \end{bmatrix} = \begin{bmatrix} W + 4V & -2V \\ -2V & V \end{bmatrix}$$

Clearly, if the weight assigned to the ordinary measurement, W, is negligible compared to V, the gain matrix will become almost singular (ill-conditioned). The same will happen in the general case, because the rows of C will not be sufficient to make the network fully observable (only in the case of the power flow, does C contain all rows).

Example 3.2:

Consider now the network shown in figure 3.2, where the ratio X_{12}/X_{01} is k.

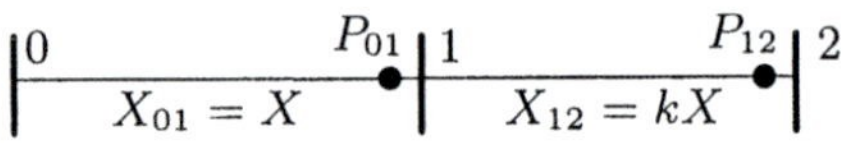

Figure 3.2. 3-bus system for example 3.2

The relevant equations in this case, assuming unit weights, are:

$$\begin{aligned} P_{01} &= \frac{-1}{X}\theta_1 \\ P_{12} &= \frac{1}{kX}(\theta_1 - \theta_2) \end{aligned}$$

$$H = \frac{1}{X} \begin{bmatrix} -1 & 0 \\ \frac{1}{k} & \frac{-1}{k} \end{bmatrix}$$

$$G = H^T H = \frac{k^2}{X^2} \begin{bmatrix} k^2 + 1 & -1 \\ -1 & 1 \end{bmatrix}$$

Note again that the gain matrix becomes ill-conditioned if the line 1-2 is much shorter than line 0-1 ($k \ll 1$).

Example 3.3:

The former example comprises two power flow measurements. When the two lines are identical ($k = 1$), the Jacobian and gain matrices become:

$$H_F = \frac{1}{X}\begin{bmatrix} -1 & 0 \\ 1 & -1 \end{bmatrix}$$

$$G_F = \frac{1}{X^2}\begin{bmatrix} 2 & -1 \\ -1 & 1 \end{bmatrix}$$

and $\text{cond}(G_F) = 6.854$.

Assume now that, instead of the two power flows, the two injections P_1, P_2 are measured. In this case, the matrix values are:

$$H_I = \frac{1}{X}\begin{bmatrix} 2 & -1 \\ -1 & 1 \end{bmatrix}$$

$$G_I = \frac{1}{X^2}\begin{bmatrix} 5 & -3 \\ -3 & 2 \end{bmatrix}$$

and cond $(G_I) = 46.98 =$ cond $(G_F)^2$. Observe that, except for a scaling factor, $G_F = H_I$, which explains the relationship between the two condition numbers.

A theoretical justification is as follows: In the DC model, the power flow vector is given by,

$$P_F = X^{-1}A^T\theta$$

where X is the diagonal primitive branch reactance matrix, A is the node-branch incidence matrix and θ is the phase angle column vector. Therefore, for a measurement set exclusively composed of (all) power flows the Jacobian is:

$$H_F = X^{-1}A^T$$

and the gain matrix (ignoring weights):

$$G_F = AX^{-2}A^T$$

Similarly, the power injection vector can be expressed as follows:

$$P_I = AX^{-1}A^T\theta$$

and, when only power injections are measured at all system buses, the resulting matrices are:

$$H_I = AX^{-1}A^T$$
$$G_I = (AX^{-1}A^T)^2$$

Consequently, according to (3.2), the gain matrix corresponding to power injection measurements is more ill-conditioned than that of power flows. For branch reactances of the same order of magnitude, it is expected that cond $(G_I) \approx$ cond $(G_F)^2$.

In the following sections, several alternative techniques which try to circumvent the shortcomings of the NE by avoiding the use of G and/or handling virtual measurements in a different manner, will be described.

3.2 Orthogonal Factorization

Any $m \cdot n$ matrix $\tilde{H}$ of full rank can be decomposed into two matrices of the form:

$$\tilde{H} = QR \tag{3.3}$$

where Q is an $m \cdot m$ orthogonal matrix ($Q^T = Q^{-1}$) and R is an $m \cdot n$ upper trapezoidal matrix (i.e., its first n rows are upper triangular while the remaining $m - n$ rows are null). The equivalent expression,

$$Q^T \tilde{H} = R$$

is the basis of well-known factorization algorithms, which obtain R as a sequence of elementary transformations on the columns (rows) of $\tilde{H}$ (see Appendix B).

Also, partitioning Q and R accordingly yields the following reduced form of this factorization:

$$\tilde{H} = \begin{bmatrix} Q_n & Q_0 \end{bmatrix} \begin{bmatrix} U \\ 0 \end{bmatrix} = Q_n U \quad \Rightarrow \quad Q_n^T \tilde{H} = U \tag{3.4}$$

It is therefore sufficient to build only the submatrix Q_n rather than the full Q.

In order to apply the orthogonal factorization to the WLS estimation problem the NE are written first in the following compact form:

$$\underbrace{\tilde{H}^T \tilde{H}}_{G} \Delta x = \tilde{H}^T \Delta \tilde{z} \tag{3.5}$$

where

$$\begin{aligned} \tilde{H} &= W^{1/2} H & (3.6) \\ \Delta \tilde{z} &= W^{1/2} \Delta z & (3.7) \end{aligned}$$

This way, the weighting factors do not appear explicitly but are embedded in the remaining terms.

Then, using the property $QQ^T = I$, Equation (3.5) can be successively transformed as follows:

$$\begin{aligned} \tilde{H}^T Q Q^T \tilde{H} \Delta x &= \tilde{H}^T \Delta \tilde{z} \\ R^T R \Delta x &= R^T Q^T \Delta \tilde{z} \\ U^T U \Delta x &= U^T Q_n^T \Delta \tilde{z} \end{aligned} \tag{3.8}$$

Furthermore, since U is a regular matrix, the last expression leads to,

$$U \Delta x = Q_n^T \Delta \tilde{z} \tag{3.9}$$

which is the key equation in the orthogonal factorization approach. Although this equation has been obtained strictly from algebraic manipulations, it can be also reached considering that the solution of the WLS estimation problem is equivalent to minimizing the euclidean norm of the residual vector, and keeping in mind that orthogonal transformations of vectors do not change their norms (geometric interpretation).

In summary, every iteration of the WLS estimation process consists of the following steps [9, 11, 10]:

1. Perform the factorization $\tilde{H} = QR$.

2. Compute the vector $\Delta z_q = Q_n^T \Delta \tilde{z}$

3. Obtain Δx from backsubstitution on $U \Delta x = \Delta z_q$

Therefore, it is not required to obtain and factorize G. Furthermore, since the QR factorization is not based on scalar pivots, it is numerically more robust than the LU factorization. Hence, the use of very large weights for virtual measurements poses no problems.

The only drawback of this scheme is the need to obtain the matrix Q which, in spite of being actually expressed as the product of elementary matrices, is much denser than G. However, clever square-root-free implementation of the Givens rotations is not computationally too expensive.

3.3 Hybrid Method

Comparing Equations (3.5) and (3.8), it can be concluded that the matrix U in the QR factorization is the same as that of the Cholesky factorization of G. In fact, different algorithms could lead to different Qs but to the same U. However, as explained in Appendix B, this may not be the case in practice due to round-off errors.

Based on this observation, a hybrid scheme can be devised as follows [7]:

1. Obtain U by orthogonal transformations on $\tilde{H}$. There is no need to keep track or save the components of Q.

2. Compute the independent vector $\Delta z_h = \tilde{H}^T \Delta \tilde{z}$

3. Obtain Δx by solving the system $U^T U \Delta x = \Delta z_h$

Hence, the NE are solved at step 3 but U is obtained by orthogonal transformations on H, rather than by Cholesky factorization of G.

Example 3.4:

The sample network of Chapter 2, repeated here for convenience, will be used to illustrate the orthogonal factorization technique.

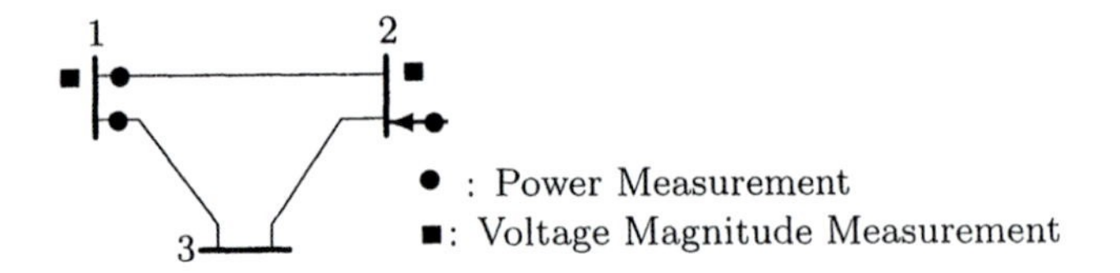

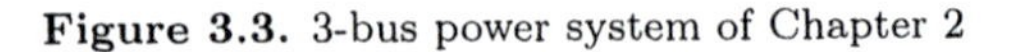

Figure 3.3. 3-bus power system of Chapter 2

Evaluating the Jacobian and the measurement residual vector at flat start, yields:

$$H = \begin{bmatrix} -30.0 & 0 & 10.0 & -10.0 & 0 \\ 0 & -17.241 & 6.8966 & 0 & -6.8966 \\ 40.959 & -10.959 & -10.0 & 14.110 & -4.1096 \\ 10.0 & 0 & 30.0 & -30.0 & 0 \\ 0 & 6.8966 & 17.241 & 0 & -17.241 \\ -14.110 & 4.1096 & -30.0 & 40.959 & -10.959 \\ 0 & 0 & 1.0 & 0 & 0 \\ 0 & 0 & 0 & 1.0 & 0 \end{bmatrix} \quad (3.10)$$

$$\Delta z = \begin{bmatrix} 0.88756 \\ 1.1739 \\ -0.50075 \\ 0.56863 \\ 0.66278 \\ -0.28578 \\ 0.0065010 \\ -0.031618 \end{bmatrix} \quad (3.11)$$

and scaling them by $W^{1/2}$:

$$\tilde{H} = \begin{bmatrix} -3750 & 0 & 1250 & -1250 & 0 \\ 0 & -2155.2 & 862.07 & 0 & -862.07 \\ 4095.9 & -1095.9 & -1000 & 1411 & -410.96 \\ 1250 & 0 & 3750 & -3750 & 0 \\ 0 & 862.07 & 2155.2 & 0 & -2155.2 \\ -1411 & 410.96 & -3000 & 4095.9 & -1095.9 \\ 0 & 0 & 250 & 0 & 0 \\ 0 & 0 & 0 & 250 & 0 \end{bmatrix} ; \Delta\tilde{z} = \begin{bmatrix} 110.95 \\ 146.73 \\ -50.075 \\ 71.079 \\ 82.847 \\ -28.578 \\ 1.625 \\ -7.905 \end{bmatrix}$$

Then, the QR factorization of $\tilde{H}$ is performed by means of Givens rotations.

The reduced factors are:

$$\tilde{H} = Q_n U \quad = \quad 10^{-2} \begin{bmatrix} -63.94 & -22.54 & 22.50 & -2.070 & -0.4446 \\ 0 & -87.91 & 14.71 & 31.60 & -4.679 \\ 69.84 & -20.08 & -18.40 & 17.44 & -1.733 \\ 21.31 & 7.514 & 67.25 & -8.329 & -1.044 \\ 0 & 35.16 & 38.96 & 77.94 & -11.62 \\ -24.06 & 8.281 & -53.65 & 48.74 & -4.773 \\ 0 & 0 & 4.485 & 9.058 & 70.02 \\ 0 & 0 & 0 & 9.599 & 70.09 \end{bmatrix} \cdot$$

$$\cdot \begin{bmatrix} 5864.5 & -864.27 & 23.359 & 0 & -23.359 \\ 0 & 2451.7 & -47.640 & 55.874 & -8.2344 \\ 0 & 0 & 5574.3 & -5260.0 & -303.03 \\ 0 & 0 & 0 & 2604.5 & -2557.9 \\ 0 & 0 & 0 & 0 & 350.28 \end{bmatrix}$$

Note that the upper triangular part of R is the same as the Choleski factors of G obtained in Chapter 2. For realistic systems, it is computationally more efficient to store and form Q as the product of elementary orthogonal matrices.

Next, the right-hand-side vector is computed as,

$$\Delta z_q = Q_n^T \Delta \tilde{z} = \begin{bmatrix} -83.891 & -111.83 & 151.25 & 79.450 & -19.902 \end{bmatrix}^T.$$

and, finally, the solution is obtained by back-substitution,

$$\Delta x = U^{-1} \Delta z_q = 10^{-3} \begin{bmatrix} -21.197 & -45.226 & 0.17702 & -25.294 & -56.816 \end{bmatrix}^T$$

Except for possible round-off errors, the state obtained after the first iteration should match the one obtained by the NE (see the table summarizing the convergence rate in Chapter 2).

The steps required by the hybrid method can be easily performed in terms of the matrices and vectors detailed above. The major difference is that $\tilde{H}$, rather than Q_n, is used to compute the right hand side term.

3.4 Method of Peters and Wilkinson

This is an alternative method, which performs an LU decomposition of $\tilde{H}$:

$$\tilde{H} = LU$$

where L is a lower trapezoidal matrix and U is upper triangular. Substituting for $\tilde{H}$ in the NE:

$$\tilde{H}^T \tilde{H} \Delta x = \tilde{H}^T \Delta \tilde{z} \tag{3.12}$$

they are successively transformed as follows:

$$\begin{aligned} U^T L^T L U \Delta x &= U^T L^T \Delta \tilde{z} \\ L^T L U \Delta x &= L^T \Delta \tilde{z} \\ (L^T L) \Delta y &= L^T \Delta \tilde{z} \end{aligned} \tag{3.13}$$

where:

$$\Delta y = U \Delta x \tag{3.14}$$

The solution procedure consists of the following steps [4]:

1. Perform the LU factorization of $\tilde{H}$.
2. Compute Δy from (3.13). This involves the Cholesky factorization of $L^T L$ followed by a forward/backward substitution.
3. Obtain Δx by backward substitution using (3.14).

The main advantage of using this scheme is the fact that $L^T L$ is less ill-conditioned than $\tilde{H}^T \tilde{H}$.

3.5 Equality-Constrained WLS State Estimation

The use of very high weights for modeling very accurate virtual measurements such as zero injections, leads to ill-conditioning of the G matrix. One way to avoid the use of high weights, is to model these measurements as explicit constraints in the WLS estimation. The Constrained WLS State Estimation problem can then be formulated as follows [2]:

$$\begin{aligned} \text{minimize} \quad & J(x) = \frac{1}{2}[z - h(x)]^T W [z - h(x)] \\ \text{subject to} \quad & c(x) = 0 \end{aligned} \tag{3.15}$$

where $c(x) = 0$ represents the accurate virtual measurements such as zero-injections, which are now excluded from $h(x)$.

This problem can be solved by the Lagrangian method, where the following Lagrangian is built:

$$\mathcal{L} = J(x) - \lambda^T c(x) \tag{3.16}$$

and the first-order optimality conditions are derived:

$$\begin{aligned} \partial \mathcal{L}(x)/\partial x = 0 \Rightarrow \quad H^T W [z - h(x)] + C^T \lambda &= 0 \\ \partial \mathcal{L}(x)/\partial \lambda = 0 \Rightarrow \quad c(x) &= 0 \end{aligned} \tag{3.17}$$

where the matrix $C = \partial c(x)/\partial x$, is the Jacobian of $c(x)$.

Applying the Gauss-Newton method, the nonlinear set of equations (3.17) is solved iteratively by means of the following linear system:

$$\begin{bmatrix} H^T W H & C^T \\ C & 0 \end{bmatrix} \begin{bmatrix} \Delta x \\ -\lambda \end{bmatrix} = \begin{bmatrix} H^T W \Delta z^k \\ -c(x^k) \end{bmatrix} \tag{3.18}$$

where:
$\Delta x = x^{k+1} - x^k$
$\Delta z^k = z - h(x^k)$

Note that the W matrix no longer has large values, which eliminates one of the main sources of ill-conditioning. However, the drawback of Equation (3.18) lies in its coefficient matrix being indefinite. This means that row-pivoting to preserve numerical stability must be combined with sparsity-oriented techniques during LU factorization, destroying the initial symmetry. More sophisticated techniques, capable of resorting on-the-fly to 2x2 pivots to preserve the symmetry, have been developed to deal with indefinite matrices. Recently, other block-pivot approaches have been presented in which the pivot size is decided in advance based on available measurements (see Section 3.7).

It is worth mentioning that the condition number of the coefficient matrix in Equation (3.18) can be further improved by simply scaling the term of the Lagrangian corresponding to the objective function, yielding:

$$\mathcal{L} = \alpha J(x) - \lambda_s^T c(x) \tag{3.19}$$

It is easy to show that the scaling factor α has no influence on the estimated state and that $\lambda_s = \alpha\lambda$. The equation system that must be solved at each iteration is obtained by substituting αW for W:

$$\begin{bmatrix} \alpha H^T W H & C^T \\ C & 0 \end{bmatrix} \begin{bmatrix} \Delta x \\ -\lambda_s \end{bmatrix} = \begin{bmatrix} \alpha H^T W \Delta z^k \\ -c(x^k) \end{bmatrix} \tag{3.20}$$

Very low condition numbers are obtained when α is chosen as

$$\alpha = \frac{1}{\max W_{ii}} \qquad \text{or} \qquad \alpha = \frac{m}{\sum_{i=1}^{m} W_{ii}}$$

The reader should be aware that $\alpha = 1$ might lead to condition numbers which are even worse than that of the conventional G, because the values W_{ii} are usually very large compared to the coefficients of C.

This flexibility is not possible in the conventional approach, where scaling the objective function has no effect on cond(G). Hence, this is an added advantage of modeling virtual measurements as equality constraints.

It is also interesting to show the relationship between both formulations. To this end, let us write $J(x)$ for the conventional formulation in such a way that ordinary and virtual measurements appear separately:

$$J(x) = \frac{1}{2}[z - h(x)]^T W[z - h(x)] + \frac{\rho}{2} c(x)^T c(x) \tag{3.21}$$

where ρ is a weighting factor several orders of magnitude larger than any W_{ii}. The optimality conditions for the above scalar are

$$H^T W[z - h(x)] - \rho C^T c(x) = 0 \tag{3.22}$$

which can be rewritten as follows:

$$\begin{aligned} H^T W[z - h(x)] + C^T \lambda &= 0 \\ c(x) + \frac{1}{\rho} \lambda &= 0 \end{aligned} \tag{3.23}$$

In turn, this augmented system involves the repetitive solution of the following linear equation:

$$\begin{bmatrix} H^T W H & C^T \\ C & -1/\rho \end{bmatrix} \begin{bmatrix} \Delta x \\ -\lambda \end{bmatrix} = \begin{bmatrix} H^T W \Delta z^k \\ -c(x^k) \end{bmatrix} \tag{3.24}$$

Clearly, Equations (3.23) and (3.24) approach (3.17) and (3.18) respectively for very large values of ρ. On the other hand, eliminating λ in (3.24) leads to the NE of the unconstrained formulation, namely:

$$[H^T W H + \rho C^T C]\, \Delta x = H^T W \Delta z^k - \rho C^T c(x^k) \tag{3.25}$$

Therefore, the equality-constrained formulation is simply an augmented way of writing the NE in which high-confidence measurements become 'unsquared' (i.e., the product $C^T C$ is not carried out).

At the optimum, the Lagrange multipliers are given by

$$\lambda = -\rho\, c(\hat{x})$$

That is, the larger ρ the smaller the residuals of the virtual measurements, but their product tend to the Lagrange multiplier vector. This interpretation allows Lagrange multipliers to be handled in the same manner as residuals for bad data analysis (see Chapter 5).

3.6 Augmented Matrix Approach

Similar to the virtual measurements, regular measurement equations can be written as equality constraints if the associated residuals are retained as explicit variables. In this approach, the WLS problem can be restated as [3, 6]:

$$\begin{aligned} \text{minimize} \quad & J(x) = \frac{1}{2} r^T W r \\ \text{subject to} \quad & c(x) = 0 \\ & r - z + h(x) = 0 \end{aligned} \tag{3.26}$$

The resulting Lagrangian will have two sets of Lagrange multipliers:

$$\mathcal{L} = J(x) - \lambda^T c(x) - \mu^T (r - z + h(x)) \tag{3.27}$$

and the optimality conditions will be given by:

$$\begin{array}{lll} \partial \mathcal{L}(x)/\partial x = 0 \Rightarrow & C^T \lambda + H^T \mu & = 0 \\ \partial \mathcal{L}(x)/\partial \lambda = 0 \Rightarrow & c(x) & = 0 \\ \partial L(x)/\partial r = 0 \Rightarrow & W r - \mu & = 0 \\ \partial \mathcal{L}(x)/\partial \mu = 0 \Rightarrow & r - z + h(x) & = 0 \end{array} \tag{3.28}$$

The third equation allows r (or μ) to be eliminated ($r = R\mu$). Linearizing the remaining three, the following system of equations will be obtained:

$$\begin{bmatrix} R & H & 0 \\ H^T & 0 & C^T \\ 0 & C & 0 \end{bmatrix} \begin{bmatrix} \mu \\ \Delta x \\ \lambda \end{bmatrix} = \begin{bmatrix} \Delta z^k \\ 0 \\ -c(x^k) \end{bmatrix} \tag{3.29}$$

The coefficient matrix in Equation (3.29) is called the Hachtel's matrix. Note that, Equation (3.29) will become identical to (3.18) if μ is eliminated. Hence, this is the most primitive or augmented formulation and, according to the theory and examples discussed above, lower condition numbers are expected. On the other hand, since the Hachtel's matrix is very sparse, solving the above enlarged system is not particularly expensive in terms of arithmetic operations, but a more involved logic is needed to control and track the required row pivoting.

Besides, as in the case of Equation (3.18), the condition number of the Hachtel's matrix can be further improved if the residual weights are properly scaled. This is achieved simply by using αW instead of W in Equation (3.29):

$$\begin{bmatrix} \alpha^{-1} R & H & 0 \\ H^T & 0 & C^T \\ 0 & C & 0 \end{bmatrix} \begin{bmatrix} \mu_s \\ \Delta x \\ \lambda_s \end{bmatrix} = \begin{bmatrix} \Delta z^k \\ 0 \\ -c(x^k) \end{bmatrix} \tag{3.30}$$

where μ_s and λ_s are the scaled Lagrange multipliers.

Example 3.5:

The augmented matrix method will be illustrated with the help of the 3-bus system used so far (see example 3.4). As there are no virtual measurements in this case, the augmented matrix is simply:

$$\left[\begin{array}{cccccccc|c} 6.4e-5 & 0 & 0 & 0 & 0 & 0 & 0 & 0 & \\ 0 & 6.4e-5 & 0 & 0 & 0 & 0 & 0 & 0 & \\ 0 & 0 & 1e-4 & 0 & 0 & 0 & 0 & 0 & \\ 0 & 0 & 0 & 6.4e-5 & 0 & 0 & 0 & 0 & H \\ 0 & 0 & 0 & 0 & 6.4e-5 & 0 & 0 & 0 & \\ 0 & 0 & 0 & 0 & 0 & 1e-4 & 0 & 0 & \\ 0 & 0 & 0 & 0 & 0 & 0 & 1.6e-5 & 0 & \\ 0 & 0 & 0 & 0 & 0 & 0 & 0 & 1.6e-5 & \\ \hline & & & H^T & & & & & \mathbf{0}_{5\times 5} \end{array}\right]$$

and the right-hand-side vector:

$$\left[\begin{array}{c|ccccc} \Delta z^T & 0 & 0 & 0 & 0 & 0 \end{array} \right]^T$$

where the values of H and Δz are given in (3.10) and (3.11) respectively.

Solving this equation system provides, as expected, the same values for Δx as obtained before and the following Lagrange multipliers:

$$\mu = \left[\begin{array}{cccccccc} -47.84 & 16.25 & -30.07 & 257.3 & -124.4 & 196.8 & 395.2 & -395.2 \end{array} \right]^T$$

The condition number of the augmented matrix is $4.4e+6$. Such a high value, which compares badly in this case with that of the conventional gain matrix ($1.5e+3$), is due to the poor scaling of the elements of R with respect to those of H. Fortunately, using a scaling factor $\alpha^{-1} = 62500$, as indicated by equation (3.30), reduces the condition number of the Hachtel's matrix to 161. Note that $J(x)$ and the elements of μ would be scaled accordingly.

3.7 Blocked Formulation

The former sections describe two somewhat extreme cases from the point of view in which ordinary measurements are dealt with. While in one case the product $\tilde{H}^T\tilde{H}$ is formed for every measurement, the whole matrix H remains unsquared in the other.

Among the several possibilities lying in between, the one described below has attracted most interest [1, 8]. It is based on the following observations regarding injection measurements:

- Once virtual measurements are excluded from H and handled as equality constraints, injections are the main source of ill-conditioning in $\tilde{H}^T\tilde{H}$.

- Only injections give rise to second-neighbor adjacency in $\tilde{H}^T\tilde{H}$.

Consequently, the method first divides the set of measurements into two sets, namely the injection set, denoted by the index I, and the remaining measurements (flows and voltage magnitudes) denoted by F. The 'tableau' equation system becomes:

$$\left[\begin{array}{cc|c|c} R_F & 0 & H_F & 0 \\ 0 & R_I & H_I & 0 \\ \hline H_F^T & H_I^T & 0 & C^T \\ \hline 0 & 0 & C & 0 \end{array}\right] \left[\begin{array}{c} \mu_F \\ \mu_I \\ \hline \Delta x \\ \hline \lambda \end{array}\right] = \left[\begin{array}{c} \Delta z_F \\ \Delta z_I \\ \hline 0 \\ \hline -c(x_k) \end{array}\right]$$

Now, only μ_F is eliminated, yielding:

$$\begin{bmatrix} R_I & H_I & 0 \\ H_I^T & -H_F^T W_F H_F^T & C^T \\ 0 & C & 0 \end{bmatrix} \begin{bmatrix} \mu_I \\ \Delta x \\ \lambda \end{bmatrix} = \begin{bmatrix} \Delta z_I \\ -H_F^T W_F \Delta z_F \\ -c(x_k) \end{bmatrix} \tag{3.31}$$

(3.32)

The resulting model is a hybrid between Hachtel's and the conventional equality-constrained method, and comprises the following variables:

- μ_I: Multipliers associated with active-reactive injection measurements
- λ: Multipliers associated with active-reactive null injections
- x: Bus voltage magnitudes and phase angles

The key observation is that all those unknowns are exclusively related to buses (μ_I and λ refer to disjoint sets of nodes). Therefore, symmetrical row/column permutations can be carried out so that the variables corresponding to each node appear consecutively and can be arranged as a block. From the above development it is clear that the following block sizes are possible:

- 4x4 when injection or virtual measurements exist (2x2 for every subproblem of the decoupled formulation)
- 2x2 otherwise (1x1 in the decoupled formulation)

Since eliminating the set of measurements F does not cause second-neighbor elements, it follows that the resulting blocked matrix has the same topology as the bus admittance matrix, except for irrelevant branches which are missing in any formulation. Hence, solving the resulting equation system by means of block arithmetic would be rather simple if the diagonal blocks were always appropriate pivots because, in such a case, the conventional minimum degree ordering adopted in the load flow problem and other routines could be applied. Unfortunately, there are situations in which, in spite of the network being observable, a block pivot may be singular when the block ordering is performed exclusively to preserve sparsity. As will be clear from the following example, this happens when a bus is made observable by means of an adjacent injection whose block is eliminated later, and requires that the measured bus be reordered before the singular one to avoid the singularity. In more complex cases, several adjacent buses may be involved in this problem by the 'domino' effect, which makes the block-based approach less elegant and simple than expected. An alternative to bus reordering consists of adding a small number to the appropriate diagonal of the critical block to make it non-singular, but then numerical observability analysis is complicated (see Chapter 4).

Example 3.6:

Consider the 6-bus system shown in figure 3.6, comprising 1 voltage magnitude, 4 power flows and 3 power injections (for our purposes it is not important if they are actual or virtual measurements). For simplicity, only the reactive subproblem will be pursued.

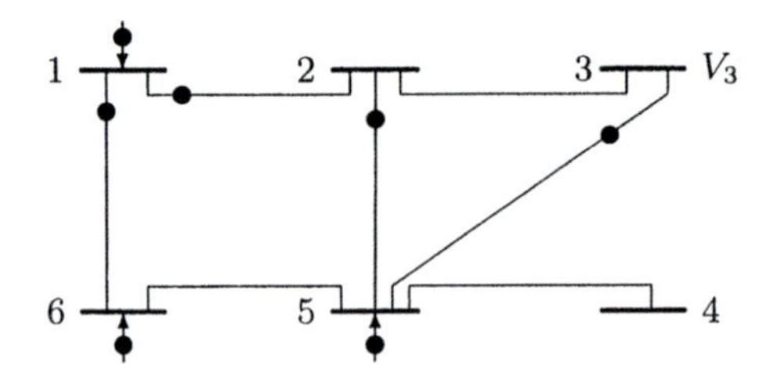

Figure 3.4. 6-bus power system to illustrate the block method

The Jacobian for this example has the following structure:

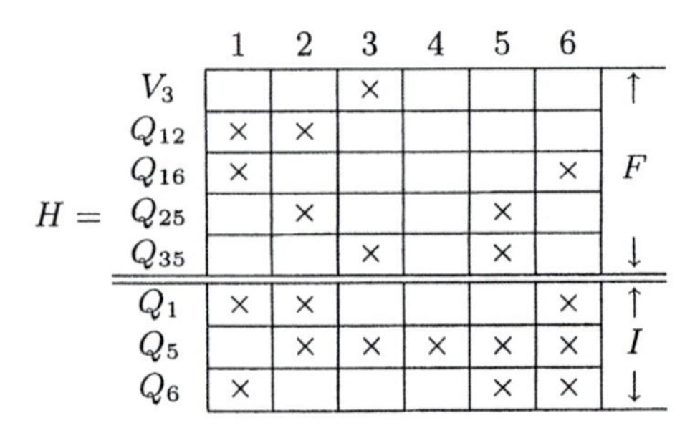

$H =$

	1	2	3	4	5	6	
V_3			×				↑
Q_{12}	×	×					
Q_{16}	×					×	F
Q_{25}		×			×		
Q_{35}			×		×		↓
Q_1	×	×				×	↑
Q_5		×	×	×	×	×	I
Q_6	×				×	×	↓

and the augmented matrix becomes

						1	5	6	1	2	3	4	5	6
	×										×			
		×							×	×				
			×						×					×
				×						×			×	
					×						×		×	
1						×			×	×				×
5							×			×	×	×	×	×
6								×	×				×	×
1		×	×			×		×						
2		×		×		×	×							
3	×				×		×							
4							×							
5				×	×		×	×						
6			×			×	×	×						

Following the steps explained above, the variables labeled with F are eliminated, yielding (the matrices corresponding to the null-injection case are shown in parentheses):

	1	5	6	1	2	3	4	5	6	
	×			×	×				×	
$W_I(0) \rightarrow$		×			×	×	×	×	×	$\leftarrow H_I(C)$
			×	×				×	×	
	×		×	•	•				•	
	×	×		•	•			•		
$H_I^T(C^T) \rightarrow$		×				•		•		$\leftarrow H_F^T W H_F$
		×								
		×	×		•	•		•		
		×	×	•					•	

As expected, the fill-in elements, represented by dots, appear only among adjacent buses. Rearranging by buses we get,

	1	1	2	3	4	5	5	6	6
1	×	×	×						×
1	×	•	•					×	•
2	×	•	•			×	•		
3				•		×	•		
4						×			
5			×	×	×	×	×		×
5			•	•		×		×	
6		×					×	×	×
6	×	•				×		×	•

whose block structure is:

	1	2	3	4	5	6
1	$\times$	$\times$				$\times$
2	$\times$	$\times$	$\square$		$\times$	
3		$\square$	$\times$		$\times$	
4				$\square$	$\times$	
5		$\times$	$\times$	$\times$	$\times$	$\times$
6	$\times$				$\times$	$\times$

The symbol $\square$ has been used above to denote null blocks which are non-null in the bus admittance matrix. The elements (2,3) and (3,2) correspond to an irrelevant branch while the null block (4,4) refers to a non-measured bus (these are the only structural differences between both matrices). The null diagonal block becomes a valid pivot only if bus 5 is eliminated beforehand. Note that node 4 would lead node 5 if the minimum degree criterion was adopted, but this may not be the case in general.

3.8 Comparison of Techniques

In this section we will first resort to the 6-bus system of example 3.6 to compare the robustness of several formulations in the presence of adverse factors, like short lines, virtual measurements, etc. As usually, only the DC state estimator will be dealt with. Then, larger benchmark networks will be used to assess the respective computational efforts.

For the first set of experiments all branches of the 6-bus system are assumed to be identical ($b_{ij} = 3$ pu), and all measurement weights are set to 1. Bus 4 is the slack bus. The following cases are considered:

(a): 7 Power flow measurements (all branches), no injections.

(b): Case (a) plus regular injection measurements at buses 2 and 6.

(c): Same as (b) except for injection 6 being null (in the NE formulation its weight is set to 1000).

(d): Same as (b) except for line 1-2 being 100 times shorter ($b_{12} = 300$ pu).

Table 3.1 shows, from left to right, the resulting condition numbers for the gain matrix of the NE, the intermediate matrix used by the blocked method (power flow measurements 'squared') and the Hachtel's matrix. It can be observed that the NE approach is significantly affected by the presence of null injections. It is also noticeable the robustness of the Hachtel's method against short lines, which is lost when power flows are 'squared'.

As all weights are 1 in this case, there is no need to normalize them with the scaling factor α introduced in the former sections.

Case	NE	Sq. Flows	Hachtel
(a) 7 power flows	48.4	48.4	20
(b) ordinary injec.	165.8	54.6	38.2
(c) null injec.	5.5e+4	57	47
(d) short line	2.6e+5	1.4e+5	1200

Table 3.1. Condition numbers for different formulations

It is worth mentioning also that the choice of the slack bus affects to some extent the condition number. For instance, for case (b), the condition number of the gain matrix ranges from 21.5 (when bus 2 is the slack) to 165.8 (when bus 4 is the slack). The Hachtel's matrix is again less sensitive to such a choice (29.5 and 38.2 for slack buses 6 and 4 respectively).

The reader should realize, however, that in mild cases, like (a) or (b), in which all line reactances and measurement weights are of the same order of magnitude, the NE's performance is quite good.

The next comparison refers to the number of flops per iteration. For this purpose, three IEEE test networks, comprising 57, 118 and 300 buses respectively, have been tested. Table 3.2 shows the results for the NE, the hybrid orthogonal method and the conventional augmented method when all power flows, but none injection, are measured. The figures shown do not include the operations required to obtain H and Δz, which are common to all formulations. Sparsity has been preserved in all cases by means of appropriate factorization techniques preceded by row/column ordering [10, 7, 6].

Network	NE	Hybrid QR	Hachtel
57	3605	3976	4431
118	8250	10634	10253
300	21265	28048	26337

Table 3.2. Flops per iteration for different formulations (only power flows)

From this table it can be concluded that the NE approach is less expensive than its competitors, although the differences are not so significant to be determinant [5, 12]. Note also that the computational effort grows almost linearly with the system size.

In order to assess the influence of injection measurements, the same experiments are repeated by adding all injections to the existing power flows. The results are shown in table 3.3.

Network	NE	Hybrid QR	Hachtel
57	13564	26822	13605
118	33166	62178	29388
300	89693	176034	82578

Table 3.3. Flops per iteration for different formulations (power flows plus injections)

Comparing this table with the former one, it can be concluded that the less sparse rows in the Jacobian corresponding to injection measurements significantly affect the computational costs. This time, in spite of its larger dimension, the Hachtel's equation system becomes the cheapest approach for realistic networks. This is due both to the second-neighbor fill-ins created by injections in the gain matrix of the NE and to the fact that no operations are involved in computing the right-hand side of the Hachtel's equation system. On the other hand, the increased number of Givens rotations required by the hybrid QR method makes it the most expensive, albeit not dramatically.

Even though the number of injections is lower than the number of power flows, the cost per iteration is 3 times as much as in the previous case for the Hachtel's method, about 4 times for the NE and over 6 times for the QR method. This means that the augmented approach is less sensitive to the presence of injection measurements.

3.9 Problems

1. The three lines of the 3-bus system shown in the figure have the same reactance, X. Considering for simplicity the DC state estimation problem, perform the following analyses:

   a) Assuming there are three, equally weighted power flow measurements, obtain the gain matrix in terms of X and compute its condition number by means of an appropriate computer tool.

   b) Repeat a) by successively adding one, two and three power injection measurements to the three power flows of the base case. Obtain the gain matrix in terms of the weight assigned to the injection measurements relative to the power flow weight. If ρ denotes this relative weight, compute the condition number for $\rho = 0.1$, $\rho = 1$ and $\rho = 10$. Draw your own conclusions regarding the influence of injection measurements on the NE ill-conditioning.

2. Same as problem 1, except that the length of line 2-3 is k times that of the remaining lines. Analyze the cases $k = 0.1$, $k = 10$ and $k = 25$.

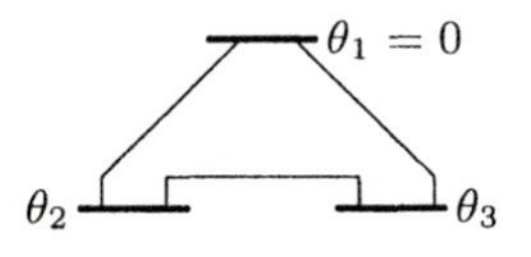

Figure 3.5. 3-bus system for problems 1, 2 and 3

3. In problem 1, case b), compute the condition number of the gain matrix for $\rho = 1000$ when only injection 3 is measured. Such a high value of ρ can be used to model perfect measurements, like null injections. Obtain also the coefficient matrix, and its condition number, when the injection 3 is modeled as an equality constraint. Do the same with Hachtel's augmented matrix.

4. Perform the LU factorization of matrix H for case a) of problem 1. Then obtain the condition number of the matrix $L^T L$ used by the Peters-Wilkinson method and compare this value with that obtained for the conventional matrix.

5. Consider the Hachtel's matrix of problem 3. Obtain its condition number for different values of the scaling factor α in Equation (3.30). Try to obtain, in terms of X, the value of α that minimizes the condition number (trial-and-error procedure).

6. From the Jacobian of example 3.6, determine the "sparse" structure of the gain matrix. Compare its number of non-zero elements with those of the alternative matrices developed in that example. Repeat the analysis when injection 5 is substituted by the power flow 5-4.

References

[1] Alvarado F., Tinney W., "State Estimation Using Augmented Blocked Matrices". *IEEE Transactions on Power Systems*, Vol. 5(3), pp. 911-921, August 1990.

[2] Aschmoneit F., Peterson N., Adrian E., "State Estimation with Equality Constraints". 10th PICA Conference Proceedings, Toronto, pp. 427-430, May 1977.

[3] Gjelsvik A., Aam S., Holten L., "Hachtel's Augmented Matrix Method - A Rapid Method Improving Numerical Stability in Power System Static State Estimation". *IEEE Transactions on Power Apparatus and Systems*, Vol. PAS-104, pp. 2987-2993, November 1985.

[4] Gu J., Clements K., Krumpholz G., Davis P., "The Solution of Ill-conditioned Power System State Estimation Problems Via the Method of Peters and Wilkinson". PICA Conference Proceedings, Houston, pp. 239-246, May 1983.

[5] Holten L., Gjelsvik A., Aam S., Wu F., Liu W., "Comparison of Different Methods for State Estimation". *IEEE Transactions on Power Systems*, Vol. 3, pp. 1798-1806, November 1988.

[6] Liu W., Wu F., Holten L., Gjelsvik A., Aam S., "Computational Issues in the Hachtel's Augmented Matrix Method for Power System State Estimation". Proceedings 9th. Power System Computation Conference, Lisbon, September 1987.

[7] Monticelli A., Murari C., Wu F., "A Hybrid State Estimator: Solving Normal Equations by Orthogonal Transformations". *IEEE Transactions on Power Apparatus and Systems*, Vol. PAS-105(2), pp. 3460-3468, December 1985.

[8] Nucera R., Gilles M., "A Blocked Sparse Matrix Formulation for the Solution of Equality-Constrained State Estimation". *IEEE Transactions on Power Systems*, Vol. 6(1), pp. 214-224, February 1991.

[9] Simes-Costa A., Quintana V., "A Robust Numerical Technique for Power System State Estimation". *IEEE Transactions on Power Apparatus and Systems*, Vol. PAS-100, pp. 691-698, February 1981.

[10] Vempati M., Slutsker I., Tinney W., "Enhancements to Givens Rotations for Power System State Estimation". *IEEE Transactions on Power Systems*, Vol. 6(2), pp. 842-849, May 1991.

[11] Wang J., Quintana V., "A Decoupled Orthogonal Row Processing Algorithm for Power System State Estimation". *IEEE Transactions on Power Apparatus and Systems*, Vol. PAS-103, pp. 2337-2344, August 1984.

[12] Wu F.F., "Power System State Estimation". *Electrical Power & Energy Systems*, Vol. 12(2), pp. 80-87, April 1990.

Chapter 4

Network Observability Analysis

Power system state estimator uses the set of available measurements in order to estimate the system state. Given a set of measurements and their locations, the network observability analysis will determine if a unique estimate can be found for the system state. This analysis may be carried out off-line during the initial phase of a state estimator installation, in order to check the adequacy of the existing measurement configuration. If the system is found not to be observable, then additional meters may have to be placed at particular locations. Observability analysis is also done on-line, prior to running the state estimator. It ensures that a state estimate can be obtained using the set of measurements received at the last measurement scan. Telecommunication errors, topology changes or meter failures may occasionally lead to cases where the state of the entire system can not be estimated. Then, the system will contain several isolated observable islands, each one having its own phase angle reference that is independent of the rest. Network observability analysis allows detection of such cases and identifies all the existing observable islands prior to the execution of the state estimator.

Observability of a given network is determined by the type and location of the available measurements as well as by the topology of the network. Thus, the analysis of network observability utilizes the graph theory as it relates to networks, their associated equations and solutions. In this chapter, a brief review of networks and graphs, their related matrices and equations will be given. This will be followed by the presentation of the methods used for analysis of network observability.

4.1 Networks and Graphs

4.1.1 Graphs

A graph is defined by a set of *nodes* $\mathcal{N}$ and a set of *edges* $\mathcal{E}$, and is denoted by

$$G = \{\mathcal{N}, \mathcal{E}\}$$

where each edge has two distinct terminal nodes. A graph is said to be fully connected if any node can be reached from any other node by tracing the edges of the graph. A *directed graph* is a graph where all edges are assigned directions. In a directed graph, the terminal nodes of each edge are designated as sending-end and receiving-end nodes. The direction of the edge is identified by an arrow from the sending-end node towards the receiving-end node of that edge.

A *tree* of a graph is defined as a set of connected edges which does not form any loops. If any node in the graph can be reached from any other node by tracing only the edges of the tree, then it will be called a *spanning tree* of the graph.

All edges that do not belong to the spanning tree, are called *links*. A network with N nodes and L edges, will have a spanning tree with $(N-1)$ edges and there will be $(L-N+1)$ links associated with this spanning tree. While the number of nodes and edges in a spanning tree are predetermined, the set of edges that form a tree, is in general not unique.

Example 4.1:

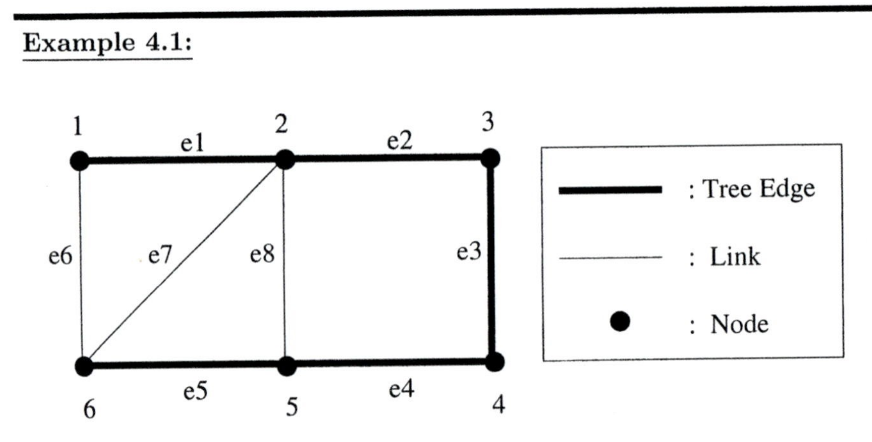

Figure 4.1. A network graph with 6 nodes

Consider the network graph given in Figure 4.1 where $N = 6$ and $L = 8$. Thus, its spanning tree will have $N - 1 = 5$ nodes and one possible choice will include the following edges:

$$T = \{e1, e2, e3, e4, e5\}$$

where e_i represents the edge number i. The graph will then have the following $(L - N + 1) = 3$ links associated with this tree:

$$\{e6, e7, e8\}$$

Note that each link forms a loop with the tree branches where no other links are present. Therefore, each link is said to belong to a *fundamental loop* of the network graph and there will be exactly $(L - N + 1)$ such fundamental loops for a given graph. The fundamental loops associated with the graph of Figure 4.1 for the chosen tree T, will be the following:

$$\ell_1 = \{e_1, e_2, e_3, e_4, e_5, e_6\} \tag{4.1}$$

$$\ell_2 = \{e_2, e_3, e_4, e_5, e_7\} \tag{4.2}$$

$$\ell_3 = \{e_2, e_3, e_4, e_8\} \tag{4.3}$$

These fundamental loops correspond to the 3 links identified in Example 4.1 above.

4.1.2 Networks

An electric network contains a collection of branches and buses where each branch has two terminal buses which may be shared by one or more other branches in the same network. If one of the terminal buses of a branch is grounded, then it will be referred to as a *shunt branch.* Branches with ungrounded terminal buses will be called *series branches.* Each electric network has a corresponding graph where the branches and buses are replaced by edges and nodes respectively.

Each branch k has an associated impedance z_k. Furthermore, if branch j and k are magnetically coupled, then there will be an associated mutual coupling impedance z_{jk} relating their terminal voltages and branch currents.

4.2 Network Matrices

For a given electric network containing L branches, the vectors of branch currents I_b and branch voltages V_b will be related through the *primitive branch impedance matrix* Z_p:

$$\underbrace{\begin{bmatrix} z_{11} & z_{12} & \cdots & z_{1L} \\ z_{21} & z_{22} & \cdots & z_{2L} \\ \vdots & \vdots & \vdots & \vdots \\ z_{L1} & z_{L2} & \cdots & z_{LL} \end{bmatrix}}_{Z_p} \underbrace{\begin{bmatrix} i_1 \\ i_2 \\ \vdots \\ i_L \end{bmatrix}}_{I_b} = \underbrace{\begin{bmatrix} v_1 \\ v_2 \\ \vdots \\ v_L \end{bmatrix}}_{V_b} \tag{4.4}$$

where z_{kk} is the self impedance of branch k and z_{km} is the mutual impedance between branches k and m. The inverse of the matrix Z_p is called the *primitive branch admittance matrix* and denoted by $Y_p = Z_p^{-1}$.

Example 4.2:

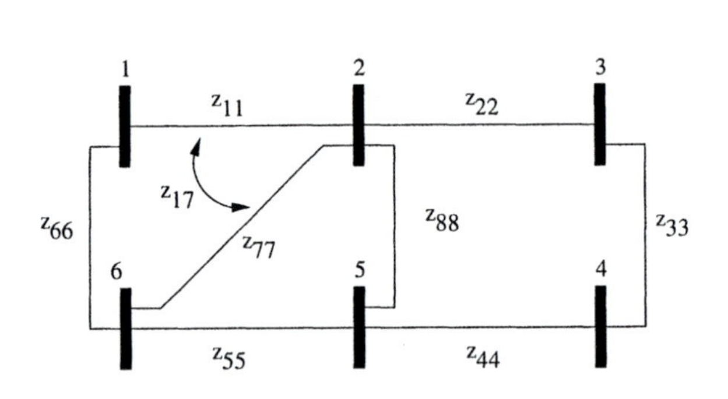

Figure 4.2. A network with 6 buses and 8 branches

Consider the electric network shown in Figure 4.2. The graph for this network is the same as given in Figure 4.1. Assume that each branch has the identical self impedance of $j1.0$ per unit, and the branches 1 and 7 are mutually coupled via a mutual impedance of $j0.2$ per unit. Then, the 8×8 primitive branch impedance matrix Z_p can be formed as follows:

$$Z_p = \begin{bmatrix} j1 & 0 & 0 & 0 & 0 & 0 & j0.2 & 0 \\ 0 & j1 & 0 & 0 & 0 & 0 & 0 & 0 \\ 0 & 0 & j1 & 0 & 0 & 0 & 0 & 0 \\ 0 & 0 & 0 & j1 & 0 & 0 & 0 & 0 \\ 0 & 0 & 0 & 0 & j1 & 0 & 0 & 0 \\ 0 & 0 & 0 & 0 & 0 & j1 & 0 & 0 \\ j0.2 & 0 & 0 & 0 & 0 & 0 & j1 & 0 \\ 0 & 0 & 0 & 0 & 0 & 0 & 0 & j1 \end{bmatrix}$$

4.2.1 Branch to Bus Incidence Matrix

Each network branch will be incident to two buses unless it is a shunt branch connecting a single bus to ground. An arbitrary direction can be assigned to each branch by designating its terminal buses as *sending* and *receiving* end terminals of the branch. The corresponding network graph will then become a directed graph. Thus, a branch-bus incidence matrix **A** can be defined based on these chosen branch directions as follows:

$$A(i,j) = \begin{cases} 1 & \text{if bus } j \text{ is the sending terminal of branch } i \\ -1 & \text{if bus } j \text{ is the receiving terminal of branch } i \\ 0 & \text{otherwise} \end{cases}$$

Each row of **A** contains two nonzero entries and they add up to zero. Thus, any one of the columns of **A** can be removed and later recovered without loss of information. The matrix formed by deleting any one of the columns of **A** is called the reduced incidence matrix and denoted by A_r.

In a fully connected network with N nodes, $(N-1)$ rows of A_r will be linearly independent. Furthermore, these $(N-1)$ rows will be linearly independent if and only if the corresponding branches form a spanning tree of the network. Assuming that a spanning tree is chosen, then the rows of A_r can be reordered and partitioned as:

$$A_r = \begin{bmatrix} A_{rT} \\ A_{rL} \end{bmatrix}$$

where the rows of A_{rT} and A_{rL} correspond to the spanning tree branches and the links respectively. The matrix A_{rT} is a $(N-1) \times (N-1)$ non-singular square matrix.

4.2.2 Fundamental Loop to Branch Incidence Matrix

As shown in section 4.1.1, each fundamental loop will include several tree branches and a single link. Each loop is assigned a direction (clockwise or counter-clockwise), which matches the direction of the corresponding link. Using the chosen direction for the branches, a fundamental loop to branch incidence matrix $\mathcal{L}$ can be defined as follows:

$$\mathcal{L}(i,j) = \begin{cases} 1 & \text{if branch } j \text{ is in loop } i \text{ and has the same direction} \\ -1 & \text{if branch } j \text{ is in loop } i \text{ and has the opposite direction} \\ 0 & \text{otherwise} \end{cases}$$

The columns of $\mathcal{L}$ can be reordered and partitioned as:

$$\mathcal{L} = \begin{bmatrix} \mathcal{L}_T & \vdots & I \end{bmatrix}$$

where the columns of $\mathcal{L}_T$ and the identity matrix I correspond to the $(N-1)$ spanning tree branches and the $(L-N+1)$ links associated with the fundamental loops, respectively.

Using the orthogonality of $\mathcal{L}$ and A_r, the following can be derived:

$$\begin{aligned} \mathcal{L} \cdot A_r &= 0 \\ \mathcal{L}_T \cdot A_{rT} + A_{rL} &= 0 \\ \mathcal{L}_T &= -A_{rL} \cdot A_{rT}^{-1} \end{aligned}$$

This provides a simple way to build $\mathcal{L}$. In practice, the inverse of A_{rT} is never explicitly calculated, since efficient methods exist for obtaining this inverse [1]. Note also that the sparsity structure of $\mathcal{L}$ depends upon the topology of the network as well as the choice of the spanning tree (see Problem 3 at the end of the chapter).

Example 4.3:

Consider the same network graph of Example 4.1. Figure 4.3 shows the arbitrarily assigned directions for the branches of this graph.

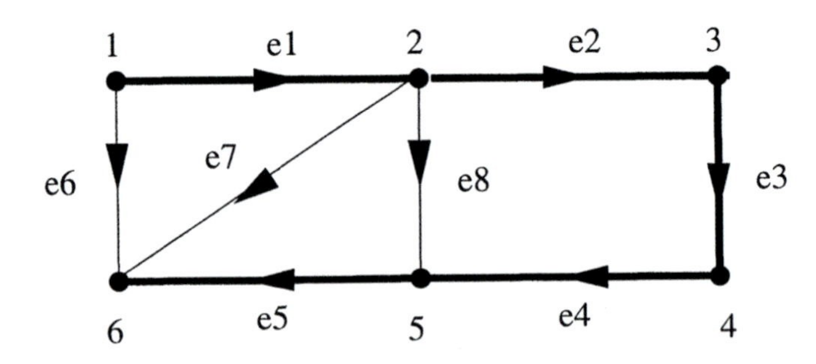

Figure 4.3. Arbitrarily assigned branch directions

Using this directed graph, a branch to bus incidence matrix A can be formed as below:

$$A = \begin{array}{c} \\ e1 \\ e2 \\ e3 \\ e4 \\ e5 \\ e6 \\ e7 \\ e8 \end{array} \begin{array}{c} \begin{array}{cccccc} 1 & 2 & 3 & 4 & 5 & 6 \end{array} \\ \left[\begin{array}{cccccc} 1 & -1 & & & & \\ & 1 & -1 & & & \\ & & 1 & -1 & & \\ & & & 1 & -1 & \\ & & & & 1 & -1 \\ 1 & & & & & -1 \\ & 1 & & & & -1 \\ & 1 & & & -1 & \end{array} \right] \end{array}$$

Furthermore, a fundamental loop to branch incidence matrix $\mathcal{L}$ corresponding to the tree in Figure 4.1 can also be built using the assigned directions for the 3 links, as follows:

$$\mathcal{L} = \begin{array}{c} \\ e6 \\ e7 \\ e8 \end{array} \begin{array}{c} \begin{array}{cccccccc} e1 & e2 & e3 & e4 & e5 & e6 & e7 & e8 \end{array} \\ \left[\begin{array}{cccccccc} -1 & -1 & -1 & -1 & -1 & 1 & & \\ & -1 & -1 & -1 & -1 & & 1 & \\ & -1 & -1 & -1 & & & & 1 \end{array} \right] \end{array}$$

Alternatively, $\mathcal{L}$ can be built by using A. Eliminating the last column of A and partitioning the rows, the reduced incidence matrix A_r is obtained as:

$$A_r = \begin{bmatrix} 1 & -1 & 0 & 0 & 0 \\ 0 & 1 & -1 & 0 & 0 \\ 0 & 0 & 1 & -1 & 0 \\ 0 & 0 & 0 & 1 & -1 \\ 0 & 0 & 0 & 0 & 1 \\ . & . & . & . & . \\ 1 & 0 & 0 & 0 & 0 \\ 0 & 1 & 0 & 0 & 0 \\ 0 & 1 & 0 & 0 & -1 \end{bmatrix} = \begin{bmatrix} A_{rT} \\ \cdots \\ A_{rL} \end{bmatrix}$$

Then, the fundamental loop to branch incidence matrix $\mathcal{L}$ will be given by:

$$\begin{aligned} \mathcal{L} &= \begin{bmatrix} -A_{rL} \cdot A_{rT}^{-1} & \vdots & I_{3\times 3} \end{bmatrix} \\ &= \begin{bmatrix} -1 & -1 & -1 & -1 & -1 & 1 & & \\ & -1 & -1 & -1 & -1 & & 1 & \\ & -1 & -1 & -1 & & & & 1 \end{bmatrix} \end{aligned}$$

4.3 Loop Equations

According to the Kirchhoff's Voltage Law, the sum of the branch voltages in a given loop should add up to zero. Given a spanning tree and the corresponding links, a linearly independent set of fundamental loop equations can be written in compact form as:

$$\mathcal{L} \cdot V_b = 0 \tag{4.5}$$

where:
$\mathcal{L}$: is the $(L - N + 1) \times L$ fundamental loop to branch incidence matrix
V_b : is the branch voltage vector.

Eq.(4.5) merely states the Kirchhoff's Voltage Law for the given network and therefore is not enough for the solution of the unknown voltages. These equations will be augmented by the measurement equations satisfying the Kirchhoff's Current Law in order to build the complete set of network equations. Measurement equations will be described in the following sections.

Note that the polarities of the branch voltages should be consistent with the assigned branch directions in building $\mathcal{L}$. Since $V_b = AV$, the two incidence matrices A and $\mathcal{L}$ should be built using the same branch direction assignments. The use of loop equations in power system analysis is discussed in more detail in [2].

4.4 Methods of Observability Analysis

Observability analysis can be carried out using the fully coupled or decoupled measurement equations. The analysis is performed on the linearized measurement model without loss of generality. Use of fully coupled model has its drawbacks, one of which is the non-uniqueness of the solution. This can be illustrated by considering a simple case of a single line, whose reactive power flow is measured at one end, along with the voltage magnitude at both end buses as shown in Figure 4.4. If the line impedance is $j0.2$ p.u., and the measurements are:

$$V_1 = 1.00 \ \text{p.u.}, \quad V_2 = 0.99 \ \text{p.u.}, \quad Q_{12} = -0.80 \ \text{p.u.}$$

then the state variable θ_2 can be estimated by solving the following equation:

$$Q_{12} = \frac{V_1 V_2}{X} \cos\theta_2 - \frac{V_1^2}{X} \tag{4.6}$$

$$-0.8 = 49.5 \cos\theta_2 - 50 \tag{4.7}$$

Equation (4.7) will be satisfied by $\theta_2 = \pm 6.31$ degrees, both of which are equally likely solutions.

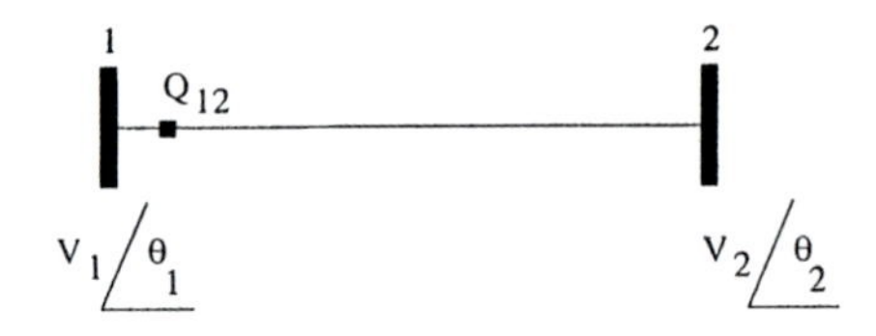

Figure 4.4. Two bus system and measurements

This example illustrates the risk involved in attempting to estimate θ using Q (or V using P) measurements. Network observability analysis methods commonly assume paired P, Q measurements and make use of the decoupled measurement model, in order to avoid such cases. Observability of θ based only on the P measurements, is analyzed using the DC power flow equations. This approach however is restricted to cases where decoupling of the measurement equations can be justified. For instance, in the presence of current magnitude measurements, the corresponding measurement equations can no longer be decoupled based on the same decoupling assumptions [3]. Hence, a coupled model has to be used along with further methods of detecting possible multiple solutions. A detailed analysis of such cases will be given in Chapter 9.

Network observability analysis can be performed using either numerical or topological approaches. Topological approaches use the decoupled measurement model and graph theory. Numerical approaches may use fully coupled or decoupled models. They are based on the numerical factorization of the measurement Jacobian or the gain matrices. These methods are formulated using either branch or nodal variables.

4.5 Numerical Method Based on the Branch Variable Formulation

This section describes a general numerical method which is developed based on branch variables [4]. Description of the new branch variables will be given first, followed by the formulation of the network observability problem using these variables.

4.5.1 New Branch Variables

Assume that branch j is connected between the sending-end bus k and the receiving-end bus m as shown in Figure 4.5.

Let the branch voltage and phase angle variables α_j and δ_j be defined as follows:

$$\alpha_j = \ln \frac{V_k}{V_m} = \ln V_k - \ln V_m \quad (4.8)$$

$$\delta_j = \theta_k - \theta_m \quad (4.9)$$

where:
V_k, V_m are the magnitude of the voltages at bus k and m.
θ_k, θ_m are the phase angles of the voltages at bus k and m. In compact

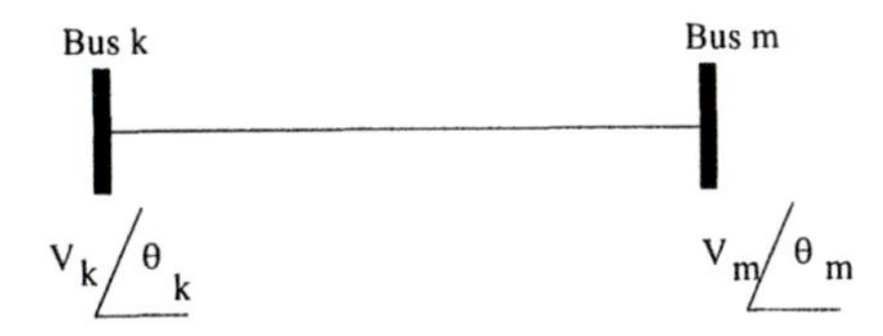

Figure 4.5. Branch j

form, the branch variable vectors X_α, X_δ can be expressed as:

$$X_\alpha = [\alpha_1 \alpha_2 \dots \alpha_L]^T = A \cdot \ln \mid V \mid \quad (4.10)$$

$$X_\delta = [\delta_1 \delta_2 \dots \delta_L]^T = A \cdot \theta \quad (4.11)$$

where:
$\mid V \mid = [V_1 V_2 \dots V_N]^T$, vector of bus voltage magnitudes
$\theta = [\theta_1 \theta_2 \dots \theta_N]^T$, vector of bus voltage phase angles
A = branch-bus incidence matrix.

4.5.2 Measurement Equations

All types of measurements can be expressed as a function of the above defined branch variables α_j, δ_j. The expressions for different measurements will be derived first, followed by their linearized analysis.

Power Flows

Real and reactive power flows along branch j of Figure 4.5 can be expressed as follows:

$$P_{km} = V_k^2 g_{km} - V_k V_m (g_{km} \cos\delta_j + b_{km} \sin\delta_j) \quad (4.12)$$

$$Q_{km} = -V_k^2 b_{km} - V_k V_m (g_{km} \sin\delta_j - b_{km} \cos\delta_j) \quad (4.13)$$

where:
$g_{km} + jb_{km}$: admittance of branch j, neglecting the line charging susceptance of the branch.
V_k : voltage magnitude at bus k,
δ_j : $\theta_k - \theta_m$, branch phase angle variable,
θ_k : voltage phase angle at bus k.

Consider the modified power flow equations where they are scaled by the squared voltage magnitude at the measured bus k:

$$P_{km}^s = \frac{P_{km}}{V_k^2} = g_{km}(1 - \frac{V_m}{V_k}\cos\delta_j) - b_{km}\frac{V_m}{V_k}\sin\delta_j \quad (4.14)$$

$$= \frac{Q_{km}}{V_k^2} = -b_{km}(1 - \frac{V_m}{V_k}\cos\delta_{km}) - g_{km}\frac{V_m}{V_k}\sin\delta_{km} \quad (4.15)$$

Using the new branch voltage variable α_j given by Equation (4.8):

$$\alpha_j = \ln\frac{V_k}{V_m}, \quad \text{or} \quad \frac{V_k}{V_m} = e^{\alpha_j} \quad (4.16)$$

Equations (4.14) and (4.15) can be expressed in terms of the branch voltage and branch phase angle variables as:

$$P_{km}^s = g_{km}(1 - e^{\alpha_j}\cos\delta_j) - b_{km}e^{\alpha_j}\sin\delta_j \quad (4.17)$$

$$Q_{km}^s = -b_{km}(1 - e^{\alpha_j}\cos\delta_j) - g_{km}e^{\alpha_j}\sin\delta_j \quad (4.18)$$

Power Injections

Scaled power injections can be expressed as linear combinations of scaled power flows that are originating from the bus where the injection is measured. Therefore, scaled power injections can be similarly expressed in terms of α_j and δ_j variables.

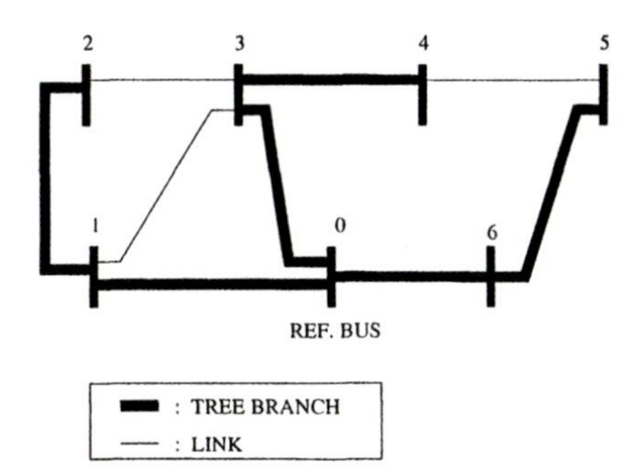

Figure 4.6. Illustration of rooted tree structure.

Voltage Magnitudes

Let us assume that there is at least one voltage magnitude measurement in the system. Let us choose one of these measured buses as the reference bus. Then, a network tree that is rooted at this reference bus, can be constructed and denoted by T. The ratio of the voltage magnitude at any bus to the voltage magnitude of the reference bus (V_{ref}) can then be expressed as:

$$V_k^s = \frac{V_k}{V_{ref}} = \frac{V_k}{V_m} \cdot \frac{V_m}{V_\ell} \cdots \frac{V_j}{V_{ref}} \tag{4.19}$$

$$= e^{\alpha_{j1}} \cdot e^{\alpha_{j2}} \cdots e^{\alpha_{jr}} \tag{4.20}$$

Taking the logarithm of both sides:

$$\ln V_k^s = \alpha_{j1} + \alpha_{j2} + \cdots \alpha_{jr} \tag{4.21}$$

where $\{j1, j2, ..., jr\}$ form the set of branches all of which belong to the tree T rooted at the reference bus and form a continuous path from bus k to the reference bus ref. This is illustrated in Figure 4.6.

Current Magnitudes

Square of the magnitude of line current through branch j connecting bus k to bus m can be expressed as:

$$I_{km}^2 = \frac{P_{km}^2 + Q_{km}^2}{V_k^2} = V_k^2[(P_{km}^s)^2 + (Q_{km}^s)^2] \tag{4.22}$$

Then, the square of the scaled line current magnitude will be:

$$(I_{km}^s)^2 = (\frac{I_{km}}{V_k})^2 = (P_{km}^s)^2 + (Q_{km}^s)^2 \tag{4.23}$$

Loop Equations

Note that, in a loop formed by the branches $j1, j2, \ldots, jL$, the branch variables will satisfy the following equations:

$$\alpha_{j1} + \alpha_{j2} + \cdots + \alpha_{jL} = 0 \tag{4.24}$$

$$\delta_{j1} + \delta_{j2} + \cdots + \delta_{jL} = 0 \tag{4.25}$$

Hence, the fundamental loop equations derived in Equation (4.5) will also hold true if the branch voltage vector V_b is replaced by either of the branch voltage magnitude difference vector X_α or branch voltage phase angle difference vector X_δ, i.e.

$$\mathcal{L} \cdot \mathcal{X}_\delta = 0 \tag{4.26}$$

$$\mathcal{L} \cdot \mathcal{X}_\alpha = 0 \tag{4.27}$$

In Equations (4.26) and (4.27), the variables corresponding to the links can be expressed in terms of the rest of the tree branch variables. Hence, a model reduction is possible by eliminating the link variables from the rest of the measurement equations. However, despite the reduction in the number of variables, this may not necessarily yield a numerically more efficient solution due to the loss of sparsity in the measurement equations.

4.5.3 Linearized Measurement Model

The above described equations for the power injections, flows, voltage and current magnitudes, and loops can be expressed in compact form as a non-linear vector equation:

$$f(X) = z \tag{4.28}$$

where:
$X^T = [\delta^T, \alpha^T]$
z : Scaled measurement vector including zeros for the loop equations. Measurements are assumed to be error free.
δ : Vector of branch phase angle variables.
α : Vector of branch voltage variables.

Note the assumption in Equation (4.28) that the measurements are free of errors. This is justified, since the measurement errors have no effect on the observability analysis of the network.

First-order Taylor approximation of Equation (4.28) yields:

$$H \cdot \Delta x = z - f(X^0) = \Delta z \tag{4.29}$$

where:
$H = \frac{\partial f(X)}{\partial X}$, evaluated at some X^0.
$\Delta X = X - X^0$.

In order to ensure that real (reactive) power measurements are not used to observe voltage (phase angle) variables, power flow and injection measurements are assumed to come in pairs. Also, in the above Equation (4.29), those terms coupling real (reactive) measurements to the voltage (phase angle) variables in matrix H, are neglected. Hence, the only type of measurements whose rows contain nonzero entries in both phase angle and voltage variable columns, will be the ampere measurements.

Elements of H

The measurement jacobian H is built by evaluating the following expressions corresponding to the first derivatives of the measurement and loop equations with respect to the branch variables.

Power flows

$$\frac{\partial P^s_{km}}{\partial \alpha_j} = -g_{km}e^{\alpha_j}\cos\delta_j - b_{km}e^{\alpha_j}\sin\delta_j \approx 0 \tag{4.30}$$

$$\frac{\partial P^s_{km}}{\partial \delta_j} = g_{km}e^{\alpha_j}\sin\delta_j - b_{km}e^{\alpha_j}\cos\delta_j \tag{4.31}$$

$$\frac{\partial Q^s_{km}}{\partial \alpha_j} = b_{km}e^{\alpha_j}\cos\delta_j - g_{km}e^{\alpha_j}\sin\delta_j \tag{4.32}$$

$$\frac{\partial Q^s_{km}}{\partial \delta_j} = -b_{km}e^{\alpha_j}\sin\delta_j - g_{km}e^{\alpha_j}\cos\delta_j \approx 0 \tag{4.33}$$

Power Injections

Neglecting the shunt elements at the buses, the scaled injection at bus k, P^s_k or Q^S_k can be expressed in terms of the incident scaled branch flows:

$$P^s_k = \sum_{i \epsilon \aleph_k} P^s_{ki} \tag{4.34}$$

$$Q^s_k = \sum_{i \epsilon \aleph_k} Q^s_{ki} \tag{4.35}$$

where, $\aleph_k$ is the set of buses directly connected to bus k.

Then,

$$\frac{\partial P_k^s}{\partial \alpha_j} = \sum_{i \epsilon \aleph_k} \frac{\partial P_{ki}^s}{\partial \alpha_j} \approx 0 \tag{4.36}$$

$$\frac{\partial P_k^s}{\partial \delta_j} = \sum_{i \epsilon \aleph_k} \frac{\partial P_{ki}^s}{\partial \delta_j} \tag{4.37}$$

$$\frac{\partial Q_k^s}{\partial \alpha_j} = \sum_{i \epsilon \aleph_k} \frac{\partial Q_{ki}^s}{\partial \alpha_j} \tag{4.38}$$

$$\frac{\partial Q_k^s}{\partial \delta_j} = \sum_{i \epsilon \aleph_k} \frac{\partial Q_{ki}^s}{\partial \delta_j} \approx 0 \tag{4.39}$$

Voltage Magnitude

Using Equation (4.21):

$$\frac{\partial}{\partial \alpha_j}(\ln V_k^s) = \begin{cases} 1 & \text{if } j = j1, j2, \ldots, jr \\ 0 & \text{otherwise} \end{cases}$$

$$\frac{\partial}{\partial \delta_j}(\ln V_k^s) = 0$$

Current Magnitude

Using the Equation (4.22), derivatives of the scaled current magnitude measurements can be expressed in terms of the scaled power injections and their derivatives as calculated above. They will be given as follows:

$$\frac{\partial I_j^s}{\partial \alpha_j} = \frac{1}{I_j^s}\{P_{km}^s \frac{\partial P_{km}^s}{\partial \alpha_j} + Q_{km}^s \frac{\partial Q_{km}^s}{\partial \alpha_j}\} \tag{4.40}$$

$$\frac{\partial I_j^s}{\partial \delta_j} = \frac{1}{I_j^s}\{P_{km}^s \frac{\partial P_{km}^s}{\partial \delta_j} + Q_{km}^s \frac{\partial Q_{km}^s}{\partial \delta_j}\} \tag{4.41}$$

Loop Equations

Combining Equations (4.26) and (4.27), and taking their derivative with respect to the branch variables X_δ, X_α will yield:

$$\frac{\partial}{\partial X}\begin{bmatrix} \mathcal{L} & 0 \\ 0 & \mathcal{L} \end{bmatrix}\begin{bmatrix} X_\delta \\ X_\alpha \end{bmatrix} = \begin{bmatrix} \mathcal{L} & 0 \\ 0 & \mathcal{L} \end{bmatrix} \tag{4.42}$$

4.5.4 Observability Analysis

Equation (4.29) relates all existing measurements to the branch variables, using the first-order Taylor approximation. An estimate for Δx can be obtained as long as the column rank of H is equal to the dimension of Δx.

One way to determine this is through matrix factorization. The method of Peters-Wilkinson [12] can be used to decompose H into its triangular factors as follows:

$$H = \begin{bmatrix} L \\ M \end{bmatrix} [U] \tag{4.43}$$

where:
L and U are square lower and upper triangular matrices respectively,
M is a rectangular matrix.

During the course of the factorization, row/column pivoting may be necessary to avoid zero pivots. When the system is unobservable, such zero pivots can not be avoided despite pivoting. In such a case a non-zero value of 1.0 will be substituted for the zero pivot and the factorization will thus be continued. The factor L obtained this way, will contain one or more artificially added pivot entries, each one corresponding to a branch variable.

Rewrite Equation (4.29) by substituting Equation (4.43):

$$L \cdot U \cdot \Delta x = \Delta z^e \tag{4.44}$$

$$M \cdot U \cdot \Delta x = \Delta z^r \tag{4.45}$$

where $\Delta z = [(\Delta z^e)^T (\Delta z^r)^T]^T$.

Note that, for the purpose of analyzing observability, the set of measurements corresponding to the top N measurements (N being the number of branch variables) are sufficient, since they represent a linearly independent set of measurements.

If the system is observable, then Equation (4.44) will yield a null solution for Δx for a null vector of Δz^e. In the case of unobservable systems, the entries in Δz^e corresponding to those zero pivots of L, that are artificially set equal to 1.0 during factorization, will be set equal to arbitrary, but distinct non-zero values. It can be shown that this procedure is equivalent to adding a new measurement of the corresponding branch variable to the existing measurement set. The set of Equation (4.44) can be reordered, so that the rows and columns of the artificially introduced nonzero pivots become the last:

$$\begin{bmatrix} L_0 & 0 \\ L_e & I_u \end{bmatrix} \begin{bmatrix} U_0 & U_e \\ 0 & I_u \end{bmatrix} \begin{bmatrix} \Delta x_0 \\ \Delta x_u \end{bmatrix} = \Delta z^e = \begin{bmatrix} 0 \\ \Delta z_u^e \end{bmatrix} \tag{4.46}$$

where

L_0 : a non-singular lower triangular matrix

U_0 : a non-singular upper triangular matrix

L_e : a rectangular matrix

U_e : a rectangular matrix

I_u : an identity matrix of dimension N_u

N_u : number of zero pivots encountered and replaced by 1.0s during the factorization of H.

Δz_u^e : vector of arbitrarily assigned but distinct nonzero entries.

Δx_0, Δx_u : branch variable solution.

Note that:

$$\Delta x_u = \Delta z_u^e \tag{4.47}$$
$$U_0 \Delta x_0 = -U_e \Delta z_u^e \tag{4.48}$$

Solution of Equation (4.48) will yield Δx_0, where the nonzero entries will correspond to the unobservable branches. The set of unobservable branches will be the union of these with the previously identified branches corresponding to the entries of Δz_u. Once the unobservable branch list is thus formed, these branches can be eliminated from the system diagram, yielding the "observable islands" of the system.

Example 4.4:

Consider the 6-bus power system with the measurement configuration as shown in Figure 4.7. For simplicity, assume that only real power measurements are used and only phase angle variables are estimated. Determine the observable island(s).

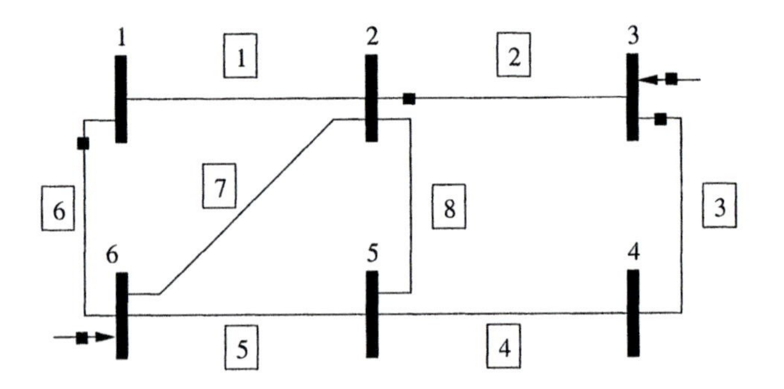

Figure 4.7. Measurement configuration of the 6-bus system

The linearized measurement jacobian with respect to the new branch variables will be given as:

$$H = \begin{array}{c} \mathcal{L}6 \\ P_3 \\ P_{34} \\ \mathcal{L}7 \\ P_6 \\ P_{16} \\ \mathcal{L}8 \\ P_{23} \end{array} \left[\begin{array}{cccccccc} -1 & -1 & -1 & -1 & -1 & 1 & 0 & 0 \\ 0 & 2 & 0 & 0 & 0 & 0 & 0 & 0 \\ 0 & 0 & 3 & 0 & 0 & 0 & 0 & 0 \\ 0 & -1 & -1 & -1 & -1 & 0 & 1 & 0 \\ 0 & 0 & 0 & 0 & 5 & 6 & 0 & 0 \\ 0 & 0 & 0 & 0 & 0 & 6 & 0 & 0 \\ 0 & -1 & -1 & -1 & 0 & 0 & 0 & 1 \\ 0 & 2 & 3 & 0 & 0 & 0 & 0 & 0 \end{array} \right]$$

Factorization of H yields one zero pivot at row/column 8, i.e. branch 8 is unobservable. This zero pivot is replaced by 1.0 and the following modified factors are obtained:

$$L = \left[\begin{array}{cccccccc} 1 & 0 & 0 & 0 & 0 & 0 & 0 & 0 \\ 0 & 1 & 0 & 0 & 0 & 0 & 0 & 0 \\ 0 & 0 & 1 & 0 & 0 & 0 & 0 & 0 \\ -1 & -1 & 0 & 1 & 0 & 0 & 0 & 0 \\ 0 & 0 & 0 & 0 & 1 & 0 & 0 & 0 \\ 0 & -1 & 0 & 1 & 0 & 1 & 0 & 0 \\ 0 & -1 & 0 & 1 & 1 & 1 & 1 & 0 \\ 0 & 1 & -1 & 0 & 0 & 0 & 0 & 1 \end{array} \right]$$

$$U = \left[\begin{array}{cccccccc} 1 & 0 & 0 & 0 & 0 & 1 & 0 & 0 \\ 0 & 1 & 1 & 0 & 0 & 0 & 0 & 0 \\ 0 & 0 & 1 & 0 & 0 & 0 & 0 & 0 \\ 0 & 0 & 0 & -1 & -1 & 2 & 0 & 0 \\ 0 & 0 & 0 & 0 & 1 & 0 & 0 & 0 \\ 0 & 0 & 0 & 0 & 0 & -2 & 1 & 0 \\ 0 & 0 & 0 & 0 & 0 & 0 & -1 & 1 \\ 0 & 0 & 0 & 0 & 0 & 0 & 0 & 1 \end{array} \right]$$

The remaining unobservable branches will be identified by partitioning the factors according to Equation (4.46) where:

$$U_0 = \left[\begin{array}{ccccccc} 1 & 0 & 0 & 0 & 0 & 1 & 0 \\ 0 & 1 & 1 & 0 & 0 & 0 & 0 \\ 0 & 0 & 1 & 0 & 0 & 0 & 0 \\ 0 & 0 & 0 & -1 & -1 & 2 & 0 \\ 0 & 0 & 0 & 0 & 1 & 0 & 0 \\ 0 & 0 & 0 & 0 & 0 & -2 & 1 \\ 0 & 0 & 0 & 0 & 0 & 0 & -1 \end{array} \right]$$

$$U_e = \left[\begin{array}{ccccccc} 0 & 0 & 0 & 0 & 0 & 0 & 1 \end{array} \right]^T$$

and solving for Δx_0:

$$\Delta x_0 = \begin{matrix} 1 \\ 2 \\ 3 \\ 4 \\ 5 \\ 6 \\ 7 \end{matrix} \begin{bmatrix} -0.5 \\ 0 \\ 0 \\ 1.0 \\ -0.5 \\ 0 \\ 0.5 \end{bmatrix}$$

Hence, in addition to branch 8, the branches 1,4,5 and 7 that correspond to the nonzero entries in Δx_0, will be declared as *unobservable branches.* Removing these branches, the observable islands shown in Figure 4.8 will be obtained.

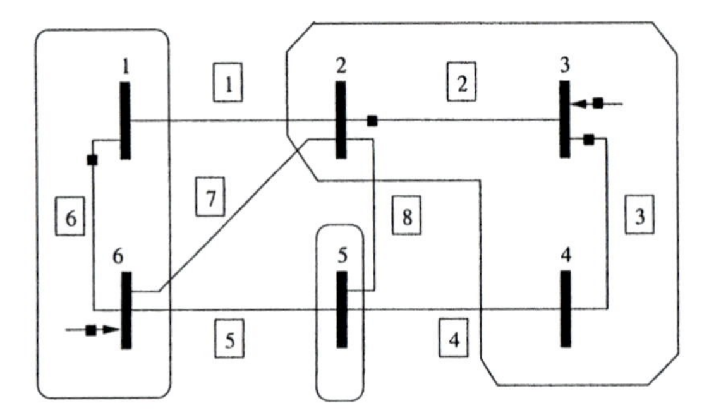

Figure 4.8. Observable islands of the 6-bus system

Before ending this section, it is noted that the use of branch variables in the formulation of the observability analysis problem has two important advantages. One is the non-iterative nature of the resulting procedure and the other is the simplification of the formulation in the absence of the slack bus concept. Although the need to build loop equations can be seen as a burden, in practice the model can be significantly reduced by direct elimination of power flow measurements and the corresponding branch variables, which is not possible in the nodal formulation to be discussed next.

4.6 Numerical Method Based on the Nodal Variable Formulation

Numerical observability analysis can also be carried out by using the nodal variables as described in [5]. Nodal variable vector is denoted by x and represents the vector of magnitude and phase angle of all bus voltages in the system. Consider the linearized measurement model, where measurement errors are ignored due to their irrelevance in the observability analysis:

$$\Delta z = H \Delta x$$

where:
$\Delta z = z - h(x_0)$, is the mismatch between the measurement vector and its calculated value at an estimate x_0.
$\Delta x = x - x_0$.
$H = \frac{\partial h(x)}{\partial x}$ evaluated at x_0.

The WLS estimate $\Delta \hat{x}$ will be given by:

$$\Delta \hat{x} = (H^T R^{-1} H)^{-1} H^T R^{-1} \Delta z$$

A unique solution for Δx can be calculated if $(H^T R^{-1} H)$ is nonsingular or equivalently if H has full column rank, i.e. rank[H] $= n$, where n is the total number of states.

Using the weak coupling between $P - V$ and $Q - \theta$, the linearized model can be decoupled as:

$$\begin{aligned} \Delta z_A &= H_{AA} \Delta \theta \\ \Delta z_R &= H_{RR} \Delta V \end{aligned}$$

where:
$\Delta z_A, \Delta z_R$ is the real and reactive power measurement mismatch vectors respectively.
$H_{AA} = \frac{\partial h_A}{\partial \theta}$ is the decoupled Jacobian for the real power measurements.
$H_{RR} = \frac{\partial h_R}{\partial V}$ is the decoupled Jacobian for the reactive power measurements.
$\Delta \theta = \theta - \theta_0$.
$\Delta V = V - V_0$.

Assuming that the P, Q measurements come in pairs, $P - \theta$ and $Q - V$ observability can be separately tested. Note that, unlike θ, the voltage solution requires a measured reference bus. Hence, following the $P - \theta$ analysis, it should be further checked to ensure that at least one voltage measurement exists per observable island.

It should be noted that the system observability is independent of the branch parameters as well as the operating state of the system. So, all system branches can be assumed to have an impedance of $j1.0$ p.u. and all bus voltages can be set equal to 1.0 p.u. for the purpose of observability analysis. Then, the d.c. power flows along these system branches can be written as:

$$P_b = A\theta \tag{4.49}$$

where: P_b is the vector of branch flows
A is the branch-bus incidence matrix
θ is the vector of bus voltage phase angles

If the estimated state $\hat{\theta}$ is zero, then all branch flows will be zero as given by Equation (4.49). Using the DC measurement model:

$$H_{AA}\theta = z_A \tag{4.50}$$

the WLS estimate for θ will be given by:

$$\hat{\theta} = (H_{AA}^T H_{AA})^{-1} H_{AA}^T z_A = G_{AA}^{-1} t_A \tag{4.51}$$

A null estimate for $\hat{\theta}$ will be obtained for an observable system when all system measurements Δz_A, i.e. flows and injection measurements, are all zero. If there exists an estimate $\hat{\theta}$ which satisfies the measurement equation:

$$H_{AA}\hat{\theta} = 0 \tag{4.52}$$

yet, yields a nonzero branch flow:

$$P_b = A\hat{\theta} \neq 0 \tag{4.53}$$

then, $\hat{\theta}$ will be called *an unobservable state.* Furthermore, those branches carrying nonzero flows, will be referred to as *unobservable branches.*

Example 4.5:

Consider the 5-bus system and its measurement configuration given in Figure 4.9. The measurement Jacobian can be formed as:

$$H_{AA} = \begin{array}{c} P_1 \\ P_5 \\ P_{12} \\ P_{13} \end{array} \overset{\begin{array}{ccccc} \theta_1 & \theta_2 & \theta_3 & \theta_4 & \theta_5 \end{array}}{\begin{bmatrix} 2 & -1 & -1 & & \\ & -1 & & -1 & 2 \\ 1 & -1 & & & \\ 1 & & -1 & & \end{bmatrix}}$$

Let $\hat{\theta}^T = [0\ 0\ 0\ 1\ 0.5]$, then $H_{AA}\hat{\theta} = 0$. Calculating the branch flows:

$$P_b = A\hat{\theta} = \begin{bmatrix} 1 & -1 & & & \\ & 1 & & & -1 \\ & & -1 & 1 & \\ -1 & & 1 & & \\ & 1 & & -1 & \end{bmatrix} \begin{bmatrix} 0 \\ 0 \\ 0 \\ 1 \\ 0.5 \end{bmatrix} = \begin{bmatrix} 0.0 \\ -0.5 \\ -0.5 \\ 1.0 \\ 0.0 \\ -1.0 \end{bmatrix}$$

Branches 2,3,4 and 6 are therefore unobservable and $\hat{\theta}$ is an unobservable state.

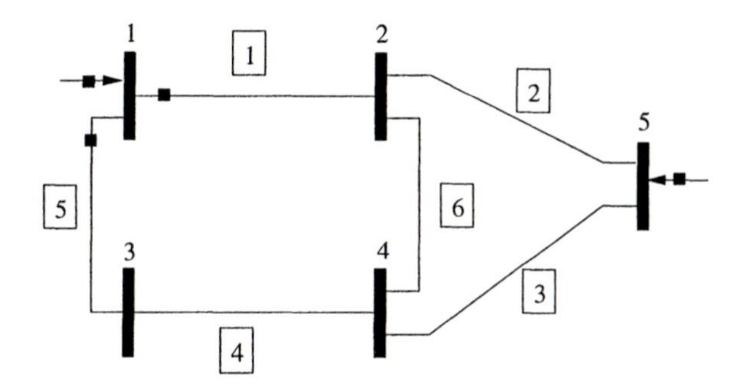

Figure 4.9. Measurement configuration of the 5-bus system

4.6.1 Determining the Unobservable Branches

If the system is found to be unobservable, then the observable islands which are separated by unobservable branches can be identified as described in [6]. Note that those branches having no incident measurements, are called *irrelevant branches* and the estimated state will be independent of the status (on/off) and parameters of these branches. Therefore, they can be disregarded when analyzing network observability.

Let us again consider the decoupled linearized model where all measurements are set equal to zero:

$$(H_{AA}^T H_{AA})\hat{\theta} = H_{AA}^T z_A = t_A = 0 \tag{4.54}$$

G_{AA} is singular, even for fully observable systems since the reference bus phase angle is included in the state vector θ. Then row/column permutations can be used to reorder and partition the matrix as follows:

$$\begin{bmatrix} G_{11} & G_{12} \\ G_{21} & G_{22} \end{bmatrix} \begin{bmatrix} \hat{\theta}_a \\ \hat{\theta}_b \end{bmatrix} = \begin{bmatrix} 0 \\ 0 \end{bmatrix}$$

where G_{11} is a nonsingular submatrix within G_{AA}. By assigning arbitrary but distinct values to $\hat{\theta}_b$ entries as $\overline{\theta_b}$, one of many possible solutions for $\hat{\theta}_a$ can be obtained as:

$$\hat{\theta}_a = -G_{11}^{-1} G_{12} \overline{\theta_b}$$

The branch flows corresponding to this solution $(\hat{\theta}_a, \overline{\theta_b}) = \hat{\theta}^*$ can then be found as:

$$A\hat{\theta}^* = P_b^*$$

Those branches i with $P_b^*(i) \neq 0$ will be identified as the *unobservable branches.*

In practice, the above procedure is carried out by triangular factorization of G_{AA} by using the Cholesky method, which is described in detail in Appendix A. Since G_{AA} is singular, at least one of the pivots will be zero during the Cholesky factorization. When a zero pivot is encountered, it will be replaced by a 1.0 and the corresponding entry of the right hand side vector will be assigned an arbitrary value. The arbitrary values assigned in this manner should be distinct from each other and this is accomplished by assigning integer numbers in increasing order, such as 0,1,2, etc. Considering the Example 4.5 again, the gain matrix will be given by:

$$G_{AA} = \begin{bmatrix} 6 & -3 & -3 & 0 & 0 \\ -3 & 3 & 1 & 1 & -2 \\ -3 & 1 & 2 & 0 & 0 \\ 0 & 1 & 0 & 1 & -2 \\ 0 & -2 & 0 & -2 & 4 \end{bmatrix}$$

Cholesky factorization of G_{AA} yields:

$$L = \begin{bmatrix} 2.4495 & 0 & 0 & 0 & 0 \\ -1.2247 & 1.2247 & 0 & 0 & 0 \\ -1.2247 & -0.4082 & 0.5774 & 0 & 0 \\ 0 & 0.8165 & 0.5774 & 1.0000 & 0 \\ 0 & -1.6330 & -1.1547 & 0 & 1.0000 \end{bmatrix}$$

where the last two zero pivots are changed to 1.0 and the right hand side vector is modified as:

$$t_A = [0\ 0\ 0\ 0\ 1]^T$$

The estimated state $\hat{\theta}$ will then be given by:

$$\hat{\theta} = (L * L^T)^{-1} t_A = \begin{bmatrix} 2.0 \\ 2.0 \\ 2.0 \\ 0.0 \\ 1.0 \end{bmatrix}$$

and the branch flow estimates can be obtained as:

$$P_b = A\hat{\theta} = \begin{bmatrix} 0 \\ 1.0 \\ 1.0 \\ -2.0 \\ 0 \\ 2.0 \end{bmatrix}$$

Hence, branches 2,3,4 and 6 that have nonzero flows, will be labelled as unobservable branches.

4.6.2 Identification of Observable Islands

The above procedure of identifying unobservable branches can be used to determine the observable islands in the system. The above procedure needs to be carried out recursively, each time eliminating the irrelevant injections until all observable islands are identified. Irrelevant injections are those that are incident to unobservable branches. The algorithm is given below:

1. Remove all irrelevant branches. These are branches that have no incident measurements.

2. Form the decoupled linearized gain matrix for the $P - \theta$ estimation problem:
$$G_{AA} = H_{AA}^T R_A^{-1} H_{AA}$$

3. Factorize G_{AA} modifying the zero pivots and the right hand side vector as described above.

4. Identify and remove all unobservable branches and all injections that are incident to these unobservable branches.

5. If no more unobservable branches are found, then determine the observable islands separated by the unobservable branches and stop. Else, go to step 2.

Example 4.6:

Consider the system and measurement configuration shown in Figure 4.10.

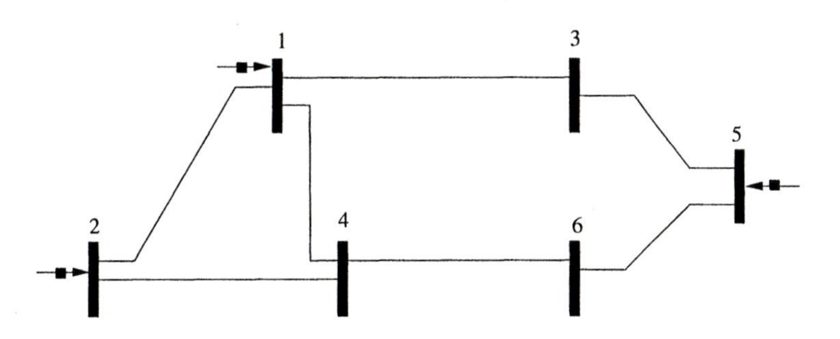

Figure 4.10. 6-bus test system and its measurements.

Note that branch 4-6 is an irrelevant branch. Removing it from the network yields the following branch-bus incidence matrix:

$$A = \begin{bmatrix} 1 & -1 & 0 & 0 & 0 & 0 \\ 1 & 0 & -1 & 0 & 0 & 0 \\ 1 & 0 & 0 & -1 & 0 & 0 \\ 0 & 1 & 0 & -1 & 0 & 0 \\ 0 & 0 & 1 & 0 & -1 & 0 \\ 0 & 0 & 0 & 0 & 1 & -1 \end{bmatrix}$$

The gain and the measurement Jacobian matrices are built as below:

$$G_{AA} = \begin{bmatrix} 10 & -5 & -3 & -2 & 0 & 0 \\ -5 & 5 & 1 & -1 & 0 & 0 \\ -3 & 1 & 2 & 1 & -2 & 1 \\ -2 & -1 & 1 & 2 & 0 & 0 \\ 0 & 0 & -2 & 0 & 4 & -2 \\ 0 & 0 & 1 & 0 & -2 & 1 \end{bmatrix}$$

$$H_{AA} = \begin{bmatrix} 3 & -1 & -1 & -1 & 0 & 0 \\ -1 & 2 & 0 & -1 & 0 & 0 \\ 0 & 0 & -1 & 0 & 2 & -1 \end{bmatrix}$$

Cholesky factorization of G yields 3 zero pivots. The upper triangular factor is shown below:

$$L^T = \begin{bmatrix} 3.1623 & -1.5811 & -0.9487 & -0.6325 & 0 & 0 \\ 0 & 1.5811 & -0.3162 & -1.2649 & 0 & 0 \\ 0 & 0 & 1.0000 & 0 & -2.0000 & 1.0000 \\ 0 & 0 & 0 & 0 & 0 & 0 \\ 0 & 0 & 0 & 0 & 0 & 0 \\ 0 & 0 & 0 & 0 & 0 & 0 \end{bmatrix}$$

Replacing the zero pivots by 1.0 and choosing the right hand side vector as:

$$t_A^T = [0\ 0\ 0\ 0\ 1\ 2]$$

the estimated state will be obtained by:

$$\hat{\theta} = (L * L^T)^{-1} t_A = \begin{bmatrix} 0.0 \\ 0.0 \\ 0.0 \\ 0.0 \\ 1.0 \\ 2.0 \end{bmatrix}$$

and the branch flow estimates can be obtained as:

$$P_b = A\hat{\theta} = \begin{bmatrix} 0 \\ 0 \\ 0 \\ 0 \\ -1.0 \\ -1.0 \end{bmatrix}$$

Hence, branches 3-5 and 5-6 are declared as unobservable and removed from the network, along with the incident injection measurement at bus 5. This results in the observable islands as shown in Figure 4.11.

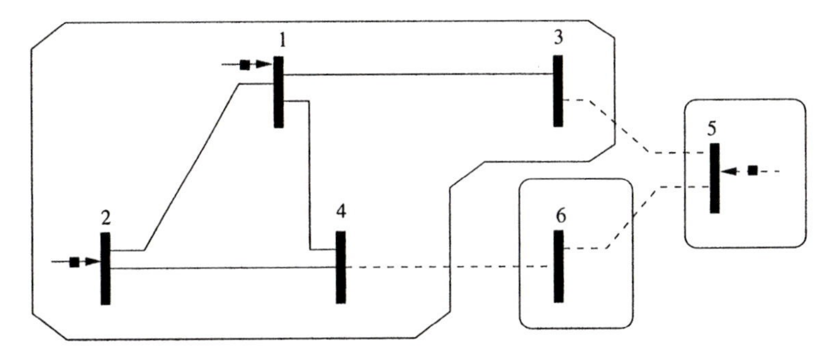

Figure 4.11. Results after the first identification cycle.

The modified network and measurement configuration will yield the following gain G_{mod} and measurement Jacobian H_{mod} matrices in the second identification cycle:

$$G_{mod} = \begin{bmatrix} 10 & -5 & -3 & -2 & 0 & 0 \\ -5 & 5 & 1 & -1 & 0 & 0 \\ -3 & 1 & 1 & 1 & 0 & 0 \\ -2 & -1 & 1 & 2 & 0 & 0 \\ 0 & 0 & 0 & 0 & 0 & 0 \\ 0 & 0 & 0 & 0 & 0 & 0 \end{bmatrix},$$

$$H_{mod} = \begin{bmatrix} 3 & -1 & -1 & -1 & 0 & 0 \\ -1 & 2 & 0 & -1 & 0 & 0 \end{bmatrix}.$$

Factorizing G_{mod}, replacing the zero pivots by 1.0:

$$L^T = \begin{bmatrix} 3.1623 & -1.5811 & -0.9487 & -0.6325 & 0 & 0 \\ 0 & 1.5811 & -0.3162 & -1.2649 & 0 & 0 \\ 0 & 0 & 1.0000 & 0 & 0 & 0 \\ 0 & 0 & 0 & 1.0000 & 0 & 0 \\ 0 & 0 & 0 & 0 & 1.0000 & 0 \\ 0 & 0 & 0 & 0 & 0 & 1.0000 \end{bmatrix}$$

and choosing the right hand side vector as:

$$r^T = [0\ 0\ 0\ 1\ 2\ 3]$$

the estimated state will be obtained by:

$$\hat{\theta} = (L_{mod} * L_{mod}^T)^{-1} r = \begin{bmatrix} 0.6 \\ 0.8 \\ 0.0 \\ 1.0 \\ 2.0 \\ 3.0 \end{bmatrix}$$

The branch flow estimates are then calculated as:

$$P_b = A_{modified}\hat{\theta} = \begin{bmatrix} 1 & -1 & 0 & 0 & 0 & 0 \\ 1 & 0 & -1 & 0 & 0 & 0 \\ 1 & 0 & 0 & -1 & 0 & 0 \\ 0 & 1 & 0 & -1 & 0 & 0 \end{bmatrix} \times \hat{\theta} = \begin{bmatrix} -0.2 \\ 0.6 \\ -0.4 \\ -0.2 \end{bmatrix}$$

Thus, all branches in the network are declared unobservable and the final result is shown in Figure 4.12. This terminates the identification procedure.

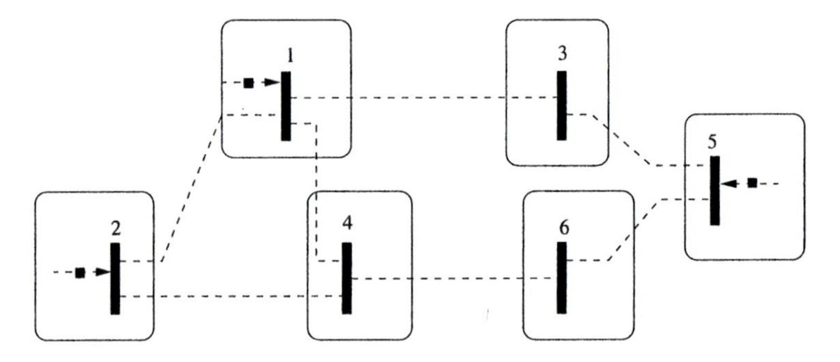

Figure 4.12. Final result of identified observable islands.

4.6.3 Measurement Placement to Restore Observability

Once the observable islands are identified, measurements can be added to merge these islands so that eventually a single observable island can be formed. The candidate measurements that can merge islands are:

- the line flows along branches that connect observable islands, and

- the injections at the boundary buses of observable islands.

Consider the gain matrix G_{AA} of Equation (4.54). The subscripts (AA) will be dropped to simplify the notation. Note that, since the slack bus is also included in the formulation, the rank of H (and G) will be at most $(n-1)$ (n being the number of buses), even for a fully observable system. Hence, the Cholesky factorization of the gain matrix G will be interrupted by at least one zero pivot. Assume that this happens after the processing of pivot i as shown below:

$$G_{red} = \begin{bmatrix} d_1 & & & & & & & \\ & d_2 & & & & & & \\ & & \ddots & & & & & \\ & & & d_i & & & & \\ & & & & 0 & 0 & \dots & 0 \\ & & & & 0 & \times & \times & \times \\ & & & & \vdots & \times & \times & \times \\ & & & & 0 & \times & \times & \times \end{bmatrix} \tag{4.55}$$

where,

$$G_{red} = L_i^{-1} L_{i-1}^{-1} \dots L_1^{-1} G L_1^{-T} \dots L_{i-1}^{-T} L_i^{-T} \tag{4.56}$$

and L_i's are elementary factors given by:

$$L_i = \begin{bmatrix} 1 & & & & \\ & \ddots & & & \\ & & 1 & & \\ & & \times & \ddots & \\ & & \times & & 1 \end{bmatrix} \tag{4.57}$$

where the i'th column has nonzero elements below its diagonal marked by $\times$'s in Equation (4.57) above.

Setting $L_{i+1} = I_{n \times n}$, the Cholesky factorization of G_{red} in Equation (4.55) can proceed with the $(i+2)$-nd column. This procedure can be repeated each time a zero pivot is detected until completion of the entire factorization. The following expression can then be written:

$$\begin{aligned} D &= L_n^{-1} L_{n-1}^{-1} \dots L_1^{-1} G L_1^{-T} \dots L_{n-1}^{-T} L_n^{-T} \\ &= L^{-1} G L^{-T} \end{aligned} \tag{4.58}$$

where D is a singular and diagonal matrix with zeros in rows corresponding to the zero pivots encountered during the factorization of G, and L is a nonsingular lower triangular matrix.

Now, consider the addition of a single candidate measurement which will contribute a new row h_k to the measurement Jacobian. Then, the new gain matrix G' can be expressed as:

$$
\begin{aligned}
G' &= G + h_k^T h_k \\
&= L(D + MM^T)L^T \\
&= LD'L^T
\end{aligned} \tag{4.59}
$$

where $D' = D + MM^T$ and $M = L^{-1}h_k^T$. It can be shown that, the rank of G' will increase by 1 if and only if, $M(i) \neq 0$ for any i such that $D(i,i) = 0$.

The value of $M(i)$ can be obtained by taking the inner product of h_k and the i'th row of L^{-1}, which is computed by a single back substitution step as shown below:

$$
L^T w^T = e_i \tag{4.60}
$$

where, e_i is a singleton array with all elements zero except for a 1.0 as its ith entry and w is the i-th row of L^{-1}, with the following structure:

$$
w = [w_1 \; w_2 \; w_3 \; \ldots \; w_{i-1} \; 1 \; 0 \; 0 \; \ldots \; 0]
$$

Let the matrix W be defined as the matrix containing only those rows of L^{-1} corresponding to the zero pivots in the diagonal matrix D. Rows of W can be obtained by repeated solution of Equation (4.60) for all i for which $D(i,i) = 0$.

The following measurement placement algorithm can then be implemented based on the above defined matrices:

1. Form the gain matrix and compute its Cholesky factors.

2. Check if D has only one zero pivot. If yes, stop, the system is observable. Else, compute the W matrix by repeated solution of Equation (4.60) for each zero pivot row.

3. Form the candidate measurement list. The list will contain flow and injection measurements incident to branches connecting observable islands, which have already been identified by the observability analysis.

4. Build the measurement Jacobian matrix H_c for the candidate measurements.

5. Compute $B = H_c W^T$ and compute the reduced echelon form E of B. The linearly independent rows of E will correspond to all the measurements required to be placed.

Detailed derivation of this multiple measurement placement method can be found in [13].

Example 4.7:

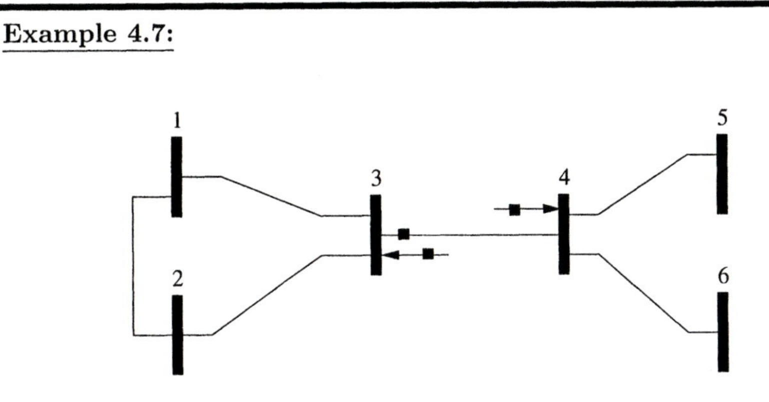

Figure 4.13. Example System for Meter Placement.

Consider the 6-bus system and its measurement configuration shown in Figure 4.13. Previous observability analysis identified 5 observable islands defined by buses [3 4], [1], [2], [5], and [6]. Place measurements in order to make the system fully observable.

Step 1 Form the Jacobian matrix H and the gain matrix G:

$$H = \begin{bmatrix} -1 & -1 & 3 & -1 & 0 & 0 \\ 0 & 0 & 1 & -1 & 0 & 0 \\ 0 & 0 & -1 & 3 & -1 & -1 \end{bmatrix}$$

$$G = \begin{bmatrix} 1 & 1 & -3 & 1 & 0 & 0 \\ 1 & 1 & -3 & 1 & 0 & 0 \\ -3 & -3 & 11 & -7 & 1 & 1 \\ 1 & 1 & -7 & 11 & -3 & -3 \\ 0 & 0 & 1 & -3 & 1 & 1 \\ 0 & 0 & 1 & -3 & 1 & 1 \end{bmatrix}$$

Triangular factors of G will be:

$$L = \begin{bmatrix} 1 & & & & & \\ 1 & 1 & & & & \\ -3 & 0 & 1 & & & \\ 1 & 0 & -2 & 1 & & \\ 0 & 0 & 0.5 & -0.5 & 1 & \\ 0 & 0 & 0.5 & -0.5 & 0 & 1 \end{bmatrix}$$

$$D = \begin{bmatrix} 1 & & & & & \\ & 0 & & & & \\ & & 2 & & & \\ & & & 2 & & \\ & & & & 0 & \\ & & & & & 0 \end{bmatrix}$$

Step 2 Form the W matrix using the 2nd, 5th and 6th rows of the inverse of L:

$$W = \begin{bmatrix} -1 & 1 & 0 & 0 & 0 & 0 \\ 1 & 0 & 0.5 & 0.5 & 1 & 0 \\ 1 & 0 & 0.5 & 0.5 & 0 & 1 \end{bmatrix}$$

Step 3 Consider only the available boundary injections as candidates. They are injections at buses 1, 2, 5, and 6. Form the candidate measurement Jacobian sub-matrix H_c, given by:

$$H_c = \begin{bmatrix} 2 & -1 & -1 & 0 & 0 & 0 \\ -1 & 2 & -1 & 0 & 0 & 0 \\ 0 & 0 & 0 & -1 & 1 & 0 \\ 0 & 0 & 0 & -1 & 0 & 1 \end{bmatrix}$$

where, the rows correspond to the candidate injection measurements at buses 1,2,5 and 6.

Step 4 Form $B = H_c W^T$:

$$B = \begin{bmatrix} -3 & 1.5 & 1.5 \\ 3 & -1.5 & -1.5 \\ 0 & 0.5 & -0.5 \\ 0 & -0.5 & 0.5 \end{bmatrix}$$

and its reduced echelon form E:

$$E = \begin{bmatrix} 1 & 0 & -1 \\ 0 & 0 & 0 \\ 0 & 1 & -1 \\ 0 & 0 & 0 \end{bmatrix}$$

Hence, the injections at buses 1 and 5, corresponding to the linearly independent first and third rows of E, should be placed in order to make the system fully observable.

4.7 Topological Observability Analysis Method

Observability analysis can also be carried out by using a topological method [7]. This method differs from the numerical methods in that it does not use any floating point calculations in the analysis. The decision is based strictly on logical operations and therefore requires the information about the network connectivity, measurement type and their location only. The actual parameters of the network elements are not used in any part of the analysis at all. Furthermore, it is assumed that the measurements come in real and reactive pairs, and therefore the real part of the decoupled (or DC) measurement model can be used for the observability analysis. In this model, the error-free real power flow and injection measurements are related to the bus voltage phase angles (excluding the slack bus) linearly as:

$$z_A = H_{AA}\theta$$

Consider a single branch with reactance $x_{km} = 1.0$ p.u. connected between buses k and m. Assuming the voltage magnitude at each terminal to be 1.0 p.u., the first order approximation around $\theta_k^0 = \theta_m^0 = 0$ of the real power flow through this branch can be written as:

$$\begin{aligned} \Delta P &= \left[\frac{V_k V_m}{x_{km}} \cos(\theta_k^0 - \theta_m^0) \right] (\theta_k - \theta_m) \\ &= (\theta_k - \theta_m) \end{aligned}$$

If a tree can be formed such that each branch of this tree contains a power flow measurement, then the phase angles at all buses can be determined, i.e. the system will be fully observable. The topological method hence starts out by assigning power flow measurements to their respective branches and tries to form a spanning tree, i.e. a tree that reaches each and every bus in the system, using these branches. If this procedure is not successful, then it will yield a forest where there are several smaller size trees. In that case, the remaining measurements which are of injection type, will be used in order to merge these trees and reduce the size of the forest. If successful, this reduction process will result in a single tree, in which case the system will be declared as observable.

4.7.1 Topological Observability Algorithm

While the implementation of the topological observability analysis can be carried out in several different ways, the essential steps of the algorithm can be summarized as follows:

1. First assign all the flow measurements to their respective branches.

2. Then, try to assign the injection measurements in order to reduce the existing forest by merging existing trees. Note that there is no way to predict the correct sequence for processing injections. Implementation of the method requires proper back-up and re-assignment of injections when necessary. Different implementation details for the topological method can be found in the literature [7, 8, 9, 10, 11].

4.7.2 Identifying the Observable Islands

After processing all the flows and injections, if a spanning tree can not be found, then the observable islands need to be identified. This can be done as follows:

1. Discard those injections that have *at least one incident branch* which does *not* form a loop with the branches of the already defined forest.
2. Update the forest accordingly and repeat step 1 until no more injections need to be removed.

Example 4.8:

Use the topological observability analysis method to solve Example 4.6.
Solution:

- Start by assigning flow measurements to the corresponding branches:
  Flow 1–2 ⇒ branch 1–2,
  Flow 3–5 ⇒ branch 3–5,
  Flow 5–6 ⇒ branch 5–6.

  They form a forest defined by the branches 3-5, 5-6, and 1-2.
- Assign the injection at bus 4 ⇒ branch 4–6 in order to merge two trees. However, the injection at bus 4 is also incident to branches 1-4 and 2-4. Neither one of these branches forms a loop with the branches of the existing forest. Therefore, the injection at bus 4 can not be used and is discarded.
- The resulting observable islands are the same as found in Example 4.6 in Figure 4.11.

4.8 Determination of Critical Measurements

Measurements can be broadly classified into two categories as critical and non-critical (or redundant). If the removal of a measurement causes an observable system to become unobservable, then this measurement is called a critical measurement. Such measurements can be identified either by topological [10] or numerical methods. State estimators that are based on the WLS method, may use the measurement error covariance matrix, which will be discussed in detail in Chapter 6, by searching its null columns in order to determine the corresponding critical measurements. Another approach which does not require explicit formation of the covariance matrix is described below.

Consider an observable power system with n states and m measurements. If the system is initially unobservable, it is assumed that proper measurements are placed to restore observability as explained in section 4.6.3. Then, a set of n measurements can be chosen out of the available m, so that the system will be observable with only these n measurements. This set of n measurements will be referred to as the "essential measurements". Such a set is in general not unique, yet will contain all of the critical measurements, since no set that excludes them can make the system observable.

Ordering the essential measurements first and partitioning the matrices, the linearized measurement equations will be:

$$\begin{bmatrix} H_1 \\ H_2 \end{bmatrix} \cdot [x] = \begin{bmatrix} z_1 \\ z_2 \end{bmatrix} \tag{4.61}$$

where the rows of H_1, z_1 and H_2, z_2 correspond to the essential and non-essential measurements respectively. Applying the Peters-Wilkinson [12] decomposition:

$$\begin{bmatrix} H_1 \\ H_2 \end{bmatrix} = \begin{bmatrix} L_1 \\ M_2 \end{bmatrix} \cdot [U] \tag{4.62}$$

where
L_1 is a $n \times n$ lower triangular matrix,
M_2 is a $(m - n) \times n$ rectangular matrix,
U is a $n \times n$ upper triangular matrix.

Substituting Equation (4.62) into (4.61):

$$z_1 = L_1 \cdot U \cdot x \tag{4.63}$$
$$z_2 = M_2 \cdot U \cdot x \tag{4.64}$$

Eliminating $U \cdot x$:

$$z_2 = M_2 \cdot L_1^{-1} \cdot z_1 \tag{4.65}$$
$$z_2 = T \cdot z_1 \tag{4.66}$$

Equation (4.66) shows the linear dependency among the non-essential and essential measurements. Hence, an element of z_1 will be critical if the corresponding column of T is null.

Example 4.9:

Consider the 6-bus power system and the measurement configuration given in Figure 4.14. Determine the critical measurements. Solution:

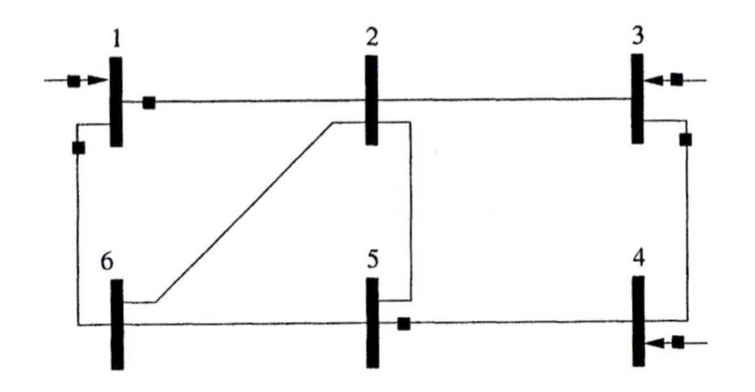

Figure 4.14. Example System Containing Critical Measurement.

First, form the measurement Jacobian H, excluding the first column for bus 1 which is chosen as the slack bus:

$$H = \begin{bmatrix} -1 & 0 & 0 & 0 & -1 \\ -1 & 2 & -1 & 0 & 0 \\ 0 & -1 & 2 & -1 & 0 \\ 0 & 1 & -1 & 0 & 0 \\ -1 & 0 & 0 & 0 & 0 \\ 0 & 0 & 0 & 0 & -1 \\ 0 & 0 & -1 & 1 & 0 \end{bmatrix}$$

Partition H into H_1, which contains the first 5 rows of essential measurements, and H_2, which contains the last 2 rows. Decompose H into LU factors using the Peters-Wilkinson decomposition method:

$$\left[\begin{array}{ccccc} 1.0 & 0 & 0 & 0 & 0 \\ 1.0 & 1.0 & 0 & 0 & 0 \\ 0 & -0.5 & 1.0 & 0 & 0 \\ 0 & 0 & -0.66 & 1.0 & 0 \\ 0 & 0 & 0 & 0 & 1.0 \\ \hline 1.0 & 0 & 0 & 0 & -1.0 \\ 0 & 0.5 & -0.33 & -1.0 & 0.0 \end{array}\right] \cdot \begin{bmatrix} -1.00 & 0 & 0 & 0 & -1.00 \\ 0 & 2.00 & -1.00 & 0 & 1.00 \\ 0 & 0 & 1.50 & -1.00 & 0.50 \\ 0 & 0 & 0 & 0.33 & 0.33 \\ 0 & 0 & 0 & 0 & -1.00 \end{bmatrix}$$

Then, the matrix T will be given by:

$$T = M_2 \cdot L_1^{-1} = \begin{bmatrix} 1.00 & 0.00 & 0.00 & 0.00 & -1.0 \\ 0.00 & 0.00 & -1.00 & -1.00 & 0.0 \end{bmatrix}$$

The second column of T is null, indicating that the measurement corresponding to the second row of H_1 is critical. In this example, this measurement is the injection at bus 3.

4.9 Measurement Design

This chapter has so far discussed various methods of analyzing network observability for a given power system and its associated measurement configuration. Methods have also been presented for placement of new measurements in order to turn an unobservable system into an observable one. However, in certain situations, considerations may go beyond just checking observability when placing new measurements. In particular, measurements may be added in order to maintain a certain level of reliability against loss of measurements or branch outages. This problem is more involved due to the additional considerations including the measurement costs. One possible formulation of this optimal measurement design problem and its solution can be found in [14].

4.10 Summary

This chapter presents the methods for network observability analysis. The methods can be broadly divided into numerical and topological categories. Both methods can be used for not only determining network observability, but also for identifying observable islands, any existing critical measurements, and placing new measurements to restore observability. These methods are discussed in sufficient detail to enable their implementation in a computer program.

4.11 Problems

1. The measurement configuration for a 14 bus test system is shown in Figure 4.15. Use the topological observability analysis method to determine the following:

   (a) All irrelevant branches.

   (b) All irrelevant injections.

   (c) All observable islands.

   (d) All unobservable branches.

   Suggest the location and type of a set of minimum number of measurements to be added to the measurement list in order to make the system observable.

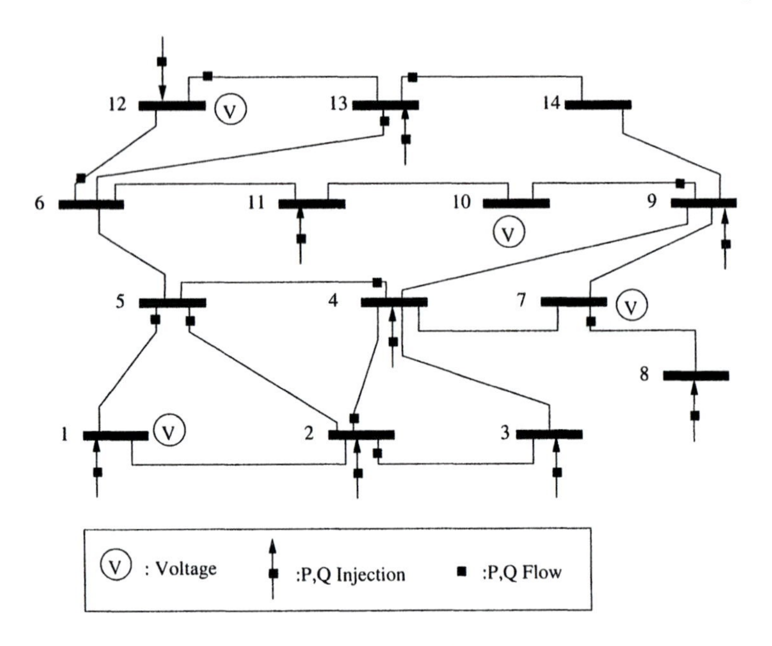

Figure 4.15. Test system for Problem 1

2. Use the numerical observability method to solve problem 1.

3. Use the numerical observability analysis method and identify the observable islands for the 10 bus test system given Figure 4.16.

4. Suggest a minimum number of measurements to be placed in the system of Problem 3, so that the system will be fully observable. Use the multiple measurement placement method.

5. Considering the system given for Problem 3. Form the $\mathcal{L}$ matrix by choosing the tree branches as:

   (a) $\tau_1 = \{1, 3, 6, 11, 15, 14, 8, 7, 10\}$

   (b) $\tau_1 = \{9, 5, 7, 10, 13, 11, 15, 2, 3\}$

   Compare the number of non-zero elements in $\mathcal{L}$ for both cases. Can you choose another tree that will yield a sparser $\mathcal{L}$?

6. Consider the system and measurement configuration given for Problem 3. Determine the observable islands using the branch variable based numerical method. Compare your results with those of Problem 3.

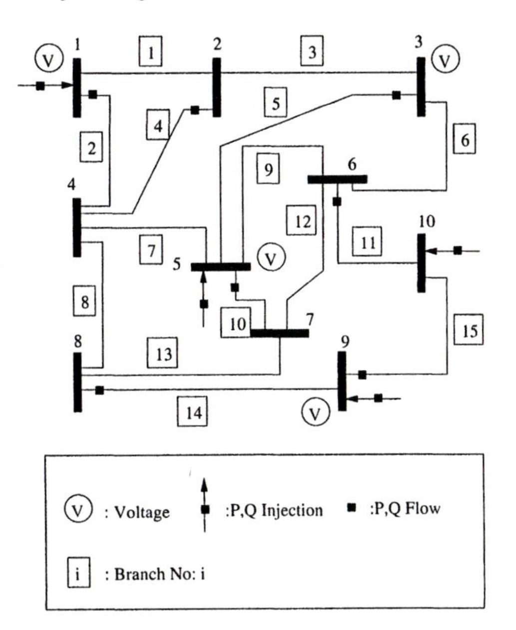

Figure 4.16. Test system for Problem 3

Design Project:

This is a project that can be carried out by two or more students as a team.

Objective:

To accomplish the most economical design that complies with the technical/cost specifications given below.

Design Specifications:

Design a measurement system for the IEEE-30 bus test system, so that the system remains observable and single bad data in any of the measurements can be detected during normal operation as well as in case of any one of the contingencies listed below:

1. Outage of line 15-23
2. Outage of line 2-5
3. Outage of line 22-24

4. Outage of line 10-20
5. Outage of transformer 27-28
6. Outage of line 16-17
7. Outage of line 12-14
8. Outage of line 12-15

Assume appropriate costs in your design for the following items:

- Remote Terminal Unit (RTU)
- Voltage magnitude meter
- P-Q meter

Note that every measurement needs to be assigned one RTU for communicating with the control center. Assume an appropriate number of available channels for each RTU.

Detailed information including the system diagram and network data for the IEEE-30 bus system can be downloaded from :
http://www.ee.washington.edu/research/pstca/pf30/pg_tca30bus.htm

Design Project Report Format

Prepare a report describing your design. The report should contain the following:

1. Executive summary (not to exceed 1 page)
2. Main body:
    - Problem statement
    - Technical approach used in the design
    - Results: tables, charts or plots
3. References: if used, should be referred to in text.
4. Appendix:
    - System diagram(s)
    - Simulation results
    - Listing of input or output files (if necessary), programs (if developed).

References

[1] H. Hale, "The Inverse of a Nonsingular Submatrix of an Incidence Matrix", *IRE Trans. on CT*, 1962, pp. 299-300.

[2] A.G. Exposito, A. Abur and E.R. Ramos, "On the Use of Loop Equations in Power System Analysis", *Proceedings of the IEEE International Symposium on Circuits and Systems*, Seattle, Washington, April 29 - May 3, 1995, Vol.2, pp. 1504-1507.

[3] A. Abur and A.G. Exposito, "Detecting Multiple Solutions in State Estimation in the Presence of Current Magnitude Measurements", *IEEE Trans. on Power Systems*, Vol.12, No.1, February 1997, pp. 370-376.

[4] A.G. Exposito and A. Abur, "Generalized Observability Analysis and Measurement Classification", *IEEE Trans. on Power Systems*, Vol.13, No.3, August 1998, pp.1090-1096.

[5] A. Monticelli and F.F. Wu, "Network Observability: Theory", *IEEE Transactions on PAS*, Vol.PAS-104, No.5, May 1985, pp.1042-1048.

[6] A. Monticelli and F.F. Wu, "Network Observability: Identification of Observable Islands and Measurement Placement", *IEEE Transactions on PAS*, Vol.PAS-104, No.5, May 1985, pp.1035-1041.

[7] G.R. Krumpholz, K.A. Clements and P.W. Davis, "Power System Observability: A Practical Algorithm Using Network Topology", *IEEE Trans. on Power Apparatus and Systems*, Vol. PAS-99, No.4, July/Aug. 1980, pp.1534-1542.

[8] K.A. Clements, G.R. Krumpholz and P.W. Davis, "Power System State Estimation with Measurement Deficiency: An Observability/Measurement Placement Algorithm", *IEEE Trans. on Power Apparatus and Systems*, Vol. 102(7), 1983, pp.2012 – 2020.

[9] K.A. Clements, G.R. Krumpholz and P.W. Davis, "Power System State Estimation with Measurement Deficiency: An Algorithm that Determines the Maximal Observable Subnetwork", *IEEE Trans. on Power Apparatus and Systems*, Vol. 101(7), 1982, pp.3044 – 3052.

[10] K.A. Clements, G.R. Krumpholz and P.W. Davis, "Power System State Estimation Residual Analysis: An Algorithm Using Network Topology", *IEEE Trans. on Power Apparatus and Systems*, Vol. 100, No.4, April 1981, pp.1779 – 1787.

[11] R. Nucera and M. Giles, "Observability Analysis: A New Topological Algorithm", *IEEE Trans. on Power Systems*, Vol.6, 1991, pp.466 – 473.

[12] G. Peters and J.H. Wilkinson, "The Least-squares Problem and Pseudo-inverses", *The Computer Journal*, Vol.13, No.4, August 1970, pp.309 – 316.

[13] Gou Bei and Abur, A., "An Improved Measurement Placement Algorithm for Network Observability", IEEE Trans. on Power Systems, Vol.16, No.4, November 2001, pp.819-824.

[14] Magnago, F.H. and Abur, A., "Unified Approach to Robust Meter Placement Against Bad Data and Branch Outages", IEEE Trans. on Power Systems, Vol.15, No.3, August 2000, pp.945-949.

Chapter 5

Bad Data Detection and Identification

One of the essential functions of a state estimator is to detect measurement errors, and to identify and eliminate them if possible. Measurements may contain errors due to various reasons. Random errors usually exist in measurements due to the finite accuracy of the meters and the telecommunication medium. Provided that there is sufficient redundancy among measurements, such errors are expected to be filtered by the state estimator. The nature of this filtering action will depend on the specific method of estimation employed.

Large measurement errors can also occur when the meters have biases, drifts or wrong connections. Telecommunication system failures or noise caused by unexpected interference also lead to large deviations in recorded measurements.

Apart from these, a state estimator may be deceived by incorrect topology information which will be subsequently interpreted as bad data by the state estimator. Such situations are more complicated to deal with and the treatment of topology errors will be separately discussed in Chapter 7.

Some bad data are obvious and can be detected and eliminated apriori state estimation, by simple plausibility checks. Negative voltage magnitudes, measurements with several orders of magnitude larger or smaller than expected values, or large differences between incoming and leaving currents at a connection node within a substation are some examples of such bad data. Unfortunately, not all types of bad data are easily detectable by such means. Hence, state estimators have to be equipped with more advanced features that will facilitate the detection and identification of any type of bad data.

Treatment of bad data depends on the method of state estimation used in the implementation. This chapter will focus on the bad data detection and identification techniques that are associated with the commonly used WLS method. Other state estimation methods such as those that will be discussed in Chapter 6, incorporate bad data processing as part of the state estimation procedure and hence their discussion will involve aspects of their treatment of bad data as well.

When using the WLS estimation method, detection and identification of bad data are done only after the estimation process by processing the measurement residuals. The analysis is essentially based on the properties of these residuals, including their expected probability distribution.

Bad data may appear in several different ways depending upon the type, location and number of measurements that are in error. They can be broadly classified as:

1. Single bad data: Only one of the measurements in the entire system will have a large error.

2. Multiple bad data: More than one measurement will be in error.

Multiple bad data may appear in measurements whose residuals are strongly or weakly correlated. Strongly correlated measurements are those whose errors affect the estimated value of each other significantly, causing the good one to also appear in error when the other contains a large error. Estimates of measurements with weakly correlated residuals are not significantly affected by the errors of each other. When measurement residuals are strongly correlated their errors may or may not be conforming. Conforming errors are those that appear consistent with each other. Multiple bad data can therefore be further classified into three groups:

1. Multiple non-interacting bad data: Bad data in measurements with weakly correlated measurement residuals.

2. Multiple interacting but non-conforming bad data: Non-conforming bad data in measurements with strongly correlated residuals.

3. Multiple interacting and conforming bad data: Consistent bad data in measurements with strongly correlated residuals.

Quantifying the degree of interaction between measurements and analysis of errors can be carried out based on the sensitivities of measurement residuals to measurement errors. Properties of the measurement residuals that are obtained by the WLS state estimation method will be reviewed below for this purpose.

5.1 Properties of Measurement Residuals

Consider the linearized measurement equations:

$$\Delta z = H\Delta x + e \tag{5.1}$$

where, $E(e) = 0$ and $cov(e) = R$, which is a diagonal matrix based on the assumption that measurement errors are not correlated. Note that measurement residuals may still be correlated even if errors are assumed independent.

Then, the WLS estimator of the linearized state vector will be given by:

$$\begin{aligned} \Delta\hat{x} &= (H^T R^{-1} H)^{-1} H^T R^{-1} \Delta z \\ &= G^{-1} H^T R^{-1} \Delta z \end{aligned} \tag{5.2}$$

and the estimated value of Δz:

$$\Delta\hat{z} = H\Delta\hat{x} = K\Delta z \tag{5.3}$$

where $K = HG^{-1}H^T R^{-1}$ and sometimes is called the *hat* matrix, for putting a hat on Δz.

A rough idea about the local measurement redundancy around a given meter can be obtained, by checking the corresponding row entries in the matrix K. A large diagonal entry relative to the off-diagonal elements in K, will imply that the estimated value corresponding to that measurement is essentially determined by its measured value, i.e. the local redundancy is poor. Furthermore, the matrix K can be shown to have the following properties:

$$K \cdot K \cdot K \cdots K = K \tag{5.4}$$

$$K \cdot H = H \tag{5.5}$$

$$(I - K) \cdot H = 0 \tag{5.6}$$

Now, the measurement residuals can be expressed as follows:

$$\begin{aligned} r &= \Delta z - \Delta\hat{z} \\ &= (I - K)\Delta z \\ &= (I - K)(H\Delta x + e) \\ &= (I - K)e \quad \text{[Substituting Equation(5.6)]} \\ &= Se \end{aligned} \tag{5.7}$$

The matrix S, called the *residual sensitivity matrix*, represents the sensitivity of the measurement residuals to the measurement errors. It has the following properties:

- It is not a symmetric matrix unless the covariance of errors are all equal, i.e. $R = kI$, where k is any scalar.
- $S \cdot S \cdot S \cdots S = S$
- $S \cdot R \cdot S^T = S \cdot R$

WLS estimation is based on the assumption that the measurement errors are distributed according to a Gaussian distribution given as below:

$$e_i \sim N(0, R_{ii}) \quad \text{for all } i$$

Using the linear relation between the measurement residuals and errors given by Equation (5.7), the mean and the covariance, and hence the probability distribution of the measurement residuals can be obtained as follows:

$$\begin{aligned} E(r) &= E(S \cdot e) = S \cdot E(e) = 0 && (5.8) \\ Cov(r) &= \Omega = E[rr^T] \\ &= S \cdot E[ee^T] \cdot S^T \\ &= SRS^T \quad \text{[See the properties of } S \text{ above.]} \\ &= SR && (5.9) \end{aligned}$$

Therefore:

$$r \sim N(0, \Omega)$$

The off-diagonal elements of the residual covariance matrix Ω can be used to identify those strongly versus weakly interacting measurements.

If $\Omega_{ij} \geq \epsilon$, then measurement i and j are said to be strongly interacting. Else, these measurements are considered as weakly interacting or non-interacting at all. The threshold ϵ depends on the network and measurement topology as well as the desired level of selectivity among measurements.

Residual covariance matrix Ω has some interesting properties which will be useful in the subsequent discussion of identification of bad data. Some of these properties are stated below:

- Ω is a real and symmetric matrix.
- $\Omega_{ij}^2 \leq \Omega_{ii} \cdot \Omega_{jj}$
- $\Omega_{ij} \leq \frac{(\Omega_{ii} + \Omega_{jj})}{2}$

Example 5.1:

Consider the 3-bus system and its measurement configuration given in Example 2 in Chapter 2. Using the DC measurement model, find the hat matrix K, the sensitivity matrix S and the covariance matrix Ω for the residuals. Verify the properties of the residual covariance matrix Ω.

Solution:

The linearized $P-\theta$ measurement equation is given by:

$$\Delta z_A = H_{AA} \cdot \theta + e$$

where

$$H_{AA} = \begin{matrix} p_{12} \\ p_{13} \\ p_2 \end{matrix} \begin{bmatrix} \theta_2 & \theta_3 \\ -33.33 & \\ & -20.0 \\ 45.8 & -12.5 \end{bmatrix}$$

Using the given measurement error covariance matrix:

$$R_{AA} = \begin{bmatrix} 0.008^2 & & \\ & 0.008^2 & \\ & & 0.01^2 \end{bmatrix}$$

Corresponding decoupled active gain matrix will then be given as:

$$G_{AA} = 10^7 \cdot \begin{bmatrix} 3.837 & -0.5729 \\ -0.5729 & 0.7812 \end{bmatrix}$$

Hat matrix K can be built as:

$$K = H_{AA} \cdot G_{AA}^{-1} \cdot H_{AA}^T \cdot R_{AA}^{-1} = \begin{bmatrix} 0.5084 & 0.2236 & -0.3577 \\ 0.2236 & 0.8983 & 0.1627 \\ -0.5589 & 0.2542 & 0.5932 \end{bmatrix}$$

and the sensitivity matrix S will be given as:

$$S = I - K = \begin{bmatrix} 0.4916 & -0.2236 & 0.3577 \\ -0.2236 & 0.1017 & -0.1627 \\ 0.5589 & -0.2542 & 0.4068 \end{bmatrix}$$

Finally, the residual covariance matrix Ω will be:

$$\Omega = S \cdot R_{AA} = 10^{-4} \begin{bmatrix} 0.3146 & -0.1431 & 0.3577 \\ -0.1431 & 0.0651 & -0.1627 \\ 0.3577 & -0.1627 & 0.4068 \end{bmatrix}$$

Note that Ω is real and symmetric. The off-diagonal entries can be verified to remain less than both the arithmetic and geometric mean of the corresponding diagonal entries. Two of them are illustrated below:

$$\Omega_{1,3}^2 = 1.2796 \times 10^{-9} \quad \leq \quad \Omega_{1,1} \cdot \Omega_{3,3} = 1.2796 \times 10^{-9}$$

$$\Omega_{1,2} = -1.4309 \times 10^{-5} \quad \leq \quad \frac{\Omega_{1,1} + \Omega_{2,2}}{2} = 1.8984 \times 10^{-5}$$

5.2 Classification of Measurements

Power systems may contain various types of measurements spread out in the system with no apparent topological pattern. These measurements will exhibit different properties and affect the outcome of the state estimation accordingly, depending upon not only their values but also their location. Therefore, they may belong to one or more of the following categories [7]:

Critical measurement: A critical measurement is the one whose elimination from the measurement set will result in an unobservable system. The column of the residual covariance matrix Ω, corresponding to a critical measurement will be identically equal zero. Furthermore, the measurement residual of a critical measurement will always be zero.

Redundant measurement: A redundant measurement is a measurement which is not critical. Only redundant measurements may have nonzero measurement residuals.

Critical pair: Two redundant measurements whose simultaneous removal from the measurement set will make the system unobservable.

Critical k-tuple: A critical k-tuple contains k redundant measurements, where removal of all of them will cause the system to become unobservable. None of these k measurements belong to a critical tuple of lower order. Those k columns of the residual covariance matrix Ω, corresponding to the members of a critical k-tuple, will be linearly dependent.

5.3 Bad Data Detection and Identifiability

Detection refers to the determination of whether or not the measurement set contains any bad data. Identification is the procedure of finding out which specific measurements actually contain bad data. Detection and identifiability of bad data depends on the configuration of the overall measurement set in a given power system.

Bad data can be detected if removal of the corresponding measurement does not render the system unobservable. In other words, bad data appearing in critical measurements can not be detected.

A single measurement containing bad data can be identified if and only if:

- it is not critical **and**
- it does not belong to a critical pair.

Bad data processing logic should be able to recognize the above inherent limitations of detection and single bad data identification. Provided that the above conditions are observed, single bad data can be detected and identified by the methods outlined next. The case of multiple bad data is more difficult to handle and will be discussed later in sections 5.7.2 and 5.8.

5.4 Bad Data Detection

One of the methods used for detecting bad data is the *Chi-squares* test. Once bad data are detected, they need to be identified and eliminated or corrected, in order to obtain an unbiased state estimate. Bad data identification methods will be discussed later in section 5.6.

5.4.1 Chi-squares χ^2 Distribution

Consider a set of N independent random variables $X_1, X_2, \ldots X_N$, where each X_i is distributed according to the Standard Normal distribution:

$$X_i \sim N(0,1)$$

Then, a new random variable Y defined by:

$$Y = \sum_{i=1}^{N} X_i^2$$

will have a χ^2 distribution with N degrees of freedom, i.e.

$$Y \sim \chi_N^2$$

The degrees of freedom N, represents the number of independent variables in the sum of squares. This value will decrease if any of the X_i variables form a linearly dependent subset.

Now, let us consider the function $f(x)$, written in terms of the measurement errors:

$$f(x) = \sum_{i=1}^{m} R_{ii}^{-1} e_i^2 = \sum_{i=1}^{m} \left(\frac{e_i}{\sqrt{R_{ii}}}\right)^2 = \sum_{i=1}^{m} (e_i^N)^2 \tag{5.10}$$

where e_i is the ith measurement error, R_{ii} is the diagonal entry of the measurement error covariance matrix and m is the total number of measurements. Assuming that e_i's are all Normally distributed random variables with zero mean and R_{ii} variance, e_i^N's will have a Standard Normal distribution, i.e.

$$e_i^N \sim N(0,1)$$

Then, $f(x)$ will have a χ^2 distribution with at most $(m-n)$ degrees of freedom. In a power system, since at least n measurements will have to satisfy the power balance equations, at most $(m-n)$ of the measurement errors will be linearly independent. Thus, the largest degree of freedom can be $(m-n)$, i.e. the difference between the total number of measurements and the system states.

5.4.2 Use of χ^2 Distribution for Bad Data Detection

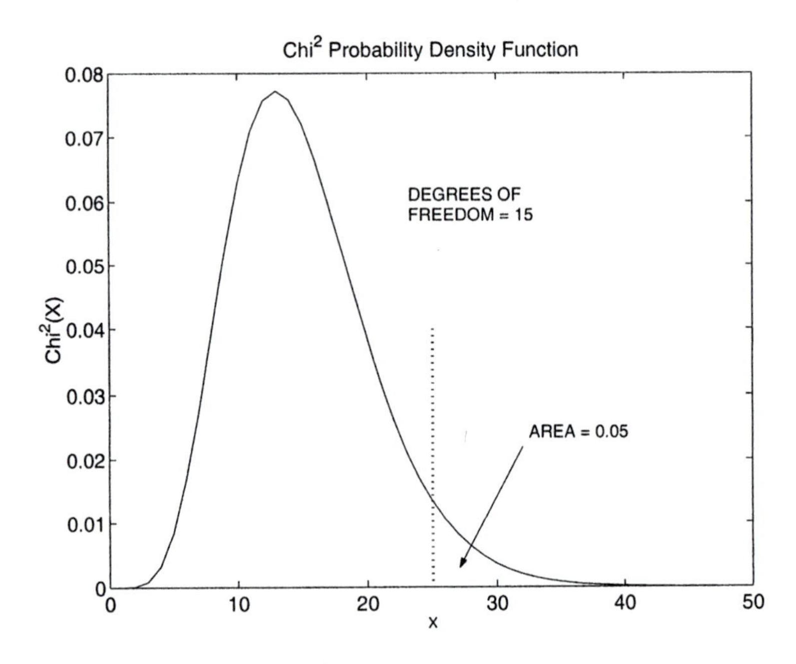

Figure 5.1. χ^2 Probability Density Function

A plot of the χ^2-probability density function (p.d.f) is shown in Figure 5.1. The area under the p.d.f. represents the probability of finding X in the corresponding region, for example:

$$Pr\{X \geq x_t\} = \int_{x_t}^{\infty} \chi^2(u) \cdot du \tag{5.11}$$

represents the probability of X being larger than a certain threshold x_t. This probability decreases with increasing values of x_t, due to the decaying tail of the distribution. Choosing a probability of error, such as 0.05, the threshold x_t can be set such that:

$$Pr\{X \geq x_t\} = 0.05$$

In Figure 5.1, this threshold corresponds to $x_t = 25$ indicated by the vertical dotted line. The threshold represents the largest acceptable value for X that will not imply any bad data. If the measured value of X exceeds this threshold, then with 0.95 probability, the measured X will not have a χ^2 distribution, i.e. presence of bad data will be suspected.

Tables containing Chi squares cumulative distribution function values for different degrees of freedom can be found in various statistical publications. Alternatively, Matlab's statistical toolbox can be used to evaluate specific values, as illustrated in the following example.

Example 5.2:

Consider 5 independent measurements of a quantity, given as follows:

Measured variable	x_1	x_2	x_3	x_4	x_5
Measured value	0.5	-1.2	0.80	0.20	-3.1

Assume that the measurements are taken from a sample which is known to have a Standard Normal distribution, i.e.

$$X_i \sim N(0,1) \quad \text{for all } i$$

Use χ^2 distribution to check for bad data with 99% confidence.

Solution:
Let us form the sum of squares of the measured variables:

$$y = \sum_{i=1}^{5} x_i^2 = 11.98$$

The probability of obtaining this value (11.98) when Y indeed has a χ_4^2 distribution, can be found by using the Matlab Statistical Toolbox function CHI2CDF(Y,DF). DF is the degrees of freedom, which is 4 for this example. This probability, denoted by P will be obtained as:

$$\text{P} = \text{CHI2CDF}(11.98, 4) = 0.9825$$

Since $0.9825 < 0.99$, bad data will not be suspected with 99% confidence.

Alternatively, the test threshold at the 99% confidence level can be obtained by using another one of the Matlab functions called CHI2INV(P,DF) where P is the confidence probability level, which is 0.99 for this example. Execution of this function yields the corresponding threshold y_t, which represents the largest acceptable value for y, without suspecting any bad data with 99% confidence:

$$y_t = \text{CHI2INV}(0.99, 4) = 13.28$$

Again, since $y_t = 13.28 > y = 11.98$, bad data will not be suspected for this example.

5.4.3 χ^2-Test for Detecting Bad Data in WLS State Estimation

The WLS state estimation objective function $J(x)$ can be used to approximate the above function $f(x)$ and a bad data detection test, referred to as the Chi-squares test for bad data, can be devised based on the properties of the χ^2 distribution.

The steps of the Chi-squares χ^2-test are given as follows:

- Solve the WLS estimation problem and compute the objective function:

$$J(\hat{x}) = \sum_{i=1}^{m} \frac{(z_i - h_i(\hat{x}))^2}{\sigma_i^2}$$

  where:
  $\hat{x}$: estimated state vector of dimension n.
  $h_i(\hat{x})$: estimated measurement i.
  z_i : measured value of the measurement i.
  $\sigma_i^2 = R_{ii}$: variance of the error in measurement i.
  m : number of measurements.

- Look up the value from the Chi-squares distribution table corresponding to a detection confidence with probability p (e.g. 95%) and $(m-n)$ degrees of freedom. Let this value be $\chi^2_{(m-n),p}$.

  Here $p = \Pr\ (\text{J}(\hat{\text{x}}) \leq \chi^2_{(m-n),p})$.

- Test if $J(\hat{x}) \geq \chi^2_{(m-n),p}$.
  If yes, then bad data will be suspected.
  Else, the measurements will be assumed to be free of bad data.

Example 5.3:

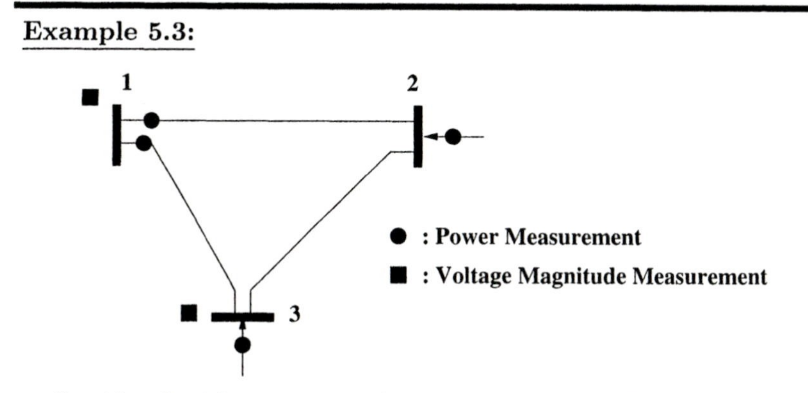

Consider the 3-bus system and its measurement configuration shown in the figure. The corresponding network data are given below:

Line		Resistance	Reactance	Total Susceptance
From Bus	To Bus	R (pu)	X (pu)	$2b_s$ (pu)
1	2	0.01	0.03	0.0
1	3	0.02	0.05	0.0
2	3	0.03	0.08	0.0

The number of state variables, n for this system is 5, made up of three bus voltage magnitudes and two bus voltage phase angles, slack bus phase angle being excluded from the state list. There are altogether $m = 10$ measurements, i.e. 2 voltage magnitude measurements, 2 pairs of real/reactive flows and 2 pairs of real/reactive injections. Therefore, the degrees of freedom for the approximate χ^2-distribution of the objective function $J(\hat{x})$ will be:

$$m - n = 10 - 5 = 5$$

Measurements are generated by solving the base case power flow and then adding Gaussian distributed errors. One of the measurements, P_2, is then changed intentionally, to simulate bad data. The state estimation solution and the objective function values that are obtained for both cases, are shown in the tables below.

Bus No:	Estimated State			
	No Bad Data		One Bad Data	
	V	θ°	V	θ°
1	1.0000	0	1.0000	0.00
2	0.9886	−0.84	0.9886	−0.67
3	0.9834	−1.19	0.9834	−1.20

Measurement No:	Measurement Type	Measured Value	
		No Bad Data	One Bad Data
1	V_1	1.0065	1.0065
2	V_3	0.9769	0.9769
3	P_2	−0.4007	**−0.3507**
4	P_3	−0.4857	−0.4857
5	Q_2	−0.3052	−0.3052
6	Q_3	−0.3850	−0.3850
7	P_{12}	0.4856	0.4856
8	P_{13}	0.4054	0.4054
9	Q_{12}	0.3821	0.3821
10	Q_{13}	0.3367	0.3367
$J(\hat{x})$		6.1	22.8

The test threshold at 95% confidence level is obtained by Matlab function CHI2INV as:

$$y_t = \text{CHI2INV}(0.95, 5) = 11.1$$

In the first case, since $J(\hat{x}) = 6.1 < 11.1$, bad data will not be suspected. However, the test will detect bad data for the second case, since the corresponding value of $J(\hat{x}) = 22.8$ exceeds the χ^2-test threshold of 11.1.

5.4.4 Use of Normalized Residuals for Bad Data Detection

As described above, the χ^2-test is inaccurate due to the approximation of errors by residuals in Equation (5.10). Therefore it may fail to detect bad data for certain cases. A more accurate test for detecting bad data can be devised by using the normalized residuals. Normalized value of the residual for measurement i can be obtained by simply dividing its absolute value by the corresponding diagonal entry in the residual covariance matrix:

$$r_i^N = \frac{\mid r_i \mid}{\sqrt{\Omega_{ii}}} = \frac{\mid r_i \mid}{\sqrt{R_{ii} S_{ii}}} \tag{5.12}$$

The normalized residual vector r^N will then have a Standard Normal distribution, i.e.

$$r_i^N \sim N(0,1)$$

Thus, the largest element in r^N can be compared against a statistical threshold to decide on the existence of bad data. This threshold can be chosen based on the desired level of detection sensitivity.

5.5 Properties of Normalized Residuals

It can be shown that if there is a single bad data in the measurement set (provided that it is neither a critical measurement nor a member of a critical pair) the largest normalized residual will correspond to the erroneous measurement. This property may hold true even for certain multiple bad data cases, where bad measurements have very weak correlation, i.e. they are essentially non-interacting.

Consider the case where the only bad data occurs in measurement k, i.e. $e_k \neq 0$ and all the remaining measurements are free of errors, $e_j = 0, j \neq k$. Using Equation (5.7), the normalized residual for the erroneous measurement k, can be shown to be the largest among all other error free measurements:

$$
\begin{aligned}
r_j &= S_{jk} \cdot e_k \qquad j = 1, \ldots, m. \\
r_j^N &= \frac{S_{jk} \cdot e_k}{\sqrt{R_{jj}}\sqrt{S_{jj}}} = \frac{\Omega_{jk} \cdot e_k}{\sqrt{\Omega_{jj}}R_{kk}} \\
&\leq \frac{\sqrt{\Omega_{jj}}\sqrt{\Omega_{kk}} \cdot e_k}{\sqrt{\Omega_{jj}}R_{kk}} \qquad \text{[Using the property} \quad \Omega_{jk}^2 \leq \Omega_{jj} \cdot \Omega_{kk}] \\
&= \frac{\sqrt{\Omega_{kk}} \cdot e_k}{R_{kk}} = \frac{S_{kk} \cdot e_k}{\sqrt{R_{kk}}\sqrt{S_{kk}}} = r_k^N
\end{aligned}
$$

The above inequality becomes a strict equality, if the measurements j and k form a critical pair, since the corresponding columns of Ω matrix will be linearly dependent. Hence, the normalized residuals of a critical pair will always be equal, making the identification of bad data impossible even though it can be detected. The same is true for any $(k-1)$ member subset of a set of measurements forming a critical k-tuple, i.e. errors associated with them can be detected but not identified.

5.6 Bad Data Identification

Upon detection of bad data in the measurement set, their identification can be accomplished by further processing of the residuals. Among the existing methods, two of them, namely the *Largest Normalized Residual* (r^N_{max}) *Test* and the *Hypothesis Testing Identification (HTI)* method, will be described here.

5.7 Largest Normalized Residual (r^N_{max}) Test

The properties of normalized residuals for a single bad data existing in the measurement set, can be used to devise a test for identifying and subsequently eliminating bad data. The test is referred to as the *Largest Normalized Residual* r^N_{max}-test and is composed of the following steps:

1. Solve the WLS estimation and obtain the elements of the measurement residual vector:

$$r_i = z_i - h_i(\hat{x}), \qquad i = 1, \ldots, m$$

2. Compute the normalized residuals:

$$r_i^N = \frac{|\, r_i \,|}{\sqrt{\Omega_{ii}}} \qquad i = 1, \ldots, m$$

3. Find k such that r_k^N is the largest among all r_i^N, $i = 1, \ldots, m$.

4. If $r_k^N > c$, then the k-th measurement will be suspected as bad data. Else, stop, no bad data will be suspected. Here, c is a chosen identification threshold, for instance 3.0.

5. Eliminate the k-th measurement from the measurement set and go to step 1.

Example 5.4:

Consider the same system studied in Example 5.3. Apply the normalized residual test to identify and eliminate the bad data for this measurement set.

The WLS state estimator results for the significant measurement residuals sorted in descending order are given in the below table.

Measurement, i	r_i^N	r_i
P_2	4.2	0.0286
P_{12}	4.1	0.0235
V_1	2.3	0.006
V_3	2.3	−0.006
P_3	1.1	0.007

The threshold of detection is assumed to be 3.0. Hence, the power injection at bus 2, is identified as bad data and eliminated from the measurement set. State estimation is repeated using the modified measurement set and the largest normalized residuals and the state estimation solution are given below.

Measurement, i	r_i^N	r_i
V_1	2.3	0.006
V_3	2.2	−0.006

Bus, i	V_i	θ_i°
1	0.9999	0
2	0.9886	−0.84
3	0.9833	−1.19

Since the largest r^N value is below the detection threshold 3.0, the test is terminated and the measurement set is declared to be free of errors.

5.7.1 Computational Issues

Implementation of the largest normalized residual test may require several identification/elimination cycles. Each cycle will involve two computationally intensive stages:

- normalized residuals are calculated using the diagonal entries of the residual covariance matrix Ω.
- identified bad measurement with the largest normalized residual will be removed from the measurement set before repeating the state estimation procedure.

Computation of the matrix Ω

Using Equation (5.7):

$$S = (I - K) = (I - HG^{-1}H^T R^{-1}) \quad (5.13)$$
$$\Omega = SR = R - HG^{-1}H^T \quad (5.14)$$

Note that only the diagonal entries of Ω are needed. Furthermore, the gain matrix G has already been decomposed into its triangular Cholesky factors L, during the state estimation iterations:

$$G = LL^T$$

Each row of H corresponds to a measurement, and can be denoted by the row vector h_i for measurement i. Then, H can be written in terms of its row vectors as follows:

$$H = \begin{bmatrix} h_1 \\ h_2 \\ \vdots \\ h_m \end{bmatrix}$$

Now, let:

$$T = G^{-1}H^T$$

The columns of the temporary matrix T can be obtained by solving the following linear sparse matrix equation for each of the m measurements:

$$LL^T T_i = h_i^T \qquad 1 \le i \le m \quad (5.15)$$

Since the sparse Cholesky factor L, is already available from the last state estimation iteration, solution of Equation (5.15) will require a forward and a backward substitution steps. Furthermore, both substitutions can be fast due to the very sparse structure of the right hand side vector h_i^T.

The diagonal entries of Ω can then be computed easily as follows:

$$\Omega_{ii} = R_{ii} - h_i \cdot T_i \qquad 1 \leq i \leq m \tag{5.16}$$

A computationally more efficient alternative is the use of sparse inverse method to obtain only the necessary elements of G^{-1}. These correspond to the locations of nonzero elements in the products $h_i^T h_i$ for all $i = 1, \ldots, m$. Details of sparse inverse method are given in Appendix B.

Removal of the identified bad data

Once bad data are identified, they have to be removed from the measurement set before the next cycle of state estimation. Actual removal of the bad measurement may be avoided by subtracting the estimated error from the bad measurement as explained below.

Assume that all measurements are error-free except for the measurement i, which can be written as:

$$z_i + e_i = z_i^{bad} \tag{5.17}$$

where, z_i^{bad} is the measured value, z_i is the true value and e_i is the gross error associated with measurement i. Using the linearized residual sensitivity relation of Equation (5.7), the bad measurement residual can be approximated by:

$$z_i^{bad} - h(\hat{x}) = r_i^{bad} \approx S_{ii} e_i \tag{5.18}$$

where, $\hat{x}$ is the state estimate based on the measurement set including the bad measurement. Hence, an approximate value for the error e_i can be computed. Subtracting this error from the bad measurement yields:

$$z_i \approx z_i^{bad} - \frac{R_{ii}}{\Omega_{ii}} r_i^{bad} \tag{5.19}$$

State estimation can be repeated after correcting the bad measurement using the above approximation. The results of this estimation will provide approximately the same state estimate that would have been obtained if the measurement were actually removed from the measurement set. Exceptions to this case exist, when the linear residual sensitivity model, fails to properly approximate the changes in the residuals for large errors. Such cases will require iterative corrections to minimize the approximation error.

5.7.2 Strengths and Limitations of the r^N_{max} Test

The largest normalized residual test will perform differently depending upon the type of bad data and its configuration. Its performance and limitations are summarized below for all possible types of bad data.

Single Bad Data

When there is single bad data, the largest normalized residual will correspond to the bad measurement, provided that it is not critical or its removal does not create any critical measurements among the remaining ones.

Multiple Bad Data

Multiple bad data may appear in 3 ways:

- Non-interacting:
  If $S_{ik} \approx 0$, then measurement i and k are said to be non-interacting. In this case, even if bad data appear simultaneously in both measurements, the largest normalized residual test can identify them sequentially, one pass at a time.

- Interacting, non-conforming:
  If S_{ik} is significantly large, then measurements i and k are said to be interacting. However, if the errors in measurement i and k are not consistent with each other, then the largest normalized residual test may still indicate the bad data correctly.

- Interacting, conforming:
  If two interacting measurements have errors that are in agreement, then the largest normalized residual test may fail to identify either one.

The following example illustrates the above situations on a rather small scale example system. However, since bad data identification is a localized function for sparsely interconnected power system buses, similar results are likely to be observed for larger, actual size systems.

Example 5.5:

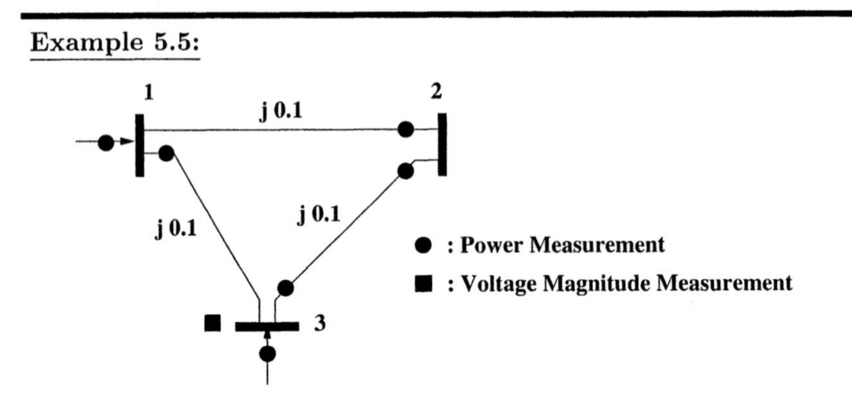

In the figure, all branches have identical impedances, z_i = j 0.1, and the voltage magnitudes are assumed to be known. State estimation is carried out for the phase angles of bus voltages using the real power measurements only. Three cases are simulated corresponding to single, multiple interacting non-conforming and multiple interacting conforming bad data present in the system. The results are summarized in the below table, where the largest normalized residual for each case is typed in boldface. Note that, the r^N_{max}-test fails when multiple interacting bad data are conforming, where errors introduced in measurements P_3 and P_{32} are consistent with each other.

Meas. Type	Bad Data					
	Single		Multiple Interacting			
			Non-conforming		Conforming	
	z_i	r_i^N	z_i	r_i^N	z_i	r_i^N
Flow 1-3	0.634	1.81	0.634	1.56	0.634	3.81
Flow 2-1	0.666	2.95	0.666	2.47	0.666	2.96
Flow 3-2	0.134	**11.1**	0.134	**14.5**	0.134	7.93
Flow 2-3	−0.034	4.57	−0.034	2.05	−0.034	**8.16**
Inj. 1	1.299	1.33	1.299	3.05	1.299	1.07
Inj. 3	−0.600	3.77	−0.700	11.16	−0.500	3.49

5.8 Hypothesis Testing Identification (HTI)

The main weakness of the r^N_{max} method is that it is based on the residuals which may be strongly correlated. Hence, in case of multiple bad data, this correlation may lead to comparable size residuals for good as well as bad measurements. A way to distinguish between good and bad measurements is by estimating the measurement errors directly, rather than devising tests based on the derived residuals. One such approach is the use of Hypothesis

Testing Identification (HTI) method [2, 3]. This method differs from the largest normalized residual test in that bad data are identified based on the computed estimates of measurement errors.

Estimation of all measurement errors using the calculated residuals is not possible, since the rank of S in Equation (5.7) is much less than the number of measurements m. In fact, the rank can not be larger than $(m-n)$ for a system with n states. Therefore, only those errors in at most $(m-n)$ measurements can be attempted to be estimated using a reduced form of the matrix S. Hence, the method's effectiveness depends upon this initial reduction, namely in the choice of an initial suspect measurement set which should include all bad data. HTI method makes use of the normalized residuals for this choice, and hence has a vulnerability due to the possibility of missing one or more bad data whose normalized residuals may appear small.

Consider that the WLS estimator is run and the normalized residuals are calculated. A set of measurements with the largest normalized residuals are picked making sure that they are linearly independent and non-critical. The number of such measurements which meet these conditions and are to be included in the initial suspect set is up to the user. The rest of the measurements are assumed to be error free. Then the sensitivity matrix S and the error covariance matrix R are partitioned according to the assumed suspect and true measurements:

$$r_s = S_{ss}e_s + S_{st}e_t \qquad (5.20)$$

$$r_t = S_{ts}e_s + S_{tt}e_t \qquad (5.21)$$

$$R = \begin{bmatrix} R_s & 0 \\ 0 & R_t \end{bmatrix} \qquad (5.22)$$

where:

r_s, r_t : is the residual vectors of suspect and true measurements
e_s, e_t : is the error vectors of suspect and true measurements
S_{ss},S_{st},S_{ts},S_{tt}, : partitioned submatrices of S
R_s, R_t : partitioned residual covariance matrices

Under the assumption that the true measurements are free of errors, i.e. $E[e_t] = 0$, an estimate for e_s can be obtained from Equation (5.20) as:

$$\hat{e}_s = S_{ss}^{-1} r_s \qquad (5.23)$$

Substituting back in Equation (5.20):

$$\hat{e}_s = e_s + S_{ss}^{-1} S_{st} e_t \qquad (5.24)$$

Note that, in order for S_{ss} to have an inverse, the suspect set should contain measurements which are independent and not critical. These are the conditions stated earlier for the selection of the suspect set. If the suspect set indeed contains all bad data in the measurement set, then the estimated errors and their expected statistical properties can be used to identify them by using the HTI algorithm. In order to devise a test for the estimated errors of the suspect set, their probability distributions must be known.

5.8.1 Statistical Properties of $\hat{\mathbf{e}}_s$

The mean and covariance of the estimated errors for the suspect set of measurements can be derived as follows:

1. Mean:
   If $E[e_t] = 0$, then $E[\hat{e}_s] = \hat{e}_s$.
   Else, $E[\hat{e}_s] \neq \hat{e}_s$.

2. Covariance:
   If $E[e_t] = 0$, then using Equation (5.24):

$$
\begin{aligned}
Cov(\hat{e}_s) &= Cov(e_s) + S_{ss}^{-1} S_{st} Cov(e_t)(S_{ss}^{-1} S_{st})^T && (5.25)\\
&= Cov(e_s) + S_{ss}^{-1} S_{st} R_t S_{st}^T S_{ss}^{-T} && (5.26)
\end{aligned}
$$

Recalling the property of residual sensitivity matrix:

$$
SRS^T = SR \tag{5.27}
$$

Substituting the partitioned form of S and R and equating the terms, will yield the following identity:

$$
S_{ss} R_s S_{ss}^T + S_{st} R_t S_{st}^T = S_{ss} R_s \tag{5.28}
$$

Substituting in Equation (5.26):

$$
\begin{aligned}
Cov(\hat{e}_s) &= Cov(e_s) + S_{ss}^{-1}(S_{ss} R_s - S_{ss} R_s S_{ss}^T) S_{ss}^{-T}\\
&= Cov(e_s) + S_{ss}^{-1}(R_s S_{ss}^T - S_{ss} R_s S_{ss}^T) S_{ss}^{-T}\\
&= Cov(e_s) + (S_{ss}^{-1} - I_s) R_s && (5.29)
\end{aligned}
$$

where I_s represents an identity matrix of order equal to the number of suspected measurements. The notation can be simplified by letting:

$$
T = S_{ss}^{-1}, \quad \sigma_i^2 = R_s(i,i)
$$

If $z_s(i)$ is good, then $e_s(i)$ is assumed to be a random variable with zero mean and $R_s(i,i)$ variance. Then $Cov(e_s)$ will cancel out R_s in Equation (5.29) and its distribution will be given by:

$$
\hat{e}_s(i) \sim N(0, T_{ii}\sigma_i^2)
$$

On the other hand, if $z_s(i)$ carries bad data, then in Equation (5.29) the variable $e_s(i)$ will be treated as an unknown but deterministic quantity [2]. Thus, its distribution will be given by:

$$\hat{e}_s(i) \sim N(e_s(i), (T_{ii} - 1)\sigma_i^2)$$

HTI method uses the above derived statistical properties of $\hat{e}_s(i)$ in order to devise a hypothesis test and its associated decision rules which are described below.

5.8.2 Hypothesis Testing

Hypothesis testing is a general method in making decisions about accepting or rejecting a statement. The statement being tested is referred to as the **null hypothesis** and denoted by H_0. Rejection of the null hypothesis implies acceptance of its complement, which is referred to as the **alternative hypothesis** and denoted by H_1.

There are two types of errors that one can make when accepting or rejecting H_0:

Type I error is rejection of H_0 when it is indeed true. The probability of making such an error is denoted by α in statistical convention. This probability is referred to as the probability of **false alarm** or the significance level of the test. Typical values for α are small such as 0.01, 0.05, or 0.10. The larger it is chosen, the more sensitive the decision will become to random errors in the observations.

Type II error is the error of rejecting the alternative hypothesis when it is indeed true. The probability of making such an error is denoted by β. It is the probability of **missing bad data** and its complement $(1 - \beta)$, which is referred to as the **power** of test.

Choosing the null H_0 and alternative H_1 hypotheses as:

H_0: measurement i is valid.
H_1: measurement i is in error.

they can be tested based on the decision rules derived below. Figure 5.2 shows an example of two distributions for $\hat{e}_{si}$ corresponding to the null H_0 and the alternative H_1 hypotheses. The distribution for H_1 has a mean of e_{si} which represents the assumed error for measurement i. Type I and type II error probabilities are also indicated in the figure for a given threshold of λ_i.

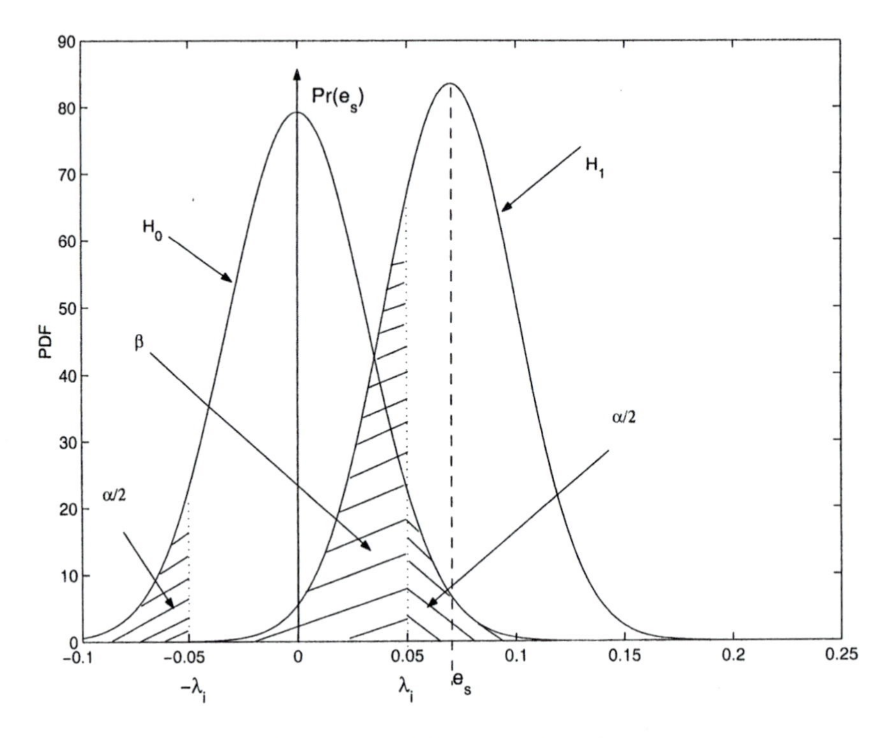

Figure 5.2. Type I and II errors in hypothesis testing

5.8.3 Decision Rules

There are two alternative strategies that can be chosen. One is to choose a fixed probability of false alarm, α and determine the corresponding threshold λ_i for deciding on the significance of $\mid \hat{e}_{si} \mid$. The other is to do the same based on a chosen fixed probability of β (or $1 - \beta$). Choice of λ_i for each strategy will be described below.

Fixed probability of false alarm, α

This type I error probability is defined as:

$$\alpha = Pr(\text{reject } H_0 \ \mid \ H_0 \text{ is true})$$

Note that if H_0 is true $\Longrightarrow \hat{e}_{si} \sim$ N (0, $\sigma_i^2 T_{ii}$). Then, λ_i should satisfy the following condition:

$$\alpha = Pr(\mid \hat{e}_{si} \mid > \lambda_i)$$

Normalizing the absolute value of the estimated error:

$$\mid \hat{e}_{si}^N \mid = \frac{\mid \hat{e}_{si} \mid}{\sigma_i \sqrt{T_{ii}}}$$

Since the normalized variable will have a standard Normal distribution, a proper cut-off value $N_{1-\frac{\alpha}{2}}$ can be looked up from the Standard Normal distribution table. Then:

$$Pr(\frac{|\hat{e}_{si}|}{\sigma_i\sqrt{T_{ii}}} > N_{1-\frac{\alpha}{2}}) = \alpha$$

Yielding the sought after threshold λ_i as:

$$\lambda_i = \sigma_i\sqrt{T_{ii}}N_{(1-\frac{\alpha}{2})}$$

Fixed probability of bad data identification, $(1-\beta)$

Similarly, one can define type II error probability and its complement as:

$$\beta \quad = \text{Pr (reject } H_1 \mid H_1 \text{ is true)}$$

$$1-\beta \quad = \text{Pr (accept } H_1 \mid H_1 \text{ is true)}$$

If H_1 is true $\Longrightarrow \hat{e}_{si} \sim \text{N} (e_{si}, \sigma_i^2(T_{ii}-1))$. Then,

$$\begin{aligned}\beta &= Pr(\hat{e}_{si} \leq \lambda_i) - Pr(\hat{e}_{si} \leq -\lambda_i) \\ \beta &\approx Pr(\hat{e}_{si} \leq \lambda_i)\end{aligned}$$

Normalizing $\hat{e}_{si}$:

$$\beta = \text{Pr} \left(\frac{\hat{e}_{si}-|e_{si}|}{\sigma_i\sqrt{T_{ii}-1}} \leq \underbrace{\frac{\lambda_i - |e_{si}|}{\sigma_i\sqrt{T_{ii}-1}}}_{N_\beta} \right)$$

N_β is the value which can be looked up from the standard Normal distribution table for the chosen value of β. Note that for the same threshold λ_i, the two cut-off values $N_{(1-\frac{\alpha}{2})}$ and N_β can be related by eliminating λ_i to yield:

$$\sigma_i N_\beta\sqrt{T_{ii}-1} = \sigma_i\sqrt{T_{ii}}N_{(1-\frac{\alpha}{2})} - |e_{si}|$$

Hence, HTI method can be implemented either under fixed α or fixed β strategy. The steps of implementing the HTI method for fixed β will be given only. Note that for small values of T_{ii}, $N_{(1-\frac{\alpha}{2})}$ may become too large increasing the risk of missing bad data. Thus, an upper bound $N_{(1-\frac{\alpha}{2})max}$ is used to limit this risk. Further details on the implementation of HTI method can be found in [3].

5.8.4 HTI Strategy Under Fixed β

The following parameters are initially set by the user:

$$\begin{aligned} \text{a} &= \tfrac{|e_{si}|}{\sigma_i}, \text{ (e.g. a = 40)} \\ N_\beta &= \text{b, (e.g. b} = -2.32 \text{ for } \beta = 0.01 \text{)} \\ N_{(1-\frac{\alpha}{2})max} &= 3.0 \end{aligned}$$

Steps of the Algorithm:

1. Select suspect set s_1 based on r^N and calculate

$$T_{s1} = S_{s1,s1}^{-1} \text{ and } \hat{e}_{s1} = T_{s1} r_{s1}$$

2. Calculate $N_{(1-\frac{\alpha}{2})i}$:

$$N_{(1-\frac{\alpha}{2})i} = \frac{|\ e_{si}\ | + \sigma_i N_\beta \sqrt{T_{ii} - 1}}{\sigma_i \sqrt{T_{ii}}}$$

   with $0 \leq N_{(1-\frac{\alpha}{2})i} \leq N_{(1-\frac{\alpha}{2})max}$.

3. Calculate the threshold for each s_{1i}:

$$\lambda_i = \sigma_i \sqrt{T_{ii}} N_{(1-\frac{\alpha}{2})i}, \quad i = 1, \ldots, s_{1i}$$

4. Select measurement s_{1i} if $|\ \hat{e}_{s_{1i}}\ | > \lambda_i$.

5. Form a shorter list of suspect measurements using those that are selected at step 4. Repeat steps 1-4 until all measurements that are suspected in the previous iteration are all selected again at step 4.

5.9 Summary

This chapter discussed the problem of bad data detection and identification when using the WLS method for state estimation. It is shown that identification of single bad data is possible by using the maximum normalized residual r_{max}^N method. Multiple bad data on the other hand, are more difficult to identify and two alternative methods are discussed for this purpose. The ability to detect and identify bad data depends also on the measurement types and their configuration. Conditions under which bad data detection or identification will fail irrespective of the method used, are also presented. Treatment of bad data can also be viewed as a robustness issue for a given estimator. Hence, rather than devising post estimation correction methods which are discussed in this chapter, alternative estimators which remain robust against bad data can be formulated. Such estimators will be discussed in detail in the next chapter.

5.10 Problems

1. Consider the following linear model:

$$z = ax + by + e$$

where, $E[e] = 0$ and $cov[e] = I$. The measurements are given as:

i	x_i	y_i	z_i
1	−0.9	0.75	−5.4
2	0.1	0.3	−0.08
3	0.2	0.5	0.02
4	0.5	0.25	2.02
5	0.4	−0.2	2.32

   (a) Find the WLS estimates for a and b.

   (b) Find the normalized residuals.

   (c) Find the estimated errors for measurements 1 and 5, assuming that the remaining measurements are error free.

2. The network diagram and its associated measurement configuration of the IEEE 14-bus test system is shown in Figure 5.3. Network data for the IEEE-14 bus system can be downloaded from : http://www.ee.washington.edu/research/pstca/pf14/pg_tca14bus.htm.

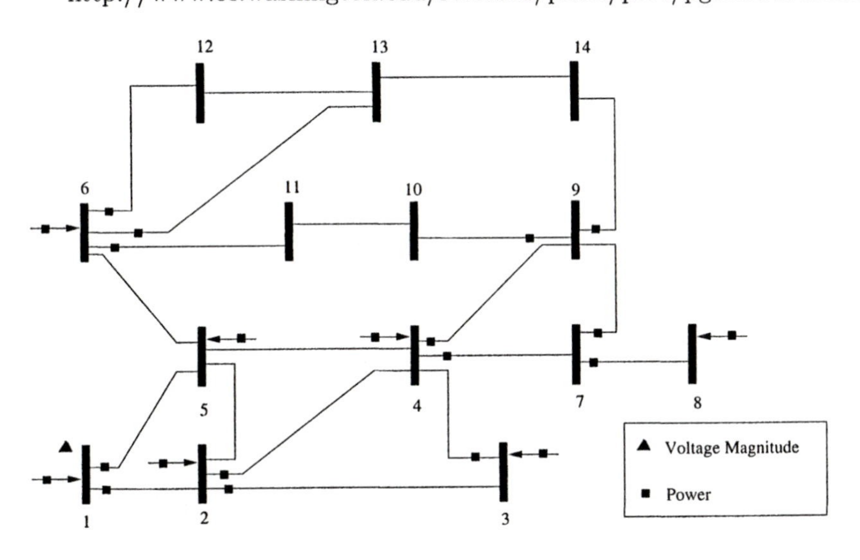

Figure 5.3. Measurement configuration for problem 2

(a) Give the results of the Chi-squares test for detecting bad data for each of the cases listed in the below table.

(b) Use the results of the normalized residual test to identify bad data for each case.

Explain the reason(s) of failure for those cases where the tests fail to detect and/or identify the bad data.

Case No:	Bad Measurement	Magnitude of Bad Data ($\times\sigma$)
1	P Flow 6-12	+30
2	P Flow 9-14	+30
3	P Flow 4-9	+20
	P Inj.at 4	+20
4	P Flow 3-4	+20

3. A 3-bus system and its measurement configuration are shown in Figure 5.4.

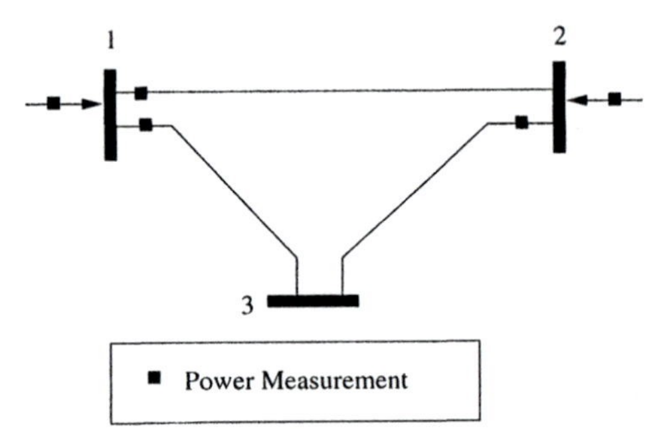

Figure 5.4. Measurement configuration for problem 3

Assume that the real power measurements can be expressed as a linear function of phase angles as below:

$$P_i = \sum_{\substack{k=1 \\ k \neq i}}^{N} g_{ik}(\theta_i - \theta_k)$$

$$P_{ik} = \theta_i - \theta_k$$

where g_{ik} is 1 if bus i and k are connected, and zero otherwise. All measurements have a standard deviation of 0.01 p.u. and they are given as:

$$P_1 = 0.53, P_2 = 0.24, P_{12} = 0.08, P_{23} = 0.42, P_{13} = 0.44.$$

(a) Use Chi-squares test to detect any bad data at 95% confidence level.

(b) Use the largest normalized residuals test to identify bad data.

References

[1] A. Monticellli and A. Garcia, "Reliable Bad Data Processing for Real-Time State Estimation", *IEEE Trans. on Power Apparatus and Systems*, Vol.PAS-102, No.3, pp.1126-1139, 1983.

[2] L. Mili, Th Van Cutsem and M. Ribbens-Pavella, "Hypothesis Testing Identification: A New Method for Bad Data Analysis in Power System State Estimation", *IEEE Trans. on Power Apparatus and Systems*, Vol.PAS-103, No.11, pp.3239-3252, 1984.

[3] L. Mili and Th. Van Cutsem, "Implementation of HTI Method in Power System State Estimation", *IEEE Trans. on Power Systems*, Vol.3, No.3, Aug. 1988, pp.887-893.

[4] E. Handschin, F.C. Schweppe, J. Kohlas and A. Fiechter, "Bad Data Analysis for Power System State Estimation", *IEEE Trans. on Apparatus and Systems*, Vol.PAS-94, No. 2, pp.329-337, 1975.

[5] Nian-de Xian, Shi-ying Wang and Er-keng Yu, "A New Approach for Detection and Identification of Multiple Bad Data in Power System State Estimation", *IEEE Trans. on Power Apparatus and Systems*, Vol.PAS-101, No.2, pp.454-462, 1982.

[6] A. Garcia, A. Monticelli and P. Abreu, "Fast Decoupled State Estimation and Bad Data Processing", *IEEE Trans. on Power Apparatus and Systems* Vol. PAS-98, pp. 1645-1652, September 1979.

[7] K. A. Clements and P. W. Davis, "Multiple Bad Data Detectability and Identifiability: a Geometric Approach", *IEEE Trans. on Power Delivery*, Vol.PWRD-1, No. 3, July 1986.

[8] U.S. Department of Energy Report, DOE/ET/29362-1, "Contribution to Power System State Estimation and Transient Stability Analysis", February 1984.

[9] H. M. Merrill and F. C. Schweppe, "Bad Data Suppression in Power System Static Estimation", *IEEE Trans. on Power Apparatus and Systems*, Vol.PAS-90, pp.2718-2725, Nov./Dec. 1971.

[10] I. W. Slutsker, "Bad Data Identification in Power System State Estimation Based on Measurement Compensation and Linear Residual Calculation", *IEEE PES Winter Meeting*, Paper No. 88 WM 210-7, 1988.

Chapter 6

Robust State Estimation

6.1 Introduction

One of the reasons for using a state estimator is to detect, identify and eliminate errors which may appear in the measurements, network model or parameters. If the estimated state remains insensitive to major deviations in a limited number of redundant measurements, then the corresponding estimator will be considered statistically robust. Unfortunately, robustness is commonly achieved at the expense of computational complexity. The topic of robust estimation is quite broad and is covered in several books and papers in the literature.

The aim of this chapter is to introduce some of the robustness issues in power system state estimation and present alternative estimation methods which are more robust as compared to the weighted least squares (WLS) estimator that has been discussed in detail so far.

The WLS state estimation problem is formulated based on certain assumptions about the measurement errors. These errors are considered to be independent random variables distributed according to the Gaussian distribution with zero mean and known variance. Measurement error variances are chosen based on the assumed metering accuracy and/or historical data on the measurement errors. However, occasional gross errors due to various telemetry noise or failure are known to occur. Post estimation processing of measurement residuals in order to detect and identify such outliers in measurements is discussed in Chapter 5. There may also be other types of outliers which strongly influence the estimated state, yet may or may not carry bad data. Robust estimators are expected to remain unbiased despite the existence of different types of outliers.

6.2 Robustness and Breakdown Points

Robustness of an estimator can be quantified by the concept of the finite-sample breakdown point [2]. Assume a set of measurements

$$z = \{z_1, z_2, \ldots, z_m\}$$

for which a given estimator yields an estimate $\hat{x}$. If one or more measurements are replaced by bad data so that a corrupted set z' is created, then the new estimate $\hat{x}'_i$ will also be biased. The maximum norm of the bias in the estimate will be:

$$b^{m_b} = \max_i \mid \hat{x}_i - \hat{x}'_i \mid \qquad i = 1, \ldots, n$$

where $\hat{x}'_i$ is estimated by replacing m_b good measurements by arbitrarily large bad data. The number of bad data m_b is increased until b^{m_b} is no longer finite. The largest ratio

$$\xi = \frac{m_b}{m}$$

for which b^{m_b} remains bounded, is called the breakdown point of this estimator.

Example 6.1:

Given $z = \{0.9, 0.95, 1.05, 1.07, 1.09\}$, find the breakdown point of the following estimators:

1. $\hat{X}_a = mean\{z_i\} = \frac{1}{5}\sum_{i=1}^{5} z_i$
2. $\hat{X}_b = median\{z_i\}, \quad i = 1, \ldots, 5$

Solution:

1. For $m_b = 1$, replace $z_5 = 1.09$ by an infinitely large number $z'_5 = \infty$. The new estimate $\hat{X}'_a$ will then be:

$$\hat{X}'_a = \frac{1}{5}\sum_{i=1}^{5} z_i = \infty$$

Since, by replacing even a single measurement by an infinitely large number, an unbounded estimate $\hat{X}'_a$ is obtained, the breakdown point for this estimator will be 0.

2. For $m_b = 1$, replace $z_5 = 1.09$ by an infinitely large number $z'_5 = \infty$. The new estimate $\hat{X}'_b$ will then be:

$$\hat{X}_b^1 = 1.05 \quad \text{(finite)}$$

For $m_b = 2$, replace both z_5 and z_4 by infinity. The new estimate $\hat{X}'_b$ will then be:

$$\hat{X}_b^2 = 1.05 \quad \text{(finite)}$$

For $m_b = 3$, replacing z_5, z_4 and z_3 by infinity yields:

$$\hat{X}_b^3 = \infty$$

Hence, the breakdown point of this estimator will be $2/5 = 0.4$.

A robust estimator will have a large breakdown point. The largest possible breakdown point will be limited by the measurement redundancy. Even the most robust estimators can not reject more than half of the redundant measurements when they are in error.

In the case of linear regression with n-unknown parameters, robustness can be quantified by a simpler concept referred to as the *exact-fit point* which is introduced in [3, 4]. An estimator is said to have the *exact-fit property* if the estimate is an n-dimensional hyperplane when the majority of the measurements lie exactly on this hyperplane. Then, the exact-fit point of an estimator will be defined as the largest fraction of bad data for which the exact-fit property continues to hold.

Prior to discussing some of the robust estimation methods that are applied to the power system state estimation problem, the concept of an outlier in state estimation will be introduced. The concept of bad data and outliers are usually used within the context of robust estimation and they are closely related. Bad data usually refers to a measurement whose value is incorrectly recorded. Since it lies away from its expected location in the measurement space, it is also an outlier in that space. A measurement, which may not contain any errors, may also appear as an outlier due to the structure of its corresponding equation. Such outliers are more difficult to identify, yet will strongly bias the state estimate when they contain errors.

6.3 Outliers and Leverage Points

Consider a linear measurement model given below:

$$z = H \cdot x + e \tag{6.1}$$

In the above model, z is the measurement vector, x is the unknown variable vector to be estimated and H is a mxn matrix. The vector e represents the measurement error which is commonly assumed to have a Normal distribution with zero mean. Denoting row i of H by H_i, the pair (z_i, H_i) will represent an observation and H_i's will lie in an n-dimensional space called *the factor space* of regression [3]. An outlier in this model can either be in the z-direction or in the factor space. An example of an outlier in the z-direction is a bad data in one of the measurements. Such an outlier can be detected and identified based on the measurement residuals as discussed in the previous chapters. When there is an outlier in the factor space, one of the rows of H, say H_i, will lie away from the rest of the factors. The corresponding measurement will then have an undue influence on the state estimate and is referred to as *the leverage point* in regression.

6.3.1 Concept of Leverage Points

Let us reconsider the simple linear regression model:

$$\tilde{z} = \widetilde{H} \cdot x + \tilde{e} \tag{6.2}$$

where $E[\tilde{e}] = 0$ and $E[\tilde{e} \cdot (\tilde{e})^T] = \text{R}$ (diagonal).

Let $z = R^{-1/2} \cdot \tilde{z}$, $H = R^{-1/2} \cdot \widetilde{H}$ and $e = R^{-1/2} \cdot \tilde{e}$, then:

$$z = H \cdot x + e \tag{6.3}$$

where $E[e] = 0$ and $E[e \cdot e^T] = I_m$, i.e., the modified measurement error vector, e, has unit covariance. Then, the least squares estimator for x can be expressed as:

$$\hat{x} = (H^T H)^{-1} H^T z \tag{6.4}$$

and the estimator for z as:

$$\hat{z} = H(H^T H)^{-1} H^T z \tag{6.5}$$

$$= K \cdot z \tag{6.6}$$

Here "K" is the "hat" matrix which is defined earlier in Chapter 5, Equation (5.3). Denoting the $i-th$ row of the matrix H by H_i, the diagonal elements of K can be expressed as follows:

$$K_{ii} = H_i (H^T H)^{-1} H_i^T \tag{6.7}$$

Since K is both symmetric ($K = K^T$) and idempotent ($K \cdot K = K$), K_{ii} can be written as:

$$K_{ii} = K_{ii}^2 + \sum_{i \neq j} K_{ij}^2 \tag{6.8}$$

It follows from the above equation that $0 \leq K_{ii} \leq 1$. The value K_{ii} represents the influence of the $i - th$ measurement z_i on its estimate $\hat{z}_i$. If this influence is high, i.e. if K_{ii} is close to 1.0, then the measurement will be likely to behave as a leverage point. Geometrically K_{ii} gives a measure of the distance of the measurement factor H_i from the bulk of the remaining $(m - 1)$ measurement factors.

The measurement residuals can also be expressed as a function of the elements of the hat matrix K:

$$r = z - \hat{z} = z - K \cdot z = (I_m - K) \cdot z \tag{6.9}$$

Thus, the residual of a measurement corresponding to a leverage point will be very small even when it is contaminated with a large error. As such, it behaves almost like a critical measurement whose residual is identically zero. However, the main difference between them will be that although the elimination of a critical measurement renders the system unobservable, leverage points can be deleted without loss of system observability.

6.3.2 Identification of Leverage Measurements

The expected value of the $i-th$ diagonal element of K can be found as:

$$E[K_{ii}] = \overline{k} = \frac{1}{m}\sum_{i=1}^{m} K_{ii} = \frac{n}{m} \tag{6.10}$$

where n and m are the number of estimated variables and measurements respectively.

This can be shown using the trace operator as follows:

$$\begin{aligned} tr[K] &= tr[H(H^T H)^{-1} H^T] && (6.11)\\ &= tr[(H^T H)^{-1} H^T H] && (6.12)\\ &= tr[I_n] = n && (6.13)\\ \overline{k} = \frac{tr[K]}{m} &= \frac{1}{m}\sum_{i=1}^{m} K_{ii} = \frac{n}{m} && (6.14) \end{aligned}$$

If any of the K_{ii}'s differs significantly from $\overline{k}$, then it is taken as an indication of a leverage point. As a rule of thumb, if :

$$K_{ii} \geq 2\frac{n}{m} \tag{6.15}$$

then the measurement i is suspected to be a leverage point.

When the leverage measurements contain bad data, identification of the bad measurement becomes very difficult by conventional methods. Residual covariances for these measurements will be numerically insignificant. Leverage measurements may appear as isolated single points or as a group. In power system state estimation, occurrence of leverage measurements can be linked to low measurement redundancy. The following conditions are known to create leverage measurements:

- An injection measurement placed at a bus which is incident to a large number of branches.

- An injection measurement placed at a bus which is incident to branches of very different impedance values.
- Flow measurements along branches whose impedances are very different from those of the other branches in the system.
- Using a very large weight for a specific measurement.

The presence of leverage measurements created as a result of the above conditions, will affect the numerical structure of the measurement Jacobian H. The rows of H corresponding to the leverage measurements will have entries of very different magnitudes compared to those of the other rows. When these rows correspond to flow measurements, the decoupled Jacobian matrix will contain only two identical entries with opposite signs. Therefore, these can easily be scaled by multiplying the entire row by an appropriate scalar. However, when the row corresponds to an injection measurement, row scaling will not necessarily work due to the large variation in the magnitudes of the entries of that row.

When there are multiple leverage points, their identification will be more difficult. Such leverage points are characterized by a set of outliers in the factor space of regression. Assuming a multi-variate Normal distribution for H_i's, the sample mean $\bar{h}$ and covariance $\bar{C}$ can be computed as:

$$\bar{h} = \frac{1}{m}\sum_{i=1}^{m} H_i \tag{6.16}$$

$$\bar{C} = \frac{1}{m-1}\sum_{i=1}^{m}(H_i - \bar{h})(H_i - \bar{h})^T \tag{6.17}$$

A given row H_i corresponding to measurement i can be checked to see if it constitutes an outlier with respect to the rest of the measurement cloud. The so-called *Mahalanobis Distance* (MD) is a simple measure defined for measurement i as:

$$MD_i^2 = (H_i - \bar{h})^T \bar{C}^{-1}(H_i - \bar{h}) \tag{6.18}$$

to indicate the distance of H_i to the rest of the cloud of H_j's. Using the Normal distribution assumption for H_i's, MD_i^2 can be shown to have a χ^2 distribution with n degrees of freedom, n being the dimension of H_i:

$$MD_i^2 \sim \chi_n^2 \tag{6.19}$$

Choosing a false alarm probability such as $\alpha = 0.025$, the measurement i will be suspected to be a leverage point if $MD_i^2 > \chi^2_{n,0.975}$. Unfortunately, this measure is not robust since leverage point clusters may be masked by causing the covariance values to increase and the sample mean to approach close to the leverage cluster.

A more robust measure of leveraging effect of a measurement is proposed by Donoho and Gasko [7] and later applied to the power system state estimation by Mili et al. [8]. This measure is called the *projection statistics* (PS_i) and defined for a measurement i as below:

$$PS_i = \max_{H_k} \frac{| H_i^T \cdot H_k |}{\beta} \qquad \text{for } k = 1, 2, \ldots, m \tag{6.20}$$

where
$\beta = \gamma \cdot \text{lomed}_i\{\text{lomed}_{j \neq i}\{| H_i^T H_k + H_j^T H_k |\}\} \quad 1 \leq i, j, k \leq m$
$\text{lomed}_i\{x\}$: low median of the m numbers in $x=\{x_1, x_2, \ldots, x_m\}$
$\gamma = 1.1926$ (see the closure of [8] for the choice of this factor)

The projection statistics PS_i can be shown to approximately behave like a chi-square random variable [8]. Furthermore, for measurement i, the degrees of freedom of the corresponding chi-square distribution is directly related to the sparsity structure of the row H_i. Hence, a measurement i will be identified as a leverage point if:

$$PS_i > \chi^2_{k,0.975} \tag{6.21}$$

where, k is the number of nonzero entries in the row H_i of the measurement Jacobian H.

Example 6.2:

Consider the 4-bus power system and its measurement configuration shown in Figure 6.1. All branches are assumed to have a reactance of $j0.1$ p.u., except for branch 1-2 which has a smaller reactance of $j0.01$ p.u.

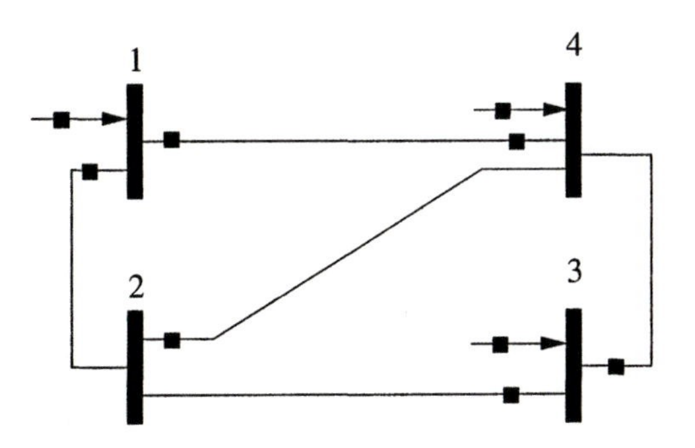

Figure 6.1. 4-bus power system with leverage points

Using the real power part of the decoupled measurement Jacobian with bus

4 chosen as the slack:

$$H = \begin{bmatrix} 110 & -100 & 0 \\ 100 & -100 & 0 \\ 10 & 0 & 0 \\ 0 & 10 & 0 \\ 0 & -10 & 20 \\ 0 & -10 & 10 \\ 0 & 0 & 10 \\ -10 & -10 & -10 \\ -10 & 0 & 0 \end{bmatrix}$$

Calculated values of MD_i are:

$$MD = \begin{bmatrix} 2.0937 \\ 1.9970 \\ 0.9072 \\ 0.9517 \\ 2.2387 \\ 1.3394 \\ 1.0792 \\ 2.2596 \\ 0.9072 \end{bmatrix}$$

Note that for a choice of a cut-off value based on $\sqrt{\chi^2_{3,0.975}} = 3.06$, none of the measurements will be identified as leverage points.

Table 6.1. Projection Statistics for the measurements

Measurement	PS	k	$\chi^2_{k,0.975}$
Flow 1-2	8.39	1	5.02
Flow 1-4	0.42	1	5.02
Flow 2-4	0.84	2	7.38
Flow 3-2	0.84	2	7.38
Flow 3-4	0.84	2	7.38
Flow 4-1	0.42	1	5.02
Inj 1	8.47	2	7.38
Inj 3	1.26	3	9.35
Inj 4	1.68	3	9.35

Now, let us compute the projection statistics for the same measurements. Table 6.1 shows the calculated PS_i values, the degrees of freedom and the corresponding cut-off $\chi^2_{k,0.975}$ values. Flow $1-2$ and Injection 1, both of which are incident to the short branch $1-2$ are identified correctly as the two leverage points.

6.4 M-Estimators

The M-estimator concept is first introduced by Huber [1] for robust estimation of the center of a distribution and is subsequently generalized to regression. In general, an M-estimator is a maximum likelihood estimator. It minimizes an objective function, which is expressed as a function of the measurement residuals $\rho(r)$, subject to the constraints given by the measurement equations:

$$\text{Minimize} \quad \sum_{i=1}^{m} \rho(r_i) \tag{6.22}$$

$$\text{Subject to } z = h(x) + r \tag{6.23}$$

where
$\rho(r_i)$ is a chosen function of the measurement residual r_i
z is the measurement vector
x is the state vector, and
$h(x)$ is the measurement function.

The influence of bad data on the estimated system state and methods of their suppression are first discussed within the context of power system state estimation by Merrill and Schweppe [5]. It is suggested that this issue can be addressed by changing the estimation algorithm in such a way that the bad measurements will be screened and suppressed during the iterative estimation process. This approach produced several M-estimators, some of which will be discussed here.

The first estimator proposed in [5] minimizes an objective function which is chosen as a non-quadratic function of the measurement residuals. Several variations of it are later introduced in [6]. These estimators are mainly designed for automatically detecting measurements with rapidly growing residuals and suppressing their influence on the state estimate. Later on, it is recognized that the structure of the measurement equations, location of meters and the network parameters lead to the creation of leverage points which have undue influence on the estimate [8]. Thus, the objective function is further modified to balance the influence of measurements irrespective of their locations and types. This estimator belongs to the class of estimators which are referred to as the generalized M-estimators or bounded influence estimators.

The function $\rho(r)$ in (6.22) should be chosen so that it has at least the following properties:

- $\rho(r) = 0$ for $r = 0$.
- $\rho(r) \geq 0$ for any r.
- $\rho(r)$ is monotonically increasing in both $+r$ and $-r$ directions.
- It is symmetric around $r = 0$, i.e. $\rho(r) = \rho(-r)$.

The following list of estimators that use such functions for $\rho(r)$, are proposed by different investigators so far. The tuning parameter a used in the definitions below, is to be specified by the users. Typical values range between 1 and 4.

- Quadratic-Constant (QC) [6]:

$$\rho(r_i) = \begin{cases} r_i^2/\sigma_i^2 & \mid r_i/\sigma_i \mid \leq a \\ a^2/\sigma_i^2 & \text{otherwise} \end{cases} \tag{6.24}$$

- Quadratic-Linear (QL) [6]:

$$\rho(r_i) = \begin{cases} r_i^2/\sigma_i^2 & \mid r_i/\sigma_i \mid \leq a \\ 2a \cdot \sigma_i \mid r_i \mid - a^2 \cdot \sigma_i^2 & \text{otherwise} \end{cases} \tag{6.25}$$

- Square Root (SR) [6]:

$$\rho(r_i) = \begin{cases} r_i^2/\sigma_i^2 & \mid r_i/\sigma_i \mid \leq a \\ 4 \cdot a^{3/2}\sqrt{r_i/\sigma_i} - 3 \cdot a^2 & \text{otherwise} \end{cases} \tag{6.26}$$

- Schweppe-Huber Generalized-M (SHGM) [8]:

$$\rho(r_i) = \begin{cases} \frac{1}{2} r_i^2/\sigma_i^2 & \mid r_i/\sigma_i \mid \leq a \cdot \omega_i \\ a \cdot \omega_i \mid r_i/\sigma_i \mid - \frac{1}{2} a^2 \cdot \omega_i^2 & \text{otherwise} \end{cases} \tag{6.27}$$

  where, ω_i is the iteratively modified weighting factor. Details of its choice will be given later in section 6.4.2.

- Least Absolute Value (LAV):

$$\rho(r_i) = \mid r_i \mid \tag{6.28}$$

The solution of the M-estimation problem defined by (6.22-6.23) can be obtained by different methods. Two alternative methods will be discussed below. One is based on the Newton's method and requires computation of the first and second derivatives of the function ρ. The second method avoids the computation of the second derivatives and is based on the iteratively re-weighted least squares method.

6.4.1 Estimation by Newton's Method

Consider the optimization problem defined by (6.22-6.23):

$$\begin{aligned} \text{Minimize} \quad J(r) &= \sum_{i=1}^{m} \rho(r_i) && (6.29) \\ \text{Subject to} \quad z &= h(x) + r && (6.30) \end{aligned}$$

This problem can be solved by forming the Lagrangian:

$$\mathcal{L}(x, r, \lambda) = \sum_{i=1}^{m} \rho(r_i) + \lambda^T (z - h(x) - r)$$

and writing the first order necessary conditions for a minimum of $J(r)$:

$$\begin{aligned} \frac{\partial \mathcal{L}}{\partial x} &= -H^T \lambda = 0 && (6.31) \\ \frac{\partial \mathcal{L}}{\partial r} &= \Upsilon - \lambda = 0 && (6.32) \\ \frac{\partial \mathcal{L}}{\partial \lambda} &= z - h(x) - r = 0 && (6.33) \end{aligned}$$

where $H = \frac{\partial h(x)}{\partial x}$ and $\Upsilon = \frac{\partial \rho(r)}{\partial r}$.

Eliminating λ from equations (6.31) and (6.32), the following set of nonlinear equations will be obtained:

$$\begin{aligned} z - h(x) - r &= 0 && (6.34) \\ H^T \Upsilon &= 0 && (6.35) \end{aligned}$$

The above non-linear equation set can be solved iteratively by substituting the following first order Taylor approximations for:

$$\begin{aligned} h(x) &\approx h(x^k) + H(x^k) \cdot \triangle x^k && (6.36) \\ \Upsilon(r) &\approx \Upsilon(r^k) + \nabla \Upsilon(r^k) \cdot (r - r^k) && (6.37) \end{aligned}$$

where:

$$\begin{aligned} \triangle x^k &= x - x^k \\ \nabla \Upsilon &= \frac{\partial \Upsilon(r^k)}{\partial r^k} \\ H(x^k) &= \frac{\partial h(x^k)}{\partial x^k} \\ r^k &= z - h(x^k) \end{aligned}$$

into (6.34) and (6.35), which yields:

$$\begin{bmatrix} \nabla\Upsilon & \overline{H} \\ \overline{H}^T & 0 \end{bmatrix} \begin{bmatrix} r \\ \triangle x^k \end{bmatrix} = \begin{bmatrix} \nabla\Upsilon \cdot r^k \\ \overline{H}^T \cdot r^k - H^T \cdot \Upsilon(r^k) \end{bmatrix} \tag{6.38}$$

where $\overline{H} = \nabla\Upsilon \cdot H(x^k)$.

Note that the above matrix equation can be further reduced to the following form by eliminating the variable r, provided that $\nabla\Upsilon$ remains non-singular:

$$H^T[\nabla\Upsilon]^{-1}H\triangle x^k = H^T\Upsilon(r^k) \tag{6.39}$$

Table 6.2. Gradient and Weighting Functions for M-estimators

Estimator	Υ	$\nabla\Upsilon$	If
QC	$\frac{2r_i}{\sigma_i^2}$ 0	$\frac{2}{\sigma_i^2}$ 0	$\mid \frac{r_i}{\sigma_i} \mid\leq a$ otherwise
QL	$\frac{2r_i}{\sigma_i^2}$ $2a\sigma_i sign(r_i)$	$\frac{2}{\sigma_i^2}$ 0	$\mid \frac{r_i}{\sigma_i} \mid\leq a$ otherwise
SR	$\frac{2r_i}{\sigma_i^2}$ $2\sqrt{\frac{a^3}{\sigma_i r_i}}$	$\frac{2}{\sigma_i^2}$ $-\sqrt{\frac{a^3}{\sigma_i r_i^3}}$	$\mid \frac{r_i}{\sigma_i} \mid\leq a$ otherwise
SHGM	$\frac{r_i}{\sigma_i^2}$ $\frac{a\omega_i}{\sigma_i} sign(r_i)$	$\frac{1}{\sigma_i^2}$ 0	$\mid \frac{r_i}{\sigma_i\omega_i} \mid\leq a$ otherwise
LAV	$sign(r_i)$	0	$r_i \neq 0$

Table 6.2 gives the functions Υ and $\nabla\Upsilon$ for different estimators. The iterative solution algorithm for (6.39) will update the diagonal matrix $\nabla\Upsilon(r^k)$ at each iteration based on the current value of r^k. Depending upon the chosen function $\rho(r)$, some of the entries in the diagonal matrix $\nabla\Upsilon(r^k)$ may be zero. This means that the corresponding measurements (with zero weights)

will effectively be eliminated from (6.39). The residuals corresponding to these measurements will still be computed and checked against their chosen tuning parameter, so that they can be reinserted in the subsequent iterations if their residuals are sufficiently reduced. Simulation results of this implementation for QC and QL estimators can be found in [9].

Note that (6.39) can be used to iteratively solve the state estimation problem for any one of the M-estimators listed in section 6.4, except for the LAV estimator for which the weighting matrix $\nabla\Upsilon$ will be zero. Formulation and solution of the LAV state estimation problem will be separately discussed in section 6.5.

6.4.2 Iteratively Re-weighted Least Squares Estimation

An alternative to the above-described Newton-based solution algorithm is to use an iteratively re-weighted least squares (IRLS) method, where the use of the $\nabla\Upsilon(r^k)$ matrix is avoided. An application of the IRLS method to the Schweppe-Huber Generalized-M (SHGM) estimation is illustrated in [8], where the SHGM estimator is developed as a robust estimator which can not only suppress bad data in regular measurements, but also avoid the influence of any existing leverage points when they carry bad data. This is accomplished through a re-weighting scheme which will be summarized next.

This estimate is obtained as the solution of the following optimization problem where the objective function is expressed in terms of the residuals $\rho(r)$ as given earlier in (6.27):

$$\text{Minimize} \quad J(r) = \sum_{i=1}^{m} \rho(r_i) \tag{6.40}$$

Writing the KKT necessary conditions for a minimum of $J(r)$:

$$\frac{\partial J}{\partial x} = \frac{\partial J}{\partial r} \cdot \frac{\partial r}{\partial x} = 0 \tag{6.41}$$

$$\Rightarrow \sum_{i=1}^{m} \frac{\partial \rho}{\partial r_i} \cdot \frac{\partial r_i}{\partial x} = 0 \tag{6.42}$$

$$\Rightarrow \sum_{i=1}^{m} \Upsilon(r_i) \cdot H_i = 0 \tag{6.43}$$

$$\Rightarrow \sum_{i=1}^{m} \frac{\Upsilon(r_i)}{r_i} \cdot r_i \cdot H_i = 0 \tag{6.44}$$

$$\Rightarrow \quad \sum_{i=1}^{m} \Phi(r_i) \cdot r_i \cdot H_i = 0 \tag{6.45}$$

$$\Rightarrow \quad H^T \cdot \Phi(z - h(x)) = 0 \tag{6.46}$$

and substituting the first order Taylor approximation for $h(x) \approx h(x^k) + H \cdot \triangle x^k$, yields:

$$H^T \cdot \Phi \cdot H \triangle x^k = H^T \cdot \Phi \cdot r^k \tag{6.47}$$

where
$r^k = z - h(x^k)$,
$H_i = \frac{\partial h_i}{\partial x}$,
$H^T = [h_1^T \ h_2^T \ \cdots \ h_m^T]$
$\Phi_{ii} = \frac{\Upsilon(r_i)}{r_i}$ is a diagonal weight matrix, whose elements are defined by:

$$\Phi_{ii} = \begin{cases} \frac{1}{\sigma_i^2} & \mid r_i/\sigma_i\omega_i \mid \leq a \\ \frac{a \cdot \omega_i}{r_i \sigma_i} \cdot \text{sign}(r_i) & \text{otherwise} \end{cases} \tag{6.48}$$

In (6.48), ω_i is the penalty factor chosen specifically to cancel the effect of any existing leverage points in the measurement set and defined as in [8]:

$$\omega_i = \min\{1, \left[\frac{\chi^2_{\nu,p}}{PS_i}\right]^2\} \tag{6.49}$$

where:
$\chi^2_{\nu,p}$: Chi-square statistics
ν: degrees of freedom which is equal to the number of nonzeros in H_i
p: probability, a typical value is 0.975
PS_i: projection statistics for measurement i, as given by (6.20).

Details of the solution algorithm can be found in [8]. This particular estimator will behave more like a LAV (WLS) estimator for small (large) values of the tuning parameter a used in (6.27).

6.5 Least Absolute Value (LAV) Estimation

In the case of LAV estimator, it can be shown that the problem can be formulated as a linear programming (LP) problem, which in turn can be solved by applying one of the well developed LP solution methods. The connection between the LAV estimation problem and linear programming was initially pointed out in [10]. LP-based implementation of LAV estimation of power system states is investigated first in [11] and [12].

In this section, we will first review the LAV estimation of the unknown vector in a linear regression model and the relevant properties of the LAV estimator. Then, we will illustrate how the power system LAV state estimation problem can be formulated and solved by means of two alternative solution methods, namely the simplex and the interior point methods.

6.5.1 Linear Regression

Let us consider the linear regression model given below:

$$z_i = A_i^T x + e_i \tag{6.50}$$

where, a set of observations $\{z_i, i = 1, ..., m\}$ are assumed to be linearly related to a set of vectors $\{A_i \in R^n, i = 1, \ldots, m\}$ and an unknown vector $x \in R^n$, and each observation i contains some random error e_i.

Then, the least absolute value estimate $\hat{x}$ for the unknown vector x will be given by the solution of the following optimization problem:

$$\text{Minimize} \quad c^T \mid r \mid \tag{6.51}$$

$$\text{Subject to} \quad z - Ax = r \tag{6.52}$$

where, A is a $m \times n$ matrix with A_i^T being its ith row, $c \in R^m$ is a vector with all of its entries equal to 1, and $r \in R^m$ is the vector of observation residuals. Thus, the objective function is equal to the sum of the absolute values of the observation residuals.

In the simple one-dimensional case, i.e. when $n = 1$, the LAV estimator yields the sample median. Note that, to find the sample median of a sample $\{z_1, \ldots, z_m\}$, the entries of z are ordered first in increasing values to obtain the ordered sample $\{z'_1, \ldots, z'_m\}$. Then the sample median $\hat{z}$ is given by:

$$\hat{z} = \begin{cases} z'_k, & \text{if } m \text{ is odd,} \\ \frac{(z'_k + z'_{k+1})}{2}, & \text{if } m \text{ is even.} \end{cases} \tag{6.53}$$

where, $k = \frac{m}{2} + 1$.

6.5.2 LAV Estimation as an LP Problem

It can be shown that the LAV estimation problem given in (6.51)-(6.52) can be formulated as a linear programming (LP) problem and solved using one of the well established LP methods. We will first illustrate the formulation of the LAV estimation problem within a standard LP framework and then present two different solution algorithms, one based on the simplex and the other on Interior Point methods.

Let ξ_i be defined such that:

$$\mid r_i \mid \; \leq \; \xi_i, \qquad 1 \leq i \leq m$$

and replace the above inequality by two equalities via the introduction of two non-negative slack variables ℓ_i and k_i:

$$r_i - \ell_i = -\xi_i \tag{6.54}$$

$$r_i + k_i = \xi_i \tag{6.55}$$

Let us now define 4 new nonnegative variables x_i^u, x_i^v, u_i, and v_i, such that:

$$x_i = x_i^u - x_i^v \tag{6.56}$$

$$r_i = u_i - v_i \tag{6.57}$$

$$u_i = \frac{1}{2}\ell_i \tag{6.58}$$

$$v_i = \frac{1}{2}k_i \tag{6.59}$$

and rewrite (6.50) in terms of these new variables:

$$z_i = \{\sum_{j=1}^{n}[A_{ij}x_j^u - A_{ij}x_j^v]\} + u_i - v_i \qquad 1 \leq i \leq m$$

Note that the term $\mid r_i \mid$ in the objective function in (6.51) can be replaced by ξ_i, which is given in terms of the new variables as:

$$\xi_i = u_i + v_i \tag{6.60}$$

Then, the LAV estimate for x will be given by the solution to the following linear programming problem :

$$\text{Minimize} \quad \sum_{i=1}^{m}[u_i + v_i] \tag{6.61}$$

$$\text{Subject to} \quad \sum_{j=1}^{n} A_{ij}(x_j^u - x_j^v) = -u_i + v_i + z_i, \quad 1 \leq i \leq m \tag{6.62}$$

$$x_j^u, x_j^v \geq 0, \quad 1 \leq j \leq n \tag{6.63}$$

$$u_i, v_i \geq 0, \quad 1 \leq i \leq m \tag{6.64}$$

The following theorem is about the interpolation property of the LAV estimator:

Theorem 1 *If the column rank of A is L, $(L \leq n)$, then there exists a LAV estimate which satisfies at least L of the observations exactly (with zero residuals) [13].*

This property will facilitate the identification and suppression of bad data in the measurements. The following is a simple example which illustrates this property.

Example 6.3:

Consider the simple regression model given by:

$$z_i = A_{i1}x_1 + A_{i2}x_2 + e_i \qquad i = 1, \ldots, 5$$

where A_{ij} represents the jth entry in vector A_i as defined above. The associated observation data are given in the table below.

i	z_i	A_{i1}	A_{i2}
1	−3.01	1.0	1.5
2	3.52	0.5	−0.5
3	−5.49	−1.5	0.25
4	4.03	0.0	−1.0
5	5.01	1.0	−0.5

The LAV estimate for x is given as:

$$x^T = [3.005 \;\; ; \;\; -4.010]$$

with an objective function of 0.0525. The residuals are all relatively small as given below:

$$r^T = [0.0 \; ; \; 0.0125 \; ; \; 0.02 \; ; \; 0.02 \; ; \; 0.0]$$

The first and the fifth observations are satisfied exactly with zero residuals.

Now, let us simulate an error in observation 5 by replacing its value by 15.01. Despite the erroneous fifth observation, LAV estimator yields almost the identical estimate for x as before:

$$x^T = [3.02 \; ; \; -4.02]$$

and the corresponding residuals for the observations are given by:

$$r^T = [0.0 \; ; \; 0.0 \; ; \; 0.045 \; ; \; 0.01 \; ; \; 9.98]$$

Note the large residual associated with the erroneous observation 5. Now, the first and the second observations are satisfied exactly with zero residuals.

In the general n-dimensional case, provided that the matrix A is of full rank, there exists a hyperplane which satisfies n out of m observations exactly, where n is the dimension of the unknown vector x. Looking back at Example 6.3 above, note that, for both cases, only two out of five observations have zero residuals. In this sense, the LAV estimator tends to discard those observations that are outliers and yields an estimate that is less sensitive to bad observations. This property however does not hold true when there are leverage points. The following example illustrates such a case.

Example 6.4:

Consider again another simple regression model given by:

$$z_i = A_{i1}x_1 + A_{i2}x_2 + e_i \qquad i = 1, \ldots, 5$$

where A_{ij} represents the jth entry in vector A_i as defined above. The associated observation data are given in the table below.

i	z_i	A_{i1}	A_{i2}
1	15	1.0	1.0
2	18	2.0	1.0
3	4	3.0	1.0
4	24	4.0	1.0
5	162	50.0	1.0

Note that one of the observations z_3 is replaced by bad data. LAV estimate for the unknown vector X is found as:

$$\hat{X}^T = [3 \quad 12]$$

and the residuals r_i are given in the below table:

i	z_i	$\hat{z}_i$	r_i
1	15	15	0
2	18	18	0
3	4	21	−17
4	24	24	0
5	162	162	0

As expected, the LAV estimator yields an unbiased estimate for X and rejects the bad observation z_3. Next, let us introduce bad data in observation 5 instead of 3:

$$z_3 = 21, \quad z_5 = 20$$

Then, the same LAV estimator will yield the following biased solution:

$$\hat{X}^T = [0.0417 \quad 17.917]$$

and the corresponding residuals r_i given in the table below:

i	z_i	$\hat{z}_i$	r_i
1	15	17.958	−2.958
2	18	18.000	0.000
3	21	18.042	2.958
4	24	18.083	5.917
5	162	20.000	0.000

Here, observation 5 constitutes a leverage point and the bad data can not be rejected by the LAV estimator. This is one of the shortcomings of the LAV estimator and must be considered when applying the method to power system state estimation. Different methods that modify the structure of the measurement equations in order to eliminate leverage points, have been proposed in [17, 16].

Two alternative methods for solving the LP problem, namely the simplex and the interior point methods have been applied to the power system state estimation problem so far. Their implementations require different considerations and they will be discussed separately.

6.5.3 Simplex Based Algorithm

Several variations of the well known simplex method of solving LP problems can be applied to the LAV estimation problem. In applying the simplex method, the special structure of the LAV estimation problem can be exploited for computational efficiency. This can be accomplished both at the initialization and the actual optimization stages of the algorithm.

Substituting (6.28) into (6.22), LAV estimation problem can be stated as:

$$\text{Minimize} \quad \sum_{i=1}^{m} | r_i | \tag{6.65}$$

$$\text{Subject to} \quad z_i = h_i(x) + r_i, \quad 1 \leq i \leq m \tag{6.66}$$

where z_i is the i'th measurement, $h_i(x)$ is the nonlinear function relating the state vector x to this measurement and r_i represents its residual.

Assuming an initial solution x^0 for the state and using the first-order approximation of $h_i(x)$ around x^0, the problem can be transformed into a successive set of linear programming (LP) problems, each one minimizing the objective function given below:

$$J(x^k) = \sum_{i=1}^{m} (u_i^k + v_i^k) \tag{6.67}$$

where $u^k - v^k = z - h(x^k) - H(x^k) \cdot \triangle x = \triangle z^k - H(x^k) \cdot \triangle x^k$, is the measurement residual vector at the k-th state estimation iteration.

Dropping the superscript k for simplicity of notation, the optimization problem to be solved at iteration k can be formulated as below:

$$\text{Minimize} \quad \sum_{i=1}^{m} (u_i + v_i) \tag{6.68}$$

$$\text{Subject to} \quad H \cdot \triangle x_u - H \cdot \triangle x_v + u - v = \triangle z \tag{6.69}$$

$$\triangle x_u, \triangle x_v, u, v \geq 0 \tag{6.70}$$

where $\Delta x = \Delta x_u - \Delta x_v$.

This can be written in compact form as a standard LP problem:

$$\text{minimize} \quad c^T \cdot Y \tag{6.71}$$

$$\text{subject to} \quad A \cdot Y = b \tag{6.72}$$

$$Y \geq 0 \tag{6.73}$$

where:

$$\begin{aligned}
c^T &= [0_n, 0_n, 1_m, 1_m], \\
0_n &= [0, \ldots, 0], \text{ a zero vector of order } n, \\
1_m &= [1, \ldots, 1], \text{ a vector of order } m, \text{ where all the entries are equal to } 1, \\
b &= \Delta z, \\
Y^T &= [\Delta x_u^T \Delta x_v^T u^T v^T], \\
A &= \begin{bmatrix} H & -H & I_m & -I_m \end{bmatrix} \\
I_m &= \text{identity matrix of order } m.
\end{aligned}$$

This LP problem can now be solved using the simplex method. The overall state estimation solution will be obtained by successively solving these LP problems until $\|\Delta x\|$ is less than a chosen threshold.

A closer look at the special structure of the coefficient matrix A and the properties of the LAV estimator, reveals that the simplex procedure can be further simplified computationally [18]. The interpolation property of the LAV estimator implies that it will select the best n measurements out of the m available ones and will yield an estimate that will satisfy the selected n measurement equations exactly. Therefore, at each state estimation iteration, the measurements can be classified into two sets:

- Set N: essential measurements with zero residuals.
- Set B: the remaining ones with non-zero residuals.

The letters N and B stand for the nonbasic and basic variables respectively. The measurement array and its Jacobian (H), can be partitioned according to these designated sets as follows:

$$H = \begin{bmatrix} H_n \\ H_b \end{bmatrix}, \Delta z = \begin{bmatrix} \Delta z_n \\ \Delta z_b \end{bmatrix}$$

where :
H_n : $n \times n$ Jacobian for those measurements in set N.
H_b : $(m - n) \times n$ Jacobian for those measurements in set B.
Δz_n : measurement mismatch vector for set N.
Δz_b : measurement mismatch vector for set B.

Simplex method involves two stages:

1. Initialization:

   Given the $(m \times (2m + 2n))$ matrix A, choose a basis B, which is an $m \times m$ nonsingular sub-matrix formed by m linearly independent columns of A. Then, partition A and Y as follows:

$$\begin{aligned} A &= [\, B \;\; D \,] \\ Y^T &= [\, Y_B^T \;\; Y_D^T \,] \end{aligned}$$

   It is observed that an initial basis can be readily chosen and a basic feasible solution can be found for the LP problem. The initial basis will contain a diagonal matrix with entries $\pm$ 1, of dimension m where the diagonal entries will be assigned the same signs as the corresponding right hand side entries b_i. This ensures the feasibility of the initial solution which will be given by:

$$\begin{aligned} Y_B &= B^{-1}b = |\, b \,| \geq 0 \\ Y_D &= 0 \end{aligned}$$

2. Iterative optimization:

   Simplex solution can then proceed by applying the simplex rules [21] in exchanging the columns of B and D until no further reduction in the objection function can be achieved. During the first n simplex iterations, the choice of the columns to enter the basis should be restricted to the columns corresponding to the state variables Δx_u and Δx_v. If no such column can be found to enter the basis anytime during the initial n iterations, then this will imply that the complete unknown vector Δx can not be estimated with the given set of observations, i.e. the system is unobservable.

   After the first n iterations, the pivoting strategy will be modified so that the columns associated with the basic Δx_u and Δx_v variables, will not be allowed to leave the basis. Hence, thereafter, the column exchanges between the columns within and outside the basis will take place among those associated with the u_i and v_i variables only.

   Note also that, if initially a minimum set of n measurements can be identified through observability analysis, then this information can be used to choose an initial basis B for the LP iterations. The minimal observable set is not necessarily unique and the chosen set will be referred to as the set of *essential* measurements.

Assuming that an essential measurement set has been chosen using either one of the methods above, the constraint equation (6.69) can be rewritten as :

$$\underbrace{\begin{bmatrix} H_n & 0 \\ H_b & I_{(m-n)} \end{bmatrix}}_{B} \begin{bmatrix} \Delta x \\ s_b \end{bmatrix} + \underbrace{\begin{bmatrix} I_n \\ 0 \end{bmatrix}}_{D} \begin{bmatrix} s_n \end{bmatrix} = \begin{bmatrix} \Delta z_n \\ \Delta z_b \end{bmatrix} \tag{6.74}$$

where :
$s_n = u_n - v_n = 0$, which also implies that $u_n = v_n = 0$,
$s_b = u_b - v_b$,
$I_{(m-n)}$, I_n : are identity matrices of order $(m - n)$ and n.

Note that, (6.74) is identical to (6.72). As long as the signs of the slack variables s_b and the states Δx are properly tracked, there is no need to explicitly store H and I matrices twice, one for each sign.

The cost vectors associated with the basic solution vector $[\Delta x \; ; \; s_b]$ and the non-basic slack variables s_n, will then be:

$$c_B = \begin{bmatrix} 0 \\ c_b \end{bmatrix}, \quad c_D = [s_n]$$

and the relative cost (r_c) of rejecting one of the already selected measurements can be found as:

$$\begin{aligned} r_c^T &= c_n^T - c_B^T \cdot B^{-1} \cdot D \\ &= c_n^T - \lambda^T \cdot D. \end{aligned}$$

Due to the structure of the matrix D, only the first n entries of λ ($\lambda^T = [\lambda_n^T \; , \; \lambda_b^T]$) will have to be calculated:

$$\begin{aligned} \lambda_n &= (H_n^T)^{-1} \cdot [-H_b^T \cdot c_b] \\ r_c &= c_n - \lambda_n \end{aligned}$$

Once the relative cost vector r_c is computed, the next step is to decide on the measurement to replace the one to be rejected (with the most negative r_c). The minimum ratio given by:

$$\min_k \quad \frac{s_b(k)}{y_b(k)} > 0$$

will indicate the measurement number to be selected among the $(m - n)$ rejected measurements. Note that:

$$\begin{aligned} y_b &= -H_b \cdot y_n \\ y_n &= H_n^{-1} \cdot e_j \end{aligned}$$

where e_j is a singleton n-vector containing a 1 as its j'th entry, j being the index of the measurement with the largest negative relative cost, to leave the selected set N.

Thus, the steps of the LAV estimation algorithm will be as follows:

1. Solve $H_n \cdot \triangle x = \triangle z_n$.
   Compute $s_b = \triangle z_b - H_b \cdot \triangle x$.

2. Solve $H_n^T \cdot \lambda_n = -H_b^T \cdot c_b$.
   Compute $r_c = c_n - \lambda_n$.
   Choose $\min_j r_c(j) < 0$.

3. If all $r_c(j) \geq 0$, stop; optimal solution is reached. Else go to 4.

4. Solve $H_n \cdot y_n = e_j$.
   Compute $y_b = -H_b \cdot y_n$.
   Compute $\min_k \{\frac{s_b(k)}{y_b(k)}\} > 0$.
   A short-cut referred to as *vertex skipping* [13] can be used at this step to further accelerate convergence.

5. Update H_n^T; Replace j'th column of H_n^T by the k'th row of H_b. Terminate the LP iterations if the preset limit is exceeded, otherwise go to 1.

Details of this implementation can be found in [18]. Further modification of this method to account for inequality and equality constraints can be found in [20]. In the absence of leverage points, LAV estimators will reject bad measurements. An illustrative example is given below.

Example 6.5:

Consider the system shown in Figure 6.1 from Example 6.2 with an added voltage magnitude measurement at bus 2. Estimate the measurements and the system state using the LAV estimation method.

Solution:
Running the LAV estimator will yield the solution for the system state:

Bus No.	Voltage	Phase
1	1.0000	0.00
2	0.9865	−0.88
3	0.9409	−4.39
4	0.9612	−2.93

Estimated measurements and residuals will be:

Type	From	To	Measured	Estimated	Residual
P-Flow	2	4	0.33966	0.33966	0.0
P-Flow	3	2	−0.56915	−0.56915	0.0
P-Flow	3	4	−0.23084	−0.23085	0.00001
P-Flow	1	2	1.50882	1.50882	0.0
P-Flow	1	4	0.49119	0.49118	0.00001
P-Flow	4	1	−0.49119	−0.49119	0.0
Q-Flow	2	4	0.25583	0.25584	−0.00001
Q-Flow	3	2	−0.41177	−0.41178	−0.00001
Q-Flow	3	4	−0.18810	−0.18810	0.0
Q-Flow	1	2	1.36472	1.36469	0.0
Q-Flow	1	4	0.40104	0.40107	−0.00003
Q-Flow	4	1	−0.36083	−0.36085	0.00002
P-Inj	3		−0.80000	−0.80000	0.0
P-Inj	1		−2.00011	2.00000	−4.0
P-Inj	4		−0.60000	−0.60000	0.0
Q-Inj	3		−0.60000	−0.59988	−0.00012
Q-Inj	1		1.76576	1.76576	0.0
Q-Inj	4		−0.40000	−0.40000	0.0
V-mag	2		0.98650	0.98650	0.0

Note that the sign error in the real power injection at bus 1 is corrected by the LAV estimator, yielding a large residual for this measurement.

6.5.4 Interior Point Algorithm

A new method for solving LP problems was introduced by N.K. Karmarkar in 1984 [14]. Several variants of the original Karmarkar's algorithm [15] have since been developed. The collection of these methods constitute what is referred to in the literature as the interior point methods for linear programming. The distinguishing feature of these methods as compared to the simplex method, is the way they reach the solution. While in the simplex method, the extreme points of the feasible region are traced along its exterior, interior point methods trace a path interior to the feasible region. These methods have been successfully applied to the solution of power system LAV state estimation problem and implementation details can be found in [22, 23, 24]. Only the primal logarithmic barrier function method of [22] will be reviewed here as an example.

Let us consider the LAV estimation problem given below:

$$\text{Minimize} \quad J = \sum_{i=1}^{m}(u_i + v_i) \tag{6.75}$$

$$\begin{aligned} \text{Subject to} \quad z_i - h_i(x) - u_i + v_i &= 0, \quad 1 \leq i \leq m & (6.76) \\ u_i, v_i &\geq 0, \quad 1 \leq i \leq m & (6.77) \end{aligned}$$

The logarithmic barrier method can be employed to remove the inequality constraints on the slack variables. This is accomplished by appending a logarithmic barrier function to the objective function J as:

$$J_a = \sum_{i=1}^{m}(u_i + v_i - \mu \ln u_i - \mu \ln v_i) \tag{6.78}$$

where μ is a positive barrier parameter, which is gradually reduced to zero as the optimal solution is reached.

Form the Lagrangian $\mathcal{L}$:

$$\mathcal{L} = \sum_{i=1}^{m}[u_i + v_i - \mu \ln u_i - \mu \ln v_i - \lambda_i(z_i - h_i(x) - u_i + v_i)] \tag{6.79}$$

and apply the Karush-Kuhn-Tucker (KKT) conditions for the minimum:

$$\begin{aligned} \mathcal{L}_x = \frac{\partial \mathcal{L}}{\partial x} &= \sum_{i=1}^{m} \lambda_i \frac{\partial h_i(x)}{\partial x} = 0 & (6.80) \\ \mathcal{L}_{\lambda_i} = \frac{\partial \mathcal{L}}{\partial \lambda_i} &= z_i - h_i(x) - u_i + v_i = 0 & (6.81) \\ \mathcal{L}_{u_i} = \frac{\partial \mathcal{L}}{\partial u_i} &= 1 - \mu(u_i)^{-1} + \lambda_i = 0 & (6.82) \\ \mathcal{L}_{v_i} = \frac{\partial \mathcal{L}}{\partial v_i} &= 1 - \mu(v_i)^{-1} - \lambda_i = 0 & (6.83) \\ & & (6.84) \end{aligned}$$

The variables λ_i, u_i, v_i, and $h_i(x)$ can be replaced by their first order approximations given below:

$$\begin{aligned} \lambda_i &\approx \lambda_i^0 + \Delta\lambda_i & (6.85) \\ u_i &\approx u_i^0 + \Delta u_i & (6.86) \\ v_i &\approx v_i^0 + \Delta v_i & (6.87) \\ (u_i)^{-1} &\approx (u_i^0)^{-1} - (u_i^0)^{-2}\Delta u_i & (6.88) \\ (v_i)^{-1} &\approx (v_i^0)^{-1} - (v_i^0)^{-2}\Delta v_i & (6.89) \\ h(x) &\approx h(x^0) + H \cdot \Delta x & (6.90) \end{aligned}$$

where
$H = \frac{\partial h(x^0)}{\partial x^0}$,
$h^T(x) = [h_1(x)h_2(x)\ldots h_m(x)]$,
$x \approx x^0 + \Delta x$.

Letting $\lambda^T = [\lambda_1\lambda_2\ldots\lambda_m]$, (6.80)-(6.83) can be expressed as:

$$\begin{aligned}
\mathcal{L}_x &= H^T\lambda_i^0 + H^T\Delta\lambda = 0 &(6.91)\\
\mathcal{L}_\lambda &= z - h(x^0) - H\Delta x - u^0 + v^0 - \Delta u^0 + \Delta v^0 = 0 &(6.92)\\
\mathcal{L}_{u_i} &= 1 - \mu(u_i^0)^{-1} + \mu(u_i^0)^{-2}\Delta u_i + \lambda_i^0 + \Delta\lambda_i = 0 &(6.93)\\
\mathcal{L}_{v_i} &= 1 - \mu(v_i^0)^{-1} + \mu(v_i^0)^{-2}\Delta v_i - \lambda_i^0 - \Delta\lambda_i = 0 &(6.94)
\end{aligned}$$

Solving for Δu_i, Δv_i from (6.93)-(6.94):

$$\begin{aligned}
\Delta u_i &= -\mu^{-1}(u_i^0)^2\{1 - \mu(u_i^0)^{-1} + \lambda_i^0\} - \mu^{-1}(u_i^0)^2\Delta\lambda_i\\
&= \alpha_i^0 - \mu^{-1}(u_i^0)^2\Delta\lambda_i\\
\Delta v_i &= -\mu^{-1}(v_i^0)^2\{1 - \mu(v_i^0)^{-1} - \lambda_i^0\} - \mu^{-1}(v_i^0)^2\Delta\lambda_i\\
&= \beta_i^0 - \mu^{-1}(v_i^0)^2\Delta\lambda_i
\end{aligned}$$

and substituting into (6.92):

$$\begin{aligned}
\mathcal{L}_\lambda &= z - h(x^0) - H\Delta x - u^0 + v^0 - \alpha_i^0 + \beta_i^0 + \mu^{-1}[Q^0]\Delta\lambda = 0\\
&\quad -\mu^{-1}[Q^0]\Delta\lambda + H\Delta x = \gamma^0
\end{aligned}$$

where:
$[Q^0] = \text{diag}\ \{(u_i^0)^2 + (v_i^0)^2\}$
$\gamma^0 = z - h(x^0) - u^0 + v^0 - \alpha_i^0 + \beta_i^0$.

Thus, (6.91 - 6.94) can be reduced to the following compact form:

$$\begin{bmatrix} -\mu^{-1}[Q^0] & H \\ H^T & 0 \end{bmatrix}\begin{bmatrix} \Delta\lambda \\ \Delta x \end{bmatrix} = \begin{bmatrix} \gamma^0 \\ -H^T\lambda^0 \end{bmatrix} \tag{6.95}$$

Equation (6.95) can be further reduced by eliminating $\Delta\lambda$:

$$H^T D^{-1} H\Delta x = H^T\lambda^0 - H^T D^{-1}\gamma^0 \tag{6.96}$$

where $D = -\mu^{-1}[Q^0]$. Iterative solution of (6.95) by reducing μ at each step will yield the optimum solution. However, successful implementation of this method requires careful choice and updating of the barrier parameter μ. Also, it may not be possible to take full Newton steps given by (6.95) without violating the inequality limits. Detailed discussion of these issues along with initialization of the solution procedure can be found in [22].

6.6 Discussion

All of the estimators which are reviewed in this chapter will have some degree of robustness against bad data in the measurements. Existence of leverage points, local measurement redundancy, weights associated with measurements and network and measurement configuration all play a role in the performance of the estimators. Furthermore, implementation of these different estimators require different solution algorithms. However, looking at (6.39) of QC, QL, SR using Newton method, (6.47) of SHGM using iteratively re-weighted least squares method and (6.95) of LAV using interior point method, it can be observed that all iterative solution equations have the form of the "Normal" equations used in weighted least squares estimation. The effective weights of the measurements and the right hand side vector will have to be updated in different ways depending upon the chosen method. Hence, all of the computational techniques that are developed for the WLS estimation problem can be effectively used in implementing these estimators.

6.7 Problems

1. Use the concept of the Mahalanobis distance to illustrate why the four conditions listed in section 6.3.2 will lead to the creation of leverage points.

2. Plot the functions of Υ and its gradient $\nabla\Upsilon$ for the SHGM and LAV estimators for different values of the tuning parameter a between 1 and 5. Verify the claim that the SHGM estimator behaves more like the LAV estimator as a is reduced.

3. Repeat Example 6.4 by adding the following new observations to the existing set:

i	z_i	A_{i1}	A_{i2}
6	42	10	1.0
7	72	20	1.0
8	102	30	1.0
9	132	40	1.0

   Comment on your results.

4. In Example 6.4, show that z_5 constitutes a leverage point. Use one of the identification methods discussed in section 6.3.2.

5. Given the following linear measurement model:

$$z = Hx + e$$

where:

$$z = \begin{bmatrix} 0.6 \\ 0.1 \\ 0.5 \\ -0.4 \\ 0.6 \\ 0.2 \\ 0.1 \end{bmatrix}$$

$$H = \begin{bmatrix} -10.0 & 0.0 & -10.0 \\ -10.0 & 0.0 & 0.0 \\ 0.0 & 0.0 & 10.0 \\ -10.0 & 0.0 & 10.0 \\ 30.0 & -10.0 & -10.0 \\ -10.0 & 10.0 & 0.0 \\ 0.0 & 10.0 & -10.0 \end{bmatrix}$$

Determine the least absolute value (LAV) estimate of x assuming that all measurements z_i are equally weighted. Detect and identify any incorrect (bad) measurements. You can use the linear programming function of Matlab to solve this problem.

References

[1] Huber, P.J., "Robust Estimation of a Location Parameter", Annals of Mathematical Statistics, 35:73-101, 1964.

[2] Hampel F.R., Ronchetti E.M., Rousseeuw P.J. and Stahel W.A., "Robust Statistics: The Approach Based on Influence Functions", Wiley Series in Probability and Mathematical Statistics, John Wiley & Sons, 1986.

[3] Rousseeuw P.J. and Leroy A.M., "Robust Regression and Outlier Detection", Wiley Series in Probability and Mathematical Statistics, John Wiley & Sons, 1987.

[4] Mili L., Cheniae M.G., and Rousseeuw P.J., "Robust State Estimation of Electric Power Systems". IEEE Transactions on Circuits and Systems, Vol. 41, No. 5, May 1994, pp.349-358.

[5] Merrill H.M. and Schweppe F.C., "Bad Data Suppression in Power System Static State Estimation". IEEE Transactions on Power Apparatus and Systems, Vol. PAS-90, pp. 2718-2725, 1971.

[6] Handschin E., Schweppe F., Kohlas J., Fiechter A., "Bad Data Analysis for Power System State Estimation". IEEE Transactions on Power Apparatus and Systems, Vol. PAS-94, No. 2, pp. 329-337, March/April 1975.

[7] Donoho D.L. and Gasko M., "Breakdown Properties of Location Estimates Based on Halfspace Depth and Projected Outlyingness", The Annals of Statistics, Vol. 20, No.4, 1992, pp.1803-1827.

[8] Mili L., Cheniae M.G., Vichare N.S and Rousseeuw P.J., "Robust State Estimation Based on Projection Statistics", IEEE Transactions on Power Systems, Vol.11, No. 2, May 1996, pp.1118-1127.

[9] Baldick R., Clements K.A., Pinjo-Dzigal Z. and Davis P.W., "Implementing Nonquadratic Objective Functions for State Estimation and Bad Data Rejection", IEEE Transactions on Power Systems, Vol. 12, No. 1, February 1997, pp.376-382.

[10] Wagner H.M., "Linear Programming Techniques for Regression Analysis", Journal of American Statistical Association, 54 (1959), pp. 206-212.

[11] Irving M.R., Owen R.C. and Sterling M.J.H., "Power System State Estimation Using Linear Programming", IEE Proceedings, Part C, Vol. 125, No.9, September 1978, pp.879-885.

[12] Kotiuga W.W. and Vidyasagar M., "Bad Data Rejection Properties of Weighted Least Absolute Value Techniques Applied to Static State Estimation", IEEE Transaction on Power Apparatus and Systems, Vol.PAS-101, No.4, April 1982, pp.844-851.

[13] Barrodale I. and Roberts F.D.K., "An Improved Algorithm for Discrete ℓ_1 Linear Approximation", SIAM Journal of Numerical Analysis, Vol. 10, No.5, October 1973.

[14] Karmarkar N.K., "A New Polynomial Time Algorithm for Linear Programming", Combinatorica, 4, pp.373-395, 1984.

[15] Karmarkar N.K., "Computational Results of an Interior Point Algorithm for Large Scale Linear Programming", Mathematical Programming, 52, pp.555-586, 1991.

[16] Abur, A., Magnago, F.H., Alvarado, F.L., "Elimination of Leverage Measurements via Matrix Stretching", Electrical Power and Energy Systems, Vol.19, No.8, pp.557-562, 1997.

[17] Çelik, M.K. and Abur, A., "A Robust WLAV State Estimator Using Transformations", IEEE Trans. on Power Systems, Vol.7, No.1, Feb. 1992, pp.106-113.

[18] Abur, A. and Çelik, M.K., "A Fast Algorithm for the Weighted Least Absolute Value State Estimation", IEEE Trans. on Power Systems, Vol.6, No.2, Feb. 1991, pp.1-8.

[19] Abur, A., "A Bad Data Identification Method for the Linear Programming State Estimation", IEEE Trans. on Power Systems, Vol.5, No.3, Aug. 1990, pp.894-901.

[20] Abur, A. and Çelik, M.K., "Least Absolute Value State Estimation with Equality and Inequality Constraints", IEEE Trans. on Power Systems, Vol.8, No.2, May 1993, pp.680-688.

[21] Luenberger D.G., "Linear and Nonlinear Programming", Addison-Wesley Publishing Co., 1984.

[22] Clements K.A., Davis P.W., and Frey K.D., "An Interior Point Algorithm for Weighted Least Absolute Value State Estimation", IEEE Power Engineering Society Winter Meeting, February 3-7, 1991.

[23] Singh H. and Alvarado F.L., "Weighted Least Absolute Value State Estimation Using Interior Point Methods", IEEE Transactions on Power Systems, Vol. 9, No. 3, August 1994, pp.1478-1484.

[24] Wei H., Sasaki H., Kubokawa J. and Yokoyama R., "An Interior Point Method for Power System Weighted Nonlinear L_1 Norm Static State Estimation", IEEE Transactions on Power Systems, Vol. 13, No. 2, May 1998, pp.617-623.

Chapter 7

Network Parameter Estimation

7.1 Introduction

The redundancy, and to a certain extent the accuracy, of the measurement system depend significantly on the voltage level and importance of the monitored network. In bulk transmission systems every bus section voltage magnitude, sending and receiving branch power flows and externally injected powers are systematically measured, which leads to *full redundancy*. For instance, for a network with 1400 buses and 2000 branches the full redundancy is 4.36. This value ignores the fact that an electrical bus is usually composed of several measured bus sections and that some key circuit breaker (CB) power flows are also measured.

In such cases, the State Estimator can be enhanced with some extra features, leading to the so-called Generalized State Estimator [4]. Among these advanced features the following can be pointed out:

1. Possibility of improving the statistical model of certain suspected measurements (e.g., the bias can be computed).

2. Ability to obtain better estimates for the suspected data base values (e.g., line parameters).

3. Capability to estimate important non-telemetered variables (e.g., transformer taps).

4. Capability to determine the unknown status of CBs and to detect topological errors (i.e., wrong CB statuses).

The first functionality is an extension of the bad data analysis techniques discussed in Chapter 5 and it is mainly of interest during the commissioning and tuning phases of a SE or when remote measurement calibration is considered. The last three items will be separately discussed in this and the next chapter.

Incorrect topological information normally produces large errors in the estimated measurements and can be more easily identified. However, branch impedance errors are less evident and may not be identified for a long time leading to permanent errors in the results provided by the SE.

7.2 Influence of parameter errors on state estimation results

Branch parameter values stored in the fixed data base and tap changer positions available in real time at the Control Center may be incorrect as a result of:

- Inaccurate manufacturing data (e.g., iron losses are customarily ignored in transformer models) or poor line length estimation. For line lengths over 200km, errors exceeding 1% are expected if a lumped parameter *pi* model is employed. Differences between topographical and actual line lengths may lead to even larger errors.
- Network changes not properly updated in the data base (e.g., an overhead line section is substituted by a cable).
- Dependence on temperature (especially the series resistance) or environmental conditions (especially the shunt conductance).
- Misoperation or miscalibration of any electrical or mechanical device involved in the tap monitoring process.
- Local modification of a tap changer without informing the Control Center. This can be done manually by an operator or automatically by the voltage regulator.

These wrong taps and inaccurate parameter values may have the following consequences:

- A noticeable degradation of the results provided by the SE which, in turn, may mislead other applications like security assessment.
- Correct measurements being identified as bad data due to their inconsistencies with the incorrect network parameters.
- Loss of operator's confidence in the SE results.

Reference [18] presents a brief review of the parameter estimation problem. Most works on parameter estimation briefly refer to the importance of including this function within the SE, in order to prevent the consequences mentioned above. However, it is difficult to find systematic experimental results supporting this claim [25, 30, 31, 33, 36]. Reference [6] presents some field experiences obtained during the process of bringing the state estimator on line. A complete study about the influence of parameter errors on state estimation, in which the most important factors are analyzed, can be found in [36, 37]. The main conclusions of this analysis, performed on the IEEE 14-bus test system, will be summarized next.

To begin with, a large enough set of different network states is generated by properly interpolating a typical 24-hour load curve. For every state, a fully redundant measurement set is generated by adding random noise in accordance with the respective standard deviation, σ. The following σ values are adopted:

- Voltage measurements: $\sigma = 0.1\gamma \cdot \text{FS}$
- Power measurements: $\sigma = \gamma \cdot \text{FS}$

where γ is the precision class of the measurement device and FS refers to its full scale. A full scale value in accordance with the largest magnitude expected at the respective measuring point has been chosen (FS=1 for voltage measurements).

Unless otherwise noted, each numerical value reported below is the average of 60 experiments, for the results to be statistically significant. Hence, as the 14-bus system comprises 20 branches, every branch is involved in three different tests.

The first analysis tries to assess how far a single parameter error spreads over the network. To this end, measurements at different *distances* from a branch are considered. Measurements at *distance 1* refer to the power flows of the erroneous branch, as well as the power injections and voltages of its edge buses (the so-called adjacent set). Measurements at *distance 2* comprise the former set plus the power flows and remote injections of those branches which are incident to the edge buses of the erroneous branch, and so on. This concept is illustrated in Figure 7.1.

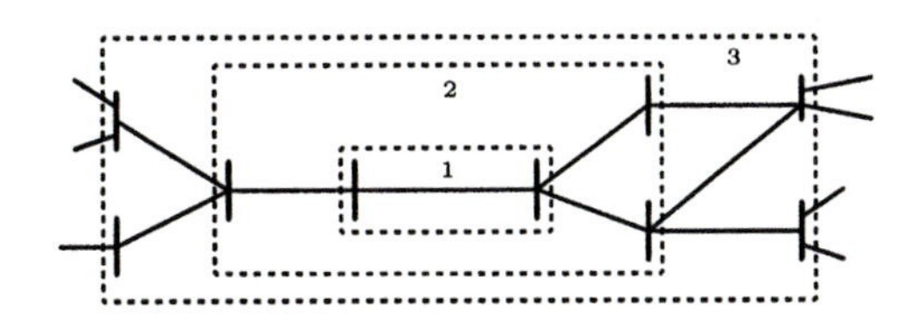

Figure 7.1. Notion of measurement distance to a branch

In simulation environments the actual measurement values are known. Consequently, estimated measurement errors ($\hat{z}_i - z_i^{true}$), rather than residuals ($\hat{z}_i - z_i$), can be used to assess the effect of a parameter error.

Figure 7.2 shows the ratio between the average estimated measurement error when a single line series susceptance is erroneous and the same average when the susceptance is correct. This ratio is computed by considering measurements at increasing distances to the erroneous susceptance. Error levels produced by class 1 transducers have been simulated.

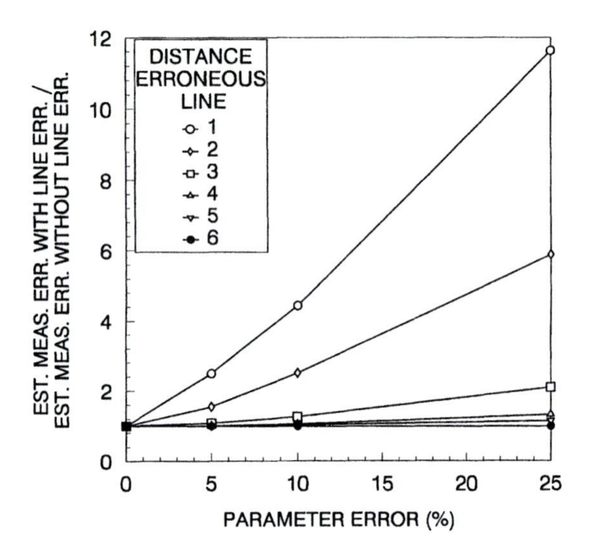

Figure 7.2. Influence of a parameter error on estimated measurement errors at different *distances* (©IEEE)

As clearly seen in this figure, the parameter error's influence is negligible at distances equal to or larger than 4. This means, reciprocally, that remote measurements are virtually useless to estimate a particular parameter value. Therefore, the parameter estimation process can be performed on small subnetworks locally surrounding the suspected branches, in order to save computational effort.

The figure also shows that, despite the high redundancy and the fact that a single parameter is erroneous, adjacent measurements (distance 1) are significantly deteriorated.

The same conclusions can be reached by an analysis of the entries of the residual sensitivity matrix presented in Chapter 5. This requires, however, that the parameter in question be included both in the state and measurement vectors (see the next sections).

The above experiments refer exclusively to measurements generated by class 1 devices, but can be repeated for different accuracies. Figure 7.3 represents, for three accuracy classes, the ratio of the average estimated measurement error to the average input measurement error, where the average is taken on adjacent measurements only. The results clearly suggest that

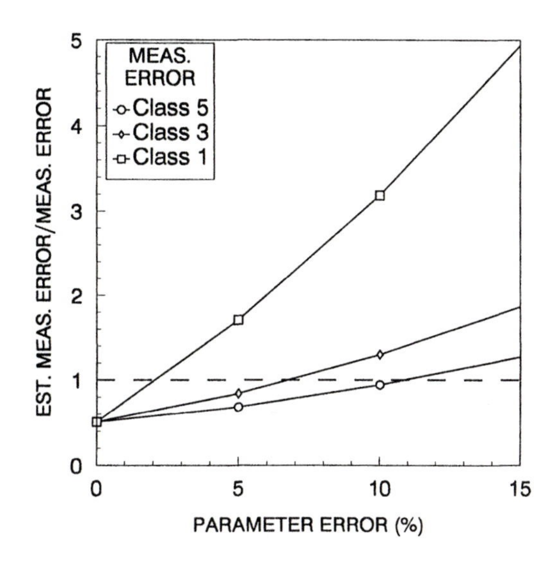

Figure 7.3. Ratio of average estimation error to average initial error versus single susceptance error

the more accurate the measurements are, the higher relative influence the parameter will have. The horizontal dashed line represents the limit beyond which no filtering is possible for the given redundancy as a consequence of the susceptance error. Consider, for instance, the case corresponding to class 1 devices; when the susceptance error is 2%, the average error of the estimated measurements is the same (2.2% for this particular case) as that of input measurements. Susceptance errors larger than 2% would make actual measurements better than estimated ones, for this precision class.

Until now full redundancy is assumed. However, partially redundant measurement sets may lead to rather different results, depending on the type of measurements available. To assess this factor, the following two sets are separately considered: A) power flows plus voltage magnitudes (no power injections); B) power injections plus voltage magnitudes (no power flows). Figure 7.4 shows, for two different accuracies, the average estimation error corresponding to the adjacent measurements. While the power flow measurements (set A) are significantly affected by the parameter

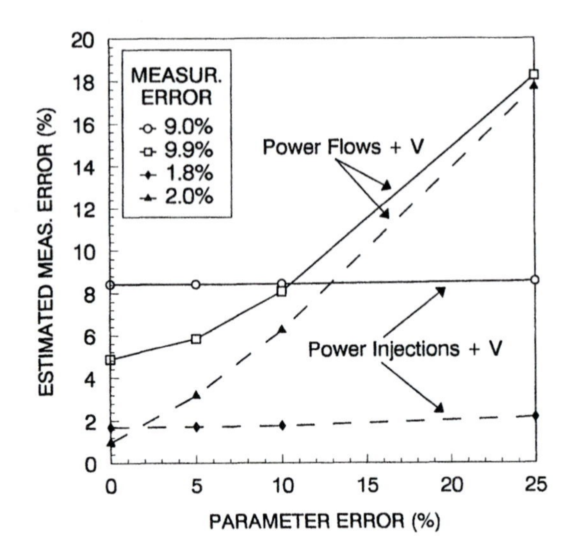

Figure 7.4. Average estimation error of adjacent measurements for incomplete measurement sets versus single susceptance error

error, the power injection measurements (set B) are hardly influenced. It should not be concluded, however, that the estimated state for the set B is more accurate. What happens in this case is that the parameter error redistributes nearby power flows in such a way that power injections remain essentially unaffected. Therefore, the presence of power flow measurements is crucial to be able to detect and estimate parameter errors. Note that the redundancy in case B is much lower than in case A.

If the line series conductance, rather than the susceptance, is erroneous, similar dependencies will be obtained, except for the fact that the resulting average errors will be much smaller. This is expected, since the power flows are more sensitive to the series susceptance than the conductance. Fortunately, it is also true that susceptances are less sensitive to weather conditions compared to the conductances. Consequently, it is reasonable to neglect possible time dependencies of line parameters.

It should be emphasized that, as every point in the above figures is the average of 60 experiments on different branches, the same or very similar diagrams and results are expected for other networks, irrespective of their size.

7.3 Identification of suspicious parameters

Although, theoretically, the parameters of all network branches can be estimated if a long series of fully redundant measurement snapshots are available, in practice this is usually not the case. Furthermore, in addition to the computational cost, there is also a concern about hitting the upper limit that is imposed by the measurement system, on the accuracy of the estimated parameters. (see sections 7.6 and 7.7). In other words, it makes no sense to estimate a parameter whose existing value is likely to be more accurate than the one provided by the SE.

Therefore, it is necessary to initially identify the candidate parameters. Sometimes, the operator's experience, or certain information provided by maintenance teams, may allow manual selection of the candidate parameters. In a majority of cases, however, an automatic procedure based on the measurement residuals is required [31].

A parameter error has the same effect on the estimated state as a set of correlated errors acting on all measurements adjacent to the erroneous branch, namely the power flows through the branch and the power injections at the edge nodes. This results from a simple manipulation of the basic measurement model [24]:

$$z_s = h_s(x,p) + e_s = h_s(x,p_0) + [h_s(x,p) - h_s(x,p_0)] + e_s \tag{7.1}$$

where p and p_0 are respectively the true and erroneous value of the network parameter, and the subscript s refers to the set of adjacent measurements only.

The term in square brackets in (7.1) is equivalent to an additional measurement error. If the parameter error is large enough, this term may lead to bad data being detected and, when this happens, the adjacent measurements will most likely have the largest residuals [23, 24, 35]. The equivalent measurement error can be linearized as:

$$h_s(x,p) - h_s(x,p_0) \approx \left[\frac{\partial h_s}{\partial p}\right] e_p \tag{7.2}$$

where $e_p = p - p_0$ is the parameter error.

Therefore, those branches whose adjacent measurements have large normalized residuals should be declared suspicious.

In reference [19] it is assumed that bad data has been previously identified and removed so that a persistent bias term in certain measurement residuals is an indication of the existence of parameter errors. The identification method proposed in [20] is essentially based on the same idea.

Finally, measurement, parameter and configuration errors on the input data are identified by means of suitable statistical tests in [1].

7.4 Classification of parameter estimation methods

In spite of the profuse literature on SE, the number of publications devoted to the parameter estimation problem is comparatively very modest. Techniques for network parameter estimation, including transformer taps, can be classified as follows [37]:

- Methods based on residual sensitivity analysis [14, 19, 22, 23, 24, 32, 35]. These methods are performed at the end of the state estimation process and resort to the same information previously used to identify suspected parameters. The main advantage of this approach is that the identification and parameter estimation procedures constitute additional and separate routines and, hence, there is no need to modify the main SE code. It is possible, and sometimes necessary, within this category of methods to carry out several iterations of the joint state-and-parameter estimation loop.

- Methods augmenting the state vector. The suspected parameters are included in the state vector and both the state and parameters are simultaneously estimated. Note, however, that a preliminary regular state estimation may be needed in order to identify which parameters should be included in the state vector. Clearly, a modification of existing SE routines is necessary under this approach. Two different but related techniques have been proposed to deal with the augmented model, namely:

  - Solution using normal equations [2, 3, 4, 8, 20, 26, 34, 36]. Except for some observability and numerical issues (e.g., risk of Jacobian singularity at flat start) this approach is a straightforward extension of the conventional SE model. Several snapshots can be resorted to, either simultaneously or sequentially, in order to increase the local redundancy around suspected parameters.

  - Solution based on Kalman filter theory [5, 9, 11, 12, 17, 29, 30, 31]. Under this approach, several measurement samples are sequentially processed in order to recursively improve existing parameter values. The need to update the covariance matrix of parameter errors, as well as other related overheads, make this approach more cumbersome and costly, especially when the number of parameters is high.

The methods comprising the above classification will be separately discussed in the following sections.

7.5 Parameter estimation based on residual sensitivity analysis

As stated above, this approach makes use of the conventional state vector and takes advantage of the results provided by the SE to perform the parameter estimation process.

The technique presented in [23, 24, 35] is based on the sensitivity relationship between residuals and measurement errors [16]:

$$r = S \cdot e \tag{7.3}$$

where S is the residual sensitivity matrix, given by (see Chapter 5):

$$S = I - HG^{-1}H^TW \tag{7.4}$$

and

$$G = H^TWH \tag{7.5}$$

is the gain matrix. Combining (7.1)-(7.3) a linear relationship can be established between the residuals of adjacent measurements, r_s, and the parameter error e_p:

$$r_s = \left(S_{ss}\frac{\partial h_s}{\partial p}\right) e_p + \overline{r}_s \tag{7.6}$$

where S_{ss} is the s x s submatrix of S corresponding to the s involved measurements and $\overline{r}_s$ is the residual vector that would be obtained when the parameter is correct.

Equation (7.6) provides a linear model linking a given vector of measurement residuals, r_s, and an unknown parameter error, e_p, in the presence of a 'noise' vector, $\overline{r}_s$. Therefore, determining e_p can be interpreted and performed as a local estimation problem in which every residual should be weighted according to its $N(0,\Omega)$ distribution (see Chapter 5 for the statistical properties of the residuals). Using $\mho$ to denote the inverse of the diagonal of Ω, the optimal value in the least squares sense, $\hat{e}_p$, is computed from,

$$\hat{e}_p = \left[\left(\frac{\partial h_s}{\partial p}\right)^T S_{ss}^T \mho_s S_{ss}\left(\frac{\partial h_s}{\partial p}\right)\right]^{-1}\left(\frac{\partial h_s}{\partial p}\right)^T S_{ss}^T \mho_s r_s \tag{7.7}$$

A simpler approximate expression is developed in [24]:

$$\hat{e}_p = \left[\left(\frac{\partial h_s}{\partial p}\right)^T W_s S_{ss}\left(\frac{\partial h_s}{\partial p}\right)\right]^{-1}\left(\frac{\partial h_s}{\partial p}\right)^T W_s r_s \tag{7.8}$$

Once the parameter error is estimated, an improved parameter value is obtained,

$$\hat{p} = p_0 + \hat{e}_p$$

Such a model can be immediately extended to several parameter errors by letting p, e_p, etc. be vectors of appropriate size instead of scalars, and by making sure that the set s contains all relevant measurements. Eventually, the state estimation can be repeated using the updated parameter value, the parameter error re-estimated, etc. until no further improvements are obtained.

Example 7.1:

Consider the 3-bus system and measurement set of Example 2.2, whose one-line diagram is repeated in Figure 7.5 for convenience. Assume that the series susceptance value of branch 1-2 available in the data base is -35 instead of its true value -30. Obtain an improved value of this parameter by an analysis of the residual sensitivity matrix.

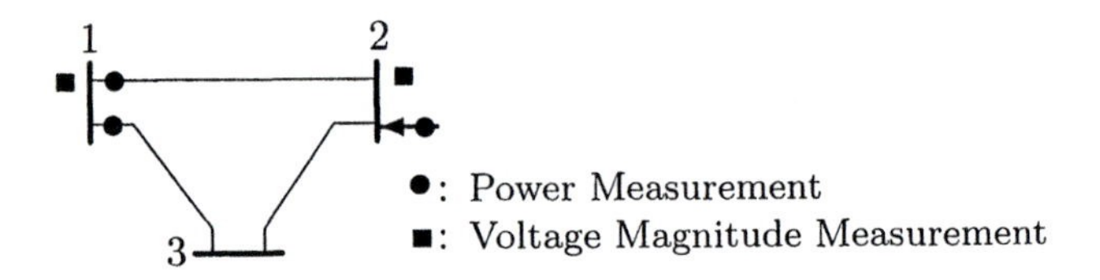

Figure 7.5. One-line diagram and measurement configuration of a 3-bus power system

The WLS algorithm converges to the following state estimation solution:

Bus i	$\hat{V}_i$ (pu)	$\hat{\theta}_i$ (degrees)
1	0.9978	0.00
2	0.9760	-1.153
3	0.9424	-2.743

The objective function is $J(\hat{x}) = 18.5$ and the normalized residuals:

meas.	p_{12}	p_{13}	p_2	q_{12}	q_{13}	q_2	V_1	V_2
r_N	-3.16	3.12	-3.15	-0.73	0.46	-0.49	2.86	-2.87

Both the objective function value and largest normalized residual confirm the presence of (perhaps multiple) bad data. The fact that the normalized residuals corresponding to p_{12}, p_{13}, p_2, V_1 and V_2 exceed, or are close to 3 suggests the possibility of something being wrong in branch 1-2.

In this case, the four measurements p_{12}, p_2, q_{12} and q_2 are directly related to the parameter b_{12}. From the expressions and values provided in Chapter 2, the following derivatives are easily obtained at the solution point:

$$\begin{aligned} \partial p_{12}/\partial b_{12} &= V_1 V_2 \sin\theta_2 = -0.0196 \\ \partial p_2/\partial b_{12} &= -V_1 V_2 \sin\theta_2 = 0.0196 \\ \partial q_{12}/\partial b_{12} &= V_1 V_2 \cos\theta_2 - V_1^2 = -0.0219 \\ \partial q_2/\partial b_{12} &= V_1 V_2 \cos\theta_2 - V_2^2 = 0.0211 \end{aligned}$$

Furthermore, the other matrices appearing in (7.8) are:

$$S_{ss} = \begin{bmatrix} 0.4640 & 0.3575 & -0.0056 & -0.0146 \\ 0.5585 & 0.4315 & 0.0216 & 0.0048 \\ -0.0056 & 0.0138 & 0.4608 & 0.3570 \\ -0.0228 & 0.0048 & 0.5579 & 0.4339 \end{bmatrix}$$

$$W_s = \begin{bmatrix} 15625 & & & \\ & 10000 & & \\ & & 15625 & \\ & & & 10000 \end{bmatrix}$$

$$r_s = [-0.0172, -0.0207, -0.0040, -0.0032]^T$$

These data yield an estimated error $\hat{e}_p = 5.372$ and a new susceptance $b_{12} = -29.63$. This new value is still inaccurate because the measurements are noisy. In this example, this is aggravated by the fact that the local redundancy is low. In general, and in absence of bad data, the higher the redundancy the more accurate the estimated parameter will be, particularly if several iterations are performed.

The procedure proposed in [19] is based both on measurement residuals and on a bias vector which combines the effect of parameter errors and the state of the system. The estimation comprises two steps: First a bias vector is estimated and then the parameter errors are obtained at the second step from a sequence of formerly computed bias vectors. The main difference between this and [23, 24, 35] is that the bias vector is expressed in terms of the line flows.

7.6 Parameter estimation based on state vector augmentation

In this class of methods, the suspected parameter, p, constitutes an additional state variable. Therefore, the objective function becomes,

$$J(x,p) = \sum_{i=1}^{m} w_i [z_i - h_i(x,p)]^2 \tag{7.9}$$

where the dependence on p affects only the set s of adjacent measurements. In general, p will be a vector containing all suspected parameters, but only the scalar case will be considered below for simplicity.

Almost always, an approximate value p_0 is available in the data base, both for line parameters and transformer taps, which can be added to the model as a pseudo-measurement. In such a case, a new term arises in the objective function,

$$J(x,p) = \sum_{i=1}^{m} w_i[z_i - h_i(x,p)]^2 + w_p(p - p_0)^2 \tag{7.10}$$

where w_p is the arbitrary weight assigned to the pseudo-measurement.

Most research works take it for granted that (7.10) rather than (7.9) should be used, to ensure the observability of p. However, this is a controversial issue to which proper attention has not been paid until recently [38]. On the one hand, if p is not observable with existing regular measurements and constraints, the new term in (7.10) is critical and, hence, useless, as the estimated value will be p_0 irrespective of the w_p value. On the other hand, if the extra term is redundant, the value assigned to w_p is critical, as it may significantly influence the estimated value $\hat{p}$. A very small w_p would be equivalent to completely neglecting the available information, p_0. In this case, $\hat{p}$ would be exclusively determined by the analog measurements. On the contrary, a very large w_p would lead to $\hat{p} \approx p_0$, irrespective of the analog measurement values. In order to assess what happens with intermediate w_p values, the series susceptance of line 7-9 (IEEE 14-bus system) is added to the state vector and estimated by means of (7.10). Figure 7.6 represents the relative error of $\hat{p}$ versus the ratio w_{av}/w_p, where w_{av} is the average of all measurement weighting factors (w_i). Different errors in p_0 have been simulated ($\pm 5\%$, $\pm 10\%$), and a measurement set characterized by an average error of 3.23% has been employed.

It is observed that, when the ratio w_{av}/w_p is smaller than 10^3 (leftmost side of the figure), the influence of the initial parameter value is dominant and there is no way to improve the estimated value. Note also that all curves merge at the rightmost side, where $w_p \approx 0$. This means that, irrespective of the initial parameter error, the estimator converges to the parameter value dictated by regular measurements. In this case, the error of the resulting parameter is positive (about 2%), but it can be negative as well for other measurement sets and/or other lines [38].

Two different situations are possible:

- The sign of the error associated to p_0 is the same as that of $\hat{p}$ when $w_p \approx 0$ (positive for this particular experiment). In this case (upper curves), the best result is obtained when $w_p = 0$, unless p_0 is more accurate than the value provided by analog measurements, which is not logical for a suspected line.

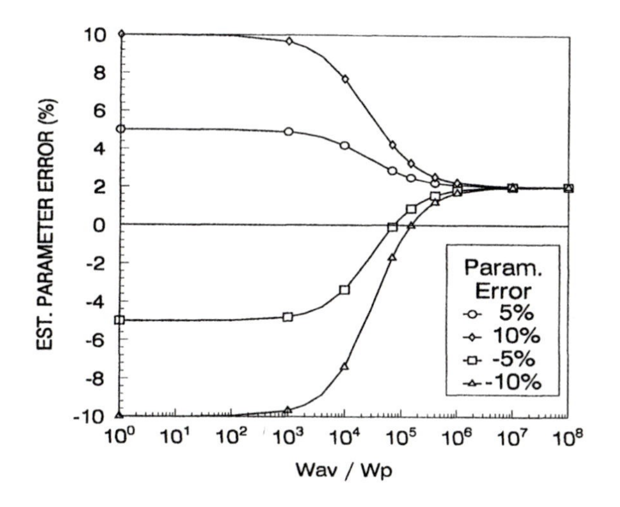

Figure 7.6. Error in the estimated susceptance value versus the relative parameter weight (©IEEE).

- The two signs are different (lower curves). Then, there must be a certain w_p (in this case $w_{av}/w_p \approx 10^5$) for which the exact parameter is estimated. However, if this optimal parameter weight is not chosen, the estimated parameter is likely to be worse than the one obtained with $w_p = 0$.

Consequently, since the sign of the initial parameter error can not be predicted in advance and, in any case, the optimal parameter weight is unknown, the pseudo-measurement p_0 should not be added to the model. If it is added for observability purposes, a rather small value should be used for w_p. It is also not advisable to estimate a parameter whose suspected initial error is smaller than the average measurement error (about 2% in this case). This is why the identification phase is so critical to prevent existing data base values from being prematurely and incorrectly substituted by estimated values.

In order to keep the local redundancy above reasonable levels, the number of extra variables added to the state vector should remain as small as possible. Typically, the line parameters per unit length are well established and the only doubtful information is the total line length. In such cases, a single parameter, e.g., the total length normalized with the existing value, L, suffices. For a line between nodes i and j, the following admittances should be used when building the admittance matrix or computing the residual vector,

$$\begin{aligned} \text{series:} \quad & (g_{ij} + jb_{ij})/L \\ \text{shunt:} \quad & jb_{ij}^{sh} L \end{aligned} \tag{7.11}$$

where g_{ij}, b_{ij} and b_{ij}^{sh} are the admittance values assumed in the data base.

7.6.1 Solution using conventional normal equations

Almost three decades ago, the augmented state vector approach was suggested in [2, 3] to estimate *all* Y_{bus} elements in polar coordinates. Since then, this straightforward technique, combined with the preliminary identification phase discussed in section 7.3, has been successfully applied in different forms. For instance, in [20] the authors extend the state vector with the incremental power flows originated by parameter errors, rather than with the parameters themselves, in an attempt to prevent the numerical problems referred to below. Parameter errors are subsequently calculated in terms of these associated power flows.

Irrespective of the particular version adopted, the Jacobian matrix must be enlarged to accommodate as many extra columns as new state variables and as many extra rows as the new pseudo-measurements. For a single parameter this yields the following structure

$$H = \left[\begin{array}{cccc|c} & & & & 0 \\ & & & & \otimes \\ & & & & 0 \\ & & & & 0 \\ & & & & \otimes \\ & & & & 0 \\ & & & & \otimes \\ & & & & 0 \\ \hline 0 & 0 & 0 & 0 & 1 \end{array}\right] \tag{7.12}$$

where the upper leftmost block corresponds to the conventional Jacobian and the nonzero elements of the new column ($\otimes$) arise only in rows corresponding to adjacent measurements. For instance, assume that the parameter L of (7.11) is added to the state vector. Then, the Jacobian elements corresponding to the power measurements (flows & injections) at node i will be

$$\frac{\partial P_{ij}(L)}{\partial L} = \frac{\partial P_i(L)}{\partial L} = -P_{ij}/L^2 \tag{7.13}$$

$$\frac{\partial Q_{ij}(L)}{\partial L} = \frac{\partial Q_i(L)}{\partial L} = -[Q_{ij} + V_i^2 b_{ij}^{sh}(1 + L^2)]/L^2 \tag{7.14}$$

where P_{ij} and Q_{ij} refer to the power flows computed for $L = 1$. Similar expressions can be obtained for the power measurements at node j (by simply exchanging the subscripts).

Of course, the best initial guess for the new state variable is $L_0 = 1$. Note that, at flat start, the extra column of the Jacobian in (7.12) will be virtually null if, as suggested above, the pseudo-measurement $L = L_0$ is not added. This will lead to a near-singular gain matrix during the first iteration and to potential numerical problems if the normal equations are solved in this situation. Including the new variable L in the state vector starting from the second iteration is one way to solve this problem, provided that the power actually flowing through the line is not null.

Example 7.2:

Consider again the 3-bus system of Example 7.1. Assume that, in the data base, the series impedance of line 1-2 is $0.0117 + j0.0351$, instead of the true value $0.01 + j0.03$ given in Chapter 2. Estimate the line length by augmenting the state vector.

The state vector is augmented with the variable L (relative length) whose initial value is set to 1 pu. It is decided not to include this information as an extra pseudo-measurement. At the beginning of every iteration, y_{12} and the 4 affected admittance matrix entries are updated by means of (7.11). Also, according to (7.12), new Jacobian elements are required at rows corresponding to p_{12}, p_2, q_{12}, q_2.

The WLS algorithm takes 5 iterations to converge to the following solution

Bus i	$\hat{V}_i$ (pu)	$\hat{\theta}_i$ (degrees)
1	0.9998	0.00
2	0.9740	−1.2605
3	0.9440	−2.7466

The estimated relative length is $L = 0.8649$ which leads to an impedance $z_{12} = 0.0101 + 0.0304j$. The objective function and the largest normalized residual are:

$$J(\hat{x}) = 8.555 \quad ; \quad r_N^{max} = 0.2169 \times 10^{-3}$$

Compared to the results of Chapter 2, a reduced objective function and much smaller residuals are obtained. The reason is simply that one extra degree of freedom is added for the same set of measurements. The additional 2 iterations are due to the larger condition number of the gain matrix and due to the fact that the new variable L is added during the second iteration.

Note that the estimated impedance error is only 1%, in spite of the low local redundancy. In practice, it is very unlikely that the existing data base value is more accurate than the one estimated in cases like this.

7.6.2 Solution based on Kalman filter theory

The method presented in [11], intended to estimate transmission line admittances, transformer taps, measurement biases and standard deviations of measurement errors, constitutes the first attempt to systematically apply the Kalman filter to this problem.

At every time sample k, the measurements are related to the states by

$$z(k) = h(x(k), k, p) + e(k) \tag{7.15}$$

where h is made dependent on k to reflect the possibility of network changes from one time sample to the next. The parameters are assumed constant for the entire time period under consideration.

In the LS formulation, the state vector is estimated by minimizing the following objective function:

$$J_k = \sum_{i=1}^{m} \left[z_i(k) - h_i\left(x(k), k, p\right)\right]^T W \left(z_i(k) - h_i(x(k), k, p)\right] \tag{7.16}$$

Starting with the available parameter vector, p_0, the idea is to get better estimates of p at every new sample from,

$$p_{k-1} = p_k + e_p(k) \tag{7.17}$$

where the error vector $e_p(k)$ is assumed to have zero mean and covariance matrix $R_p(k)$. To do so, the objective function must be augmented with as many pseudo-measurements as suspected parameters

$$J = (p_{k-1} - p_k)^T R_p^{-1}(k)(p_{k-1} - p_k) + J_k \tag{7.18}$$

This leads to the following equation being solved at iteration i of the k-th sample,

$$G^i(k) \begin{bmatrix} \Delta x^i(k) \\ \Delta p_k^i \end{bmatrix} = \begin{bmatrix} H_x^i & H_p^i \\ 0 & I \end{bmatrix}^T \begin{bmatrix} W & 0 \\ 0 & R_p^{-1}(k-1) \end{bmatrix} \begin{bmatrix} z(k) - h(x^i(k), k, p_k^i) \\ p_{k-1} - p_k^i \end{bmatrix} \tag{7.19}$$

where the gain matrix is given by,

$$G^i(k) = \begin{bmatrix} H_x^i & H_p^i \\ 0 & I \end{bmatrix}^T \begin{bmatrix} W & 0 \\ 0 & R_p^{-1}(k-1) \end{bmatrix} \begin{bmatrix} H_x^i & H_p^i \\ 0 & I \end{bmatrix} \tag{7.20}$$

and,

$$\begin{aligned} H_x^i(k) &= \left.\frac{\partial h}{\partial x}\right|_{x^i(k), p_k^i} \\ H_p^i(k) &= \left.\frac{\partial h}{\partial p}\right|_{x^i(k), p_k^i} \end{aligned}$$

At the end of the iterative process, the parameter covariance matrix is updated from,

$$R_p(k) = \Lambda_{pp}(k) \tag{7.21}$$

where $\Lambda_{pp}(k)$ is the respective block of the inverse of the gain matrix,

$$G(k)^{-1} = \begin{bmatrix} \Lambda_{xx}(k) & \Lambda_{xp}(k) \\ \Lambda_{px}(k) & \Lambda_{pp}(k) \end{bmatrix} \tag{7.22}$$

There are efficient methods to compute small selected subsets of the inverse of a sparse matrix (see Appendix B), but the computational effort may be prohibitive for on-line application if the number of parameters is relatively large.

The algorithm is recursive in the sense that, at the k-th sample, only the vector $z(k)$ is considered, but the effect of former measurement vectors (i.e., the past history) is taken into account through updated estimates of parameters and associated covariances.

The work reported in [29]-[31] presents two important differences with respect to [11, 12]:

- The problem is localized into several small observable subnetworks containing the unknown parameters.
- Parameters are modeled as Markov processes, thereby allowing estimation of time-varying parameters.

The method assumes that the probability density functions of the parameter and measurement errors are Gaussian with zero mean, leading to an adaptive parameter estimator. The procedure starts by estimating only the parameters of a few branches where the redundancy is maximum. Once the impedances of those branches become established, they are used to extend the process to less telemetered branches, and so on. The solution will eventually include all network branches with adequate local redundancy, excluding only those for which the process can not be reliably performed.

References [5, 9] and [17], combine the residual-based identification procedure with the Kalman filter theory.

7.7 Parameter estimation based on historical series of data

Assuming network parameters are essentially time invariant, it is possible and probably more convenient to perform the parameter estimation process in batch mode by means of a sufficiently large set of recorded measurement samples. This way of proceeding offers the following advantages [25, 36]:

- Parameter values can be routinely improved on a dedicated computer without interfering with the execution of any EMS critical application. Furthermore, there is no need to modify the SE code running on line.

- Optimality of the estimated values is much more important than computation times. Hence, complex and expensive algorithms can be resorted to if they provide more accurate results.

- Filtering tests can be performed in order to get rid of those samples in which measurement redundancy is locally insufficient or the existence of bad data is suspected. This "healthy" snapshot selection process can be as sophisticated (i.e., costly) as needed. Similarly, selection of suspect parameters can be based on a longer historical series of data.

- Simultaneously using several snapshots avoids deteriorating the local redundancy because the additional parameter variables are shared by all network states. This would be quite a cumbersome process if carried out on line.

The last statement can be easily proved as follows. Let n_p and q be the number of suspected parameters and simultaneous samples respectively. As always, m and n denote the measurement and state vector sizes, but in this case they can refer also to the size of the SE subproblems localized around the suspected parameters. The base-case redundancy, i.e., the redundancy when no parameters are estimated is

$$\eta_\infty = \frac{m}{n}$$

This redundancy level decreases to

$$\eta_1 = \frac{m}{n + n_p} < \frac{m}{n}$$

when n_p parameters are included in the state vector and no pseudo-measurements are added to compensate for them. For instance, if $n_p \approx n$, then $\eta_1 \approx 0.5\eta_\infty$. However, when q samples are simultaneously employed we get

$$\eta_q = \frac{mq}{nq + n_p} = \frac{m}{n + n_p/q}$$

Clearly, for $q \to \infty$ the local redundancy approaches that of the base case, even though no pseudo-measurements reflecting existing parameter values are added. For instance, for $q = 10$, with $n_p \approx n$, we get $\eta_q = 0.9\eta_\infty$ which is quite an acceptable value.

The remaining part of this section will be devoted to showing how the computations involved in the NE-based SE can be rearranged so that several measurement samples can be simultaneously handled without incurring unacceptable overheads.

Assuming q simultaneous samples are used, the overall state and measurement vectors will become,

$$x = [x_1, x_2, ..., x_q | p]^t \tag{7.23}$$
$$z = [z_1, z_2, ..., z_q]^t \tag{7.24}$$

where x_k, z_k refer to the k-th sample vectors.

The Jacobian corresponding to this enlarged model has, therefore, the following structure,

$$H = \begin{bmatrix} H_1 & & & & h_{1p} \\ & H_2 & & & h_{2p} \\ & & \ddots & & \vdots \\ & & & H_q & h_{qp} \end{bmatrix} \tag{7.25}$$

Note that, in order to avoid the potentially perverse effect discussed in section 7.6 the parameter pseudo-measurements are not added to the model. Substituting into the normal equations,

$$H^t W H \Delta x = H^t W \Delta z \tag{7.26}$$

leads to a gain matrix given by

$$H^t W H = \begin{bmatrix} G_{11} & & & & g_{1p} \\ & G_{22} & & & g_{2p} \\ & & \ddots & & \vdots \\ & & & G_{qq} & g_{qp} \\ g_{1p}^t & g_{2p}^t & \cdots & g_{qp}^t & G_{pp} \end{bmatrix} \tag{7.27}$$

where

$$G_{ii} = H_i^t W_i H_i \qquad (i = 1, 2, ..., q) \tag{7.28}$$
$$g_{ip} = H_i^t W_i h_{ip} \qquad (i = 1, 2, ..., q) \tag{7.29}$$
$$G_{pp} = \sum_{i=1}^{q} h_{ip}^t W_i h_{ip} \tag{7.30}$$

and to the following right-hand side vector

$$H^t W \Delta z = [b_1, b_2, ..., b_q | b_p]^t \tag{7.31}$$

where

$$b_i = H_i^t W_i \Delta z_i \qquad (i = 1, 2, ..., q) \tag{7.32}$$

$$b_p = \sum_{i=1}^{q} h_{ip}^t W_i \Delta z_i \tag{7.33}$$

At first glance, it seems that the huge matrix given by (7.27) must be formed and factorized at each iteration, in order to obtain the state vector correction, Δx. This is clearly impractical for large networks, even if conventional sparsity techniques are used. In the appendix of [11], it was suggested that a block-wise iterative method could be used to circumvent the dimensionality problem of the resulting model. However, convergence of such an iterative scheme has not been tested.

Instead, the direct method described below is a safer alternative provided the number of parameters simultaneously estimated remains relatively low [36]. Assuming the matrix G_{pp} and vector b_p are initially set to zero, the following steps must be performed at each iteration:

1. Block factorization and forward elimination. For each diagonal block $(i = 1, 2, ..., q)$:

   (a) Compute

$$\hat{h}_{ip} = W_i h_{ip} \tag{7.34}$$

$$\Delta \hat{z}_i = W_i \Delta z_i \tag{7.35}$$

$$g_{ip} = H_i^t \hat{h}_{ip} \tag{7.36}$$

$$b_i = H_i^t \Delta \hat{z}_i \tag{7.37}$$

   (b) Compute G_{ii} and obtain y_i, b_i' from

$$G_{ii} = H_i^t W_i H_i \tag{7.38}$$

$$G_{ii} y_i = g_{ip} \tag{7.39}$$

$$G_{ii} b_i' = b_i \tag{7.40}$$

   (c) Update G_{pp}, b_p

$$G_{pp} = G_{pp} + h_{ip}^t \hat{h}_{ip} - g_{ip}^t y_i \tag{7.41}$$

$$b_p = b_p + h_{ip}^t \Delta \hat{z}_i - g_{ip}^t b_i' \tag{7.42}$$

2. Obtain Δp from

$$G_{pp} \Delta p = b_p \tag{7.43}$$

3. Block backward substitution. Once Δp is available, all state variables are updated independently. For each diagonal block ($i = 1, 2, ..., q$) obtain Δx_i from

$$\Delta x_i = b'_i - y_i \Delta p \tag{7.44}$$

The following clarifying comments are in order:

- Each column of h_{ip} is an extremely sparse vector, as only adjacent measurements contribute non-zero derivatives with respect to p. This fact should be taken into account to save arithmetic operations in the above procedure.

- Each diagonal block, G_{ii}, is used only once in step 1.(b), and can be discarded thereafter. Hence, only the gain matrix corresponding to a single snapshot must be formed and held simultaneously in memory. In fact, steps 1 and 3 could be performed in parallel on as many processors as available samples.

- When only a single parameter is estimated, h_{ip} is a column vector and G_{pp} is a scalar.

- As explained above, the derivative terms included in h_{ip} are negligible for the usual flat start voltage profile. So, to prevent ill-conditioning of the resulting equation system, parameters are added to the state vector only after the first iteration.

Therefore, for a low number of estimated parameters, the overhead caused by equations (7.34), (7.36), (7.39) and (7.41)-(7.44), both in terms of memory requirements and computation time, is rather modest, compared to the cost of sequentially processing the whole set of samples, which is dictated mainly by equations (7.38) and (7.40). Considering, as proved in section 7.2, that only nearby measurements appreciably influence the estimated parameter values, the above methodology can be applied in such a way that each diagonal block includes only the relevant network portion. This would reduce significantly the computational cost but still provide virtually the same estimated values.

The major difference between this and the Kalman filter approach described above lies in the fact that there is no need to update the parameter error covariance matrix, as all samples are simultaneously, rather than recursively considered and existing information about the parameter values is discarded.

Resorting again to the IEEE 14-bus test system and to the simulation environment presented in section 7.2 (full redundancy, 60 experiments per point, etc.), the advantage of using several samples to estimate line parameters can be shown. Figure 7.7 represents the estimated parameter error divided by the average estimated measurement error (and its inverse) versus the number of samples. The initial parameter error is 10% and the measurement noise corresponds to class 1 devices. The linear regression and its hyperbolic inverse clearly shows that the estimated parameter error continuously decreases with the number of samples. For instance, when 25 samples are used, the estimated parameter is almost 8 times as accurate as the estimated measurements. In the limit, the exact parameter value would be estimated, irrespective of its initial error, which is a remarkable result.

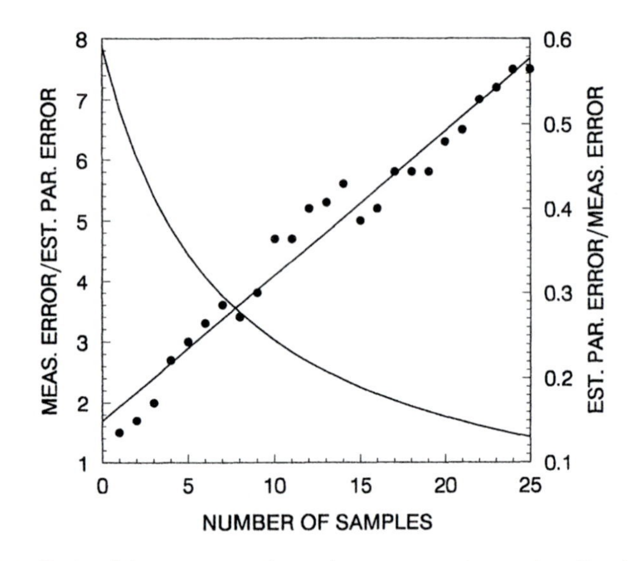

Figure 7.7. Ratio of the average estimated measurement error to estimated parameter error (left) and its inverse (right) versus the number of samples simultaneously processed

Of course, the initial noise associated with the measurements has a significant influence on the quality of the estimated parameter value. Table 7.1 presents the estimated parameter error (b) for three measurement error levels (a) when 1, 4 and 7 samples are used. Even for an unrealistically high measurement error (14%) the parameter is estimated within 3% accuracy if 7 samples are employed.

Note that the ratio a/b depends on the number of samples, but not on the measurement noise. Clearly, for a given number of samples, the estimated parameter value may be worse than that of the data base if the analog measurements are of poor quality; and vice versa. The interested reader may resort to the above figures (which are quite general in spite of being referred to the 14-bus system) in order to find out whether it is worth

Snapshots	(a)	(b)	(a/b)
1	13.53	8.73	1.5
	8.10	5.27	1.5
	2.70	1.70	1.6
4	14.20	4.38	3.2
	8.52	2.65	3.2
	2.84	.90	3.2
7	14.26	3.27	4.4
	8.56	1.95	4.4
	2.85	.65	4.4

Table 7.1. Estimated parameter error (b) for different measurement error levels (a), both in percent, and different number of samples

to estimate a certain parameter for a given redundancy and error level.

7.8 Transformer tap estimation

From an operational point of view two different classes of on-load tap changers can be identified:

- Taps of large transformers located within the bulk transmission grid or connecting the transmission networks to the sub-transmission networks. These taps are usually telemetered and remotely set by the control center operator in order to keep bus voltages within reasonable but relatively wide limits, to control reactive power flows or reactive power delivered by generators and also to reduce power losses.

- Taps of medium-size transformers connecting the sub-transmission network to the distribution feeders. In order to keep customer voltages as steady as possible against load fluctuations, these transformers are equipped with automatic controllers that sense the voltage and/or current at the feeder head and accordingly shift the tap. As there are many transformers of this type, the tap value is seldom monitored in the control center.

Unlike ordinary measurements, the telemetered tap is handled from the very beginning as a digital value and consequently no noise can be associated with this piece of information. However, malfunctioning of the mechanical or electrical equipment involved in the tap changer, albeit rare, is always possible, leading to an error of the same nature as the topological errors discussed in the next chapter. When this happens inadvertently, the transformer model used by the SE will be inaccurate and one or several nearby measurements can be wrongly identified as bad data.

Example 7.3:

Let us substitute the line 1-2 in the same 3-bus case of Example 7.1 by a voltage regulating transformer. The impedance values and measurement points are shown in Figure 7.8. Noisy measurements ($\sigma = 0.001$) are generated which are compatible with the null resistance of branch 1-2 and nominal tap setting ($a = 1$). Run the SE by assuming a wrong tap value $a = 1.05$ and perform bad data analysis.

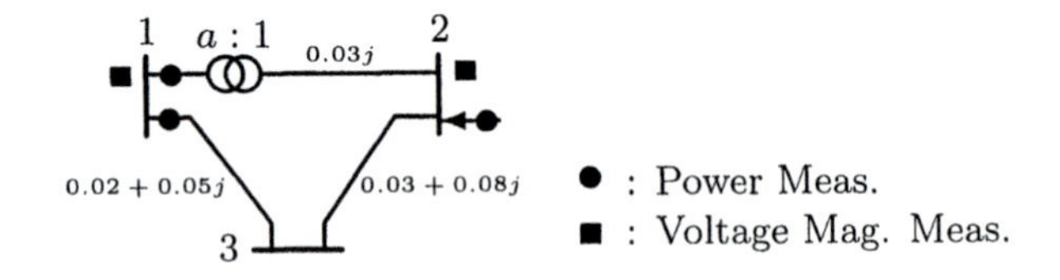

Figure 7.8. Modified 3-bus system with a transformer

The SE converges after 5 iterations to the state given in the following tables, where the estimated and actual measurements as well as the normalized residuals are also provided.

Bus i	$\hat{V}_i$ (pu)	$\hat{\theta}_i$ (degrees)
1	0.9380	0.00
2	0.8776	−1.42
3	0.8794	−3.19

Measurement	Meas. Value	Estim. Value	Norm. Res.
p_{12}	0.7135	0.6475	87.9
p_{13}	1.1390	1.1791	−120.9
p_2	−0.3230	−0.3914	121.8
q_{12}	0.6747	0.4767	285.8
q_{13}	0.5675	0.6529	−271.5
q_2	−0.4193	−0.5639	271.0
V_1	1.0000	0.9380	82.3
V_2	0.9800	0.8776	134.3

The objective function remains very high at the solution point ($J(\hat{x}) = 9.2 \cdot 10^4$), the largest normalized residual being associated with q_{12}. Note that the residuals of q_{13}, q_2 and V_2 are also very high. Removing q_{12}, the SE still detects q_{13} as bad data.

Of course, assuming $a = 1$, and repeating the experiment, the correct solution with small normalized residuals is obtained in just 3 iterations. This confirms that the presence of wrong parameters tends to deteriorate the convergence rate (divergence is also possible).

The above example illustrates that the terminal bus voltage magnitudes and reactive power through a transformer become significantly affected by an error in the tap value. Conversely, these magnitudes can not be accurately estimated if the tap value is not available, as in the second class of transformers. Therefore, there is a need to estimate transformer tap positions for the following two cases:

- The tap is telemetered but the available value is suspicious. Since tap setting errors are very infrequent, it is wasteful to include in the SE model all taps. Hence, as in the parameter estimation problem, only a few suspected taps are considered each time.

- The tap is not telemetered and the user wishes to estimate the power flow through the transformer, which requires that both terminal buses are included within the SE scope.

Obviously, transformer taps are time-varying parameters and, consequently, off-line processing can not be used. Apart from this, most of the ideas discussed in former sections for line parameter estimation are applicable for estimating transformer taps.

Residual analysis can be used both to identify suspected transformers and to obtain improved tap values. As noted above, the reactive power flow through the transformer plays an important role in this regard. In fact, it is almost mandatory for this magnitude to be measured, in addition to the terminal voltage magnitudes, if the tap setting is to be estimated reliably. Let t and f denote the tapped and fixed buses respectively. Then, ignoring the resistance, the reactive power entering bus t is given by the following equation,

$$Q_t = \frac{T^2 V_t^2}{X} - \frac{T V_t V_f}{X} \cos\theta_{tf} \tag{7.45}$$

where X is the transformer reactance and $T = 1/a$ is used to get a more linear expression. Assuming $\cos\theta_{tf} \approx 1$, the former equation can be linearized around the estimated state, yielding,

$$X \Delta Q_t = K_{tf}(V_t \Delta T + T \Delta V_t) - T V_t \Delta V_f \tag{7.46}$$

where $K_{tf} = 2TV_t - V_f$, the incremental values ΔQ_t, ΔV_t and ΔV_f refer to the respective residual (measured minus estimated magnitude), and $\Delta T = T_{\text{new}} - T_{\text{old}}$ is the inverted tap error.

Equation (7.46) provides a simple means of estimating the tap error, based on the most significant residuals related to the transformer tap,

$$\Delta T = \frac{X}{K_{tf} V_t} \Delta Q_t - \frac{T}{V_t} \Delta V_t + \frac{T}{K_{tf}} \Delta V_f \tag{7.47}$$

The new tap setting is therefore,

$$a_{\text{new}} = \frac{1}{T_{\text{new}}} = \frac{1}{1/a_{\text{old}} + \Delta T} \tag{7.48}$$

However, taking into account that $\Delta a/a = -\Delta T/T$, the following alternative expression can be used to update the tap,

$$a_{\text{new}} = a_{\text{old}} + \Delta a = a_{\text{old}}(1 - a_{\text{old}} \Delta T) \tag{7.49}$$

Note the signs of the different terms in (7.47), and the fact that the contribution of ΔV_t counteracts that of ΔV_f. The size of every term depends very much on the network topology and transformer role. For a radially connected distribution transformer, a tap error will translate into a large ΔV_f, because the values of V_t and Q_t will be mainly dictated by the measurements at the high voltage side. For a transmission regulating transformer, belonging to a meshed grid, the 3 magnitudes will be affected to a certain extent, depending on the local redundancy and relative weights (usually, ΔQ_t will be the largest residual).

A similar development can be undertaken if Q_f, rather than Q_t, is measured, but the results obtained will generally be worse.

Example 7.4:

Apply the above residual-based methodology to get a better value for the tap setting of Example 7.3.

The following data are taken directly from the results provided in that example,

$$\Delta q_{12} = 0.1980 \quad ; \quad \Delta V_1 = 0.0620 \quad ; \quad \Delta V_2 = 0.1024 \quad ;$$

Therefore, we obtain $\Delta T = 0.0513$ and,

$$a_{\text{new}} = \frac{1}{1/a_{\text{old}} + \Delta T} = 0.9963$$

The accuracy of this result depends on that of involved measurements, but also on the size of the tap deviation. The following table collects the results provided both by (7.48) and (7.49) when different tap values are assumed for the same set of measurements and the actual tap setting ($a = 1$):

a_{old}	a_{new} (7.48)	a_{new} (7.49)	# Iter.
0.875	1.0941	1.0502	7
0.9	1.0501	1.0287	6
0.925	1.0226	1.0133	5
0.95	1.0073	1.0040	4
0.975	1.0010	1.0003	3
1.0	1.0	1.0	3
1.025	0.9995	0.9988	4
1.05	0.9963	0.9934	5
1.075	0.9898	0.9825	6
1.1	0.9805	0.9659	6
1.125	0.9691	0.9440	7

The following conclusions are reached:

- The tap is overestimated (underestimated) when the assumed value is smaller (larger) than 1 pu.
- Equation (7.48) is more accurate when the assumed value is larger than 1 pu., and the opposite happens with (7.49). This is helpful to select the most appropriate expression in every case.
- The linear approximation starts to deteriorate (estimation error is larger than 1 %) for assumed tap errors exceeding 7.5 %. However, different percentages could apply when the available measurements are less accurate.

Several approximations to (7.47) have been used in an attempt to sequentially refine wrong tap settings. For instance, in [14] and [32] suspected taps are identified when the difference between the calculated and telemetered reactive power flow through a transmission transformer is higher than a preset tolerance. Then, the tap setting is raised or lowered depending on which reactive power flow is measured and the sign of the residual, ΔQ. The algorithm proposed in reference [22], intended for radial transformers, uses the estimated and measured voltages in order to generate a new tap position.

Example 7.5:

Consider the 4-bus system of Example 2.1, containing a radial transformer. This system, along with the measurement points, is shown in Figure 7.9.

A set of measurements, compatible with the actual tap value, $a = 0.98$, is generated by adding Gaussian noise ($\sigma = 0.001$) to the exact values corresponding to the state given in Chapter 2.

Assume now that the available tap value is $a = 1.05$. The following tables show the results of the WLS algorithm after 3 iterations:

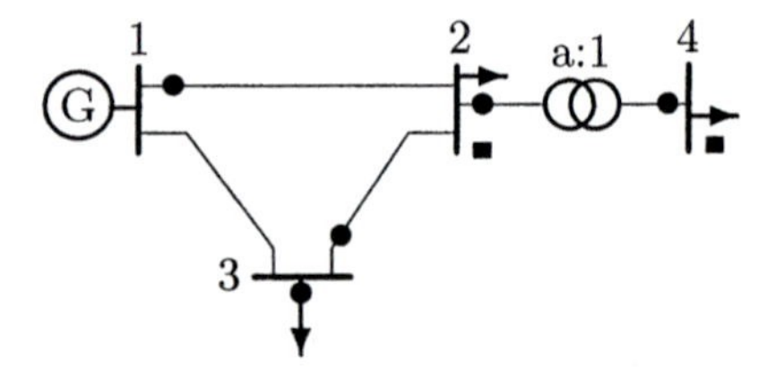

Figure 7.9. One-line diagram of a 4-bus power system

Bus i	$\hat{V}_i$ (pu)	$\hat{\theta}_i$ (degrees)
1	1.0281	0.00
2	0.9919	−2.61
3	0.9902	−3.40
4	0.9361	−3.90

Measurement	Meas. Value	Estim. Value
p_{12}	0.88726	0.88739
p_{32}	−0.11676	−0.11520
p_{24}	0.25034	0.24972
p_{42}	−0.24908	−0.24972
p_3	−1.19965	−1.20007
q_{12}	0.24140	0.23658
q_{32}	0.02923	0.04209
q_{24}	0.10645	0.10444
q_{42}	−0.09913	−0.09788
q_3	−0.79933	−0.80426
V_2	0.96290	0.99190
V_4	0.97423	0.93606

Applying (7.47) and (7.48) to the above results yields $\Delta T = 0.06615$ and $a_{\text{new}} = 0.9818$. Unlike in the former example, the term ΔQ_{24} in (7.47) is negligible in this case, due to the radial configuration.

The normalized residual test erroneously identifies V_4 as bad data ($r_N^{\max} = 50.2$). When this measurement is removed, the following results are obtained:

Bus i	$\hat{V}_i$ (pu)	$\hat{\theta}_i$ (degrees)
1	1.0002	0.00
2	0.9630	−2.76
3	0.9599	−3.58
4	0.9082	−4.14

Measurement	Meas. Value	Estim. Value
p_{12}	0.88726	0.88735
p_{32}	−0.11676	−0.11705
p_{24}	0.25034	0.24971
p_{42}	−0.24908	−0.24971
p_3	−1.19965	−1.19954
q_{12}	0.24140	0.24148
q_{32}	0.02923	0.02903
q_{24}	0.10645	0.10629
q_{42}	−0.09913	−0.099929
q_3	−0.79933	−0.79924
V_2	0.96290	0.96304
V_4	(0.97423)	0.90817

Observe that the residuals corresponding to Q_{24} and V_2 are very small now. Consequently, (7.47) reduces in this case to,

$$\Delta T \approx \frac{T}{K_{24}} \Delta V_4 = 0.06793$$

and the updated tap setting is $a_{\text{new}} = 0.9801$, which is almost the exact value.

The state vector augmentation technique is considered a more effective approach, as all surrounding measurements, not just reactive power flows or voltages magnitudes are involved in the procedure. However, this is accomplished at the expense of more complexity and computation time.

Consider, for simplicity, that a single transformer tap must be estimated. Then, the Jacobian of the augmented model has a block structure like that of (7.12). The extra row is needed only if a tap value is available whereas the nonzero elements of the extra column correspond to the sensitivities of power flows through the transformer and terminal bus power injections with respect to the tap variable.

It has recently been shown [7] that, if $T = 1/a$, rather than a, is added to the state vector, the new Jacobian elements will be readily available from existing ones by taking advantage of the symmetrical role of V_t and T in all equations. From the transformer model provided in Chapter 2 the following expressions can be written for the power flows at both sides of the transformer,

$$\begin{aligned} P_{tf} &= -TV_tV_f\left(G\cos\theta_{tf} + B\sin\theta_{tf}\right) + GT^2V_t^2 \\ Q_{tf} &= -TV_tV_f\left(G\sin\theta_{tf} - B\cos\theta_{tf}\right) - BT^2V_t^2 \\ P_{ft} &= -TV_tV_f\left(G\cos\theta_{tf} - B\sin\theta_{tf}\right) + GV_f^2 \\ Q_{ft} &= TV_tV_f\left(G\sin\theta_{tf} + B\cos\theta_{tf}\right) - BV_f^2 \end{aligned} \qquad (7.50)$$

where G and B are the transformer series conductance and susceptance respectively. Noting the systematic pairing of the variables T and V_t we can write,

$$\frac{\partial}{\partial T}\begin{bmatrix} P_{tf} \\ Q_{tf} \\ P_{ft} \\ Q_{ft} \end{bmatrix} = \frac{V_t}{T}\frac{\partial}{\partial V_t}\begin{bmatrix} P_{tf} \\ Q_{tf} \\ P_{ft} \\ Q_{ft} \end{bmatrix} \tag{7.51}$$

which allows new Jacobian elements to be obtained in a simple manner. The same idea applies to power injection measurements.

The technique proposed in [34] belongs to the state augmentation class of methods. Initially, transformer taps are modeled as continuous variables and a best fit is calculated. Then, the best fit is set to its nearest feasible discrete tap position and is removed from the state vector. The normal equations are solved again allowing changes in the state vector resulting from the tap discretization. One or several transformer taps can be estimated simultaneously.

The method described in [20], which considers incremental flows originated by parameter errors as additional state variables, can be applied also to the estimation of transformer taps.

Example 7.6:

The above ideas are applied to the 3-bus and 4-bus systems of the former examples, assuming the tap is not measured. In both cases the WLS algorithm converges in three iterations to the following states:

Bus i	$\hat{V}_i$ (pu)	$\hat{\theta}_i$ (degrees)
1	1.000	0.00
2	0.9800	−1.25
3	0.9499	−2.75
$\hat{a} = 0.9999$		

Bus i	$\hat{V}_i$ (pu)	$\hat{\theta}_i$ (degrees)
1	1.0002	0.00
2	0.9630	−2.76
3	0.9599	−3.58
4	0.9742	−3.96
$\hat{a} = 0.9801$		

Note that the estimated tap settings are slightly better than those obtained from the residuals, but probably the most noticeable effect is the fact that the convergence is not deteriorated by the presence of the wrong tap value.

7.9 Observability of network parameters

Observability of the parameters defining a particular branch depends on the nature of the measurements in the adjacent set. As defined in section 7.2, this set is composed of the branch power flows and power injections at both terminal buses. When all measurements in this set are critical to estimate the conventional state vector, none of the respective branch parameters will be observable. In fact any parameter error will remain undetected, as the respective residuals will be null. Conversely, if any member of this set, and hence all of them, are redundant then at least a single branch parameter, e.g. a line length or transformer tap, can be added to the state vector in order to be estimated [17]. Estimating several parameters requires a higher redundancy level.

Of course, any parameter can be made observable by adding a guessed value as a pseudo-measurement to the measurement set, but this is useless to improve the data base when there is no redundancy. If the added pseudo-measurement turns out to be critical when the respective parameter is included in the state vector, then such a parameter will be unobservable in the absence of the extra pseudo-measurement.

An interesting and frequent particular case arises when power flows at both ends of a branch belonging to an observable island are measured. In this case, at least two out of the four measurements can be removed without losing observability. Therefore, at least two independent branch parameters can be estimated in this situation.

From a numerical point of view, however, some care must be exercised when performing parameter estimation. Consider, for instance, the line length estimation problem discussed in section 7.6.1. The Jacobian terms corresponding to the new variable are null at flat start, but also when the power flow through the line is negligible. This means that, in the absence of the respective pseudo-measurement, the relative line length is in practice unobservable for lines whose power flow is very small, or at least the estimated value is not reliable as a consequence of numerical instabilities. For the same reason, the presence of voltage magnitude and/or reactive power measurements is of paramount importance to reliably estimate transformer taps, as the sensitivity of active power with respect to this variable is very small.

In the general case, the numerical observability methods described in Chapter 4 should be resorted to so as to decide whether a particular parameter included in the state vector is observable or not.

Nevertheless, as concluded from the results presented in former sections, assuring observability is not enough to take the decision of estimating a particular parameter. An analysis of the local redundancy and quality of available measurements is strongly recommended before running the risk of substituting a data base value by an estimated one.

7.10 Discussion

The following aspects are somewhat subject to controversy and deserve further discussion.

- *Residual analysis versus state augmentation.* If the ultimate goal is to estimate a parameter value, then the state vector augmentation technique is preferable to the one based on residual sensitivity analysis. This is due to the fact that the approach based on residuals extracts the information from a linearized model and, hence, must be applied repeatedly to provide the same accuracy. Nevertheless, the residual analysis is still necessary in the process of identifying suspect branches.

- *Kalman filter versus conventional SE.* It is evident that resorting to several snapshots, either sequentially by means of the Kalman filter or simultaneously by enlarging the conventional SE model, improves the local redundancy and provides a safer way of updating the data base. It is not clear, however, whether the recursive filter should be employed in all circumstances. The Kalman filter could be perhaps more appropriate to continuously update time-varying parameters in a localized area, while the simpler WLS approach is a better choice to estimate constant parameters. Keep in mind, anyway, that any method will provide poor estimates of parameters in the presence of persistent nearby gross errors that remain undetected.

- *On-line versus off-line processing.* Estimation of transformer tap positions, like detection of topology errors, is inherently an on-line process. However, off-line processing may be a more adequate approach to estimate those branch parameters which remain essentially constant over time, like inductance and capacitance. Fluctuations of line resistance due to temperature changes may be significant, but errors affecting this parameter have been shown to be less influential on the SE performance [33, 36]. Additionally, if representative temperatures were recorded along with every sample, then it would be possible to relate the time-varying resistance to the constant value corresponding to a reference temperature.

- *Localized versus global parameter estimation.* This issue refers to whether a few selected parameters (including transformer taps) should be estimated, or whether the estimation process should be adaptively extended to eventually include all network branches with adequate local redundancy. A tentative answer to this question can be given by looking at the results presented above. From Table 7.1 it can be concluded that, for a given redundancy, the estimated parameter errors are proportional to the average measurement error. Therefore, if the SE is fed with very accurate measurements, the estimated parameter values will most likely be better than those stored in the data base. But the opposite may also occur; an acceptable parameter value can be updated with a less accurate value if poor measurements are involved in the estimation process. Hence, deciding in advance which parameters should be estimated is not a trivial task.

7.11 Problems

1. Prove that (7.8) can be obtained from (7.7) by ignoring the off-diagonal terms in the transpose of the residual sensitivity matrix, i.e. $S_{ss}^T \approx \text{diag}(S_{ss})$.

2. Obtain $\hat{e}_p$ in Example 7.1 by means of (7.8), and compare the result with the one provided by (7.7).

3. Prove that, if (7.8) is adopted, then using the full residual vector, r, and the full matrices S, $\mho$ and $\partial h/\partial p$ will provide the same parameter value as that obtained with the subset s of adjacent measurements (Hint: reorder the rows of h into two blocks so that the first one corresponds to the set s and the second one is null).

4. Prove that resorting to the submatrices corresponding to the set s is not equivalent to using the full matrices and vectors when the exact expression (7.7) is adopted. Numerically assess the differences by applying both schemes to the data of Example 7.1.

5. In Example 7.2, add $L = 1$ as a pseudo-measurement and perform the SE. Analyze the results obtained when very different weighting factors are adopted for the new pseudo-measurement.

6. Using $\sigma = 0.01$ for all measurements, generate 4 different sets of measurements for the network of Example 7.2. Estimate separately the parameter L for the 4 cases. Next, considering simultaneously the 4 samples, obtain a single line length as explained in section 7.7. Compare the results of both experiments.

7. Repeat the computations of Examples 7.4 and 7.5 by using less accurate measurements ($\sigma = 0.01$). Assess the loss of accuracy in the estimated tap values.

8. Repeat Problem 7 for the Example 7.6.

9. Change the value of q_2 in the 3-bus network of Example 7.4 to the bad data $q_2 = 0$. Perform the SE by assuming a wrong tap $a = 1.1$.

Estimate a better tap value based on the residuals and (7.47). Obtain also the tap setting by the state augmentation technique.

10. Repeat the experiments of Problem 9, this time assuming the bad data is $q_{12} = 0$. Assess the robustness of both estimation techniques when the bad data is in the reactive power injection or in the power flow through the transformer.

11. Develop the counterpart of (7.47) when the power flow at the non-tapped bus, Q_f, is considered. Apply the resulting expression to the networks of Examples 7.4 and 7.5. Compare the accuracy of the estimated tap values with that obtained when Q_t is employed.

References

[1] Aboytes F., Cory B., "Identification of Measurement, Parameter and Configuration Errors in Static State Estimation". PICA Conference Proceedings, pp. 298-302, June 1975.

[2] Allam M., Laughton M., "A General Algorithm for Estimating Power System Variables and Network Parameters". IEEE PES 1974 Summer Meeting, Anaheim, CA, Paper C74 331-5, 1974.

[3] Allam M., Laughton M., "Static and Dynamic Algorithms for Power System Variable and Parameter Estimation". Proceedings of the 5th. Power System Computation Conference, Paper 2.3/11, Cambridge, United Kingdom, September 1975.

[4] Alsac O., Vempati N., Stott B., Monticelli A., "Generalized State Estimation". IEEE Transactions on Power Systems, Vol. 13(3), pp. 1069-1075, August 1998.

[5] Arafeh S., Schinzinger R., "Estimation Algorithms for Large-Scale Power Systems". IEEE Transactions on Power Apparatus and Systems, Vol. PAS-98, No. 6, pp. 1968-1977, Nov./Dec. 1979.

[6] Assadian M., Goddard R., Wayne H., French D., "Field Operational Experiences with On Line State Estimator". IEEE Transactions on Power Systems, Vol. 9(1), pp. 50-58, February 1994.

[7] Castrejón F.G., Gómez A., "Modeling Transformer Taps in Block-based State Estimation". Proceedings of the IEEE Powertech Conference, September 2001, Porto.

[8] Clements K., Denison O., Ringlee R., "The Effects of Measurement Non-Simultaneity, Bias and Parameter Uncertainty on Power System State Estimation". PICA Conference Proceedings, pp. 327-331, June 1973.

[9] Clements K., Ringlee R., "Treatment of Parameter Uncertainty in Power System State Estimation". IEEE Transactions on Power Apparatus and Systems, Anaheim, Cal., Paper C74 311-7, July 1974.

[10] Couch G., Sullivan A., Dembecki J., "Results from a Decoupled State Estimator with Transformer Ratio Estimation for a 5 GW Power System in the the Presence of Bad Data". Proceedings of the 5th. Power System Computation Conference, Paper 2.3/3, Cambridge, United Kingdom, Sept. 1975.

[11] Debs A., "Estimation of Steady-State Power System Model Parameters". IEEE Transactions on Power Apparatus and Systems, Vol. PAS-93, No. 5, pp. 1260-1268, 1974.

[12] Debs A., Litzenberger W., "The BPA State Estimator Project: Tuning of Network Model". IEEE Transactions on Power Systems, Paper A 75 448-1, July 1975.

[13] Do Coutto, M., Leite A., Falcão D., "Bibliography on Power System State Estimation (1968 - 1989)". IEEE Transactions on Power Systems, Vol. 5(3), pp. 950-961, August 1990.

[14] Fletcher D., Stadlin W., "Transformer Tap Position Estimation". IEEE Transactions on Power Apparatus and Systems, Vol. PAS-102, No. 11, pp. 3680-3686, November 1983.

[15] Habiballah I., Quintana V., "Efficient Treatment of Parameter Errors in Power System State Estimation". Electric Power Systems Research, No. 24, pp. 105-109, 1992.

[16] Handschin E., Schweppe F., Kohlas J., Fiechter A., "Bad Data Analysis for Power System State Estimation". IEEE Transactions on Power Apparatus and Systems, Vol. PAS-94, No. 2, pp. 329-337, March/April 1975.

[17] Handschin E., Kliokys E., "Transformer Tap Position Estimation and Bad Data Detection Using Dynamic Signal Modelling". IEEE Transactions on Power Systems, Vol. 10(2), pp. 810-817, May 1995.

[18] Koglin H., Neisius T., Beissler G., Schmitt K., "Bad Data Detection and Identification". Electrical Power & Energy Systems, Vol. 12(2), pp. 94-103, April 1990.

[19] Liu W., Wu F., Lun S., "Estimation of Parameter Errors from Measurement Residuals in State Estimation". IEEE Transactions on Power Systems, Vol. 7(1), pp. 81-89, February 1992.

[20] Liu W., Lim S., "Parameter Error Identification and Estimation in Power System State Estimation". IEEE Transactions on Power Systems, Vol. 10(1), pp. 200-209, February 1995.

[21] Mili L., Van Cutsem T., Ribbens-Pavella M., "Hypothesis Testing Identification: A New Method for Bad Data Analysis in Power System State Estimation". IEEE Transactions on Power Apparatus and Systems, Vol. PAS-103, No. 11, pp. 3239-3252, November 1984.

[22] Mukherjee B., Fuerst G., Hanson S., Monroe C., "Transformer Tap Estimation - Field Experience". IEEE Transactions on Power Apparatus and Systems, Vol. PAS-103, No. 6, pp. 1454-1458, June 1984.

[23] Quintana V., Van Cutsem T., "Real-Time Processing of Transformer Tap Positions". Canadian Electrical Engineering Journal, Vol. 12(4), pp. 171-180, 1987.

[24] Quintana V., Van Cutsem T., "Power System Network Parameter Estimation". Optimal Control Applications & Methods, Vol. 9, pp. 303-323, 1988.

[25] Reig A., Alvarez C., "Influence of Network Parameter Errors in State Estimation Results". Proceedings IASTED Power High Tech '89, pp. 199-204, Valencia, Spain, 1989.

[26] Reig A., Alvarez C., "Off-Line Parameter Estimation Techniques for Network Model Data Tuning". Proceedings IASTED Power High Tech '89, pp. 205-210, Valencia, Spain, 1989.

[27] Schweppe F., Douglas B., "Power System Static-State Estimation". IEEE Transactions on Power Apparatus and Systems, Vol. PAS-89, pp. 120-135, 1970.

[28] Schweppe F., Handschin E., "Static State Estimation in Electric Power Systems". Proceedings IEEE, Vol. 62, pp. 972-983, July 1974.

[29] Slutsker I., Mokhtari S., Clements K., "On-line Parameter Estimation in Energy Management Systems". American Power Conference, Chicago, Illinois, Paper 169, April 1995.

[30] Slutsker I., Mokhtari S., "Comprehensive Estimation in Power Systems: State, Topology and Parameter Estimation". American Power Conference, Chicago, Illinois, Paper 170, April 1995.

[31] Slutsker I., Clements K., "Real Time Recursive Parameter Estimation in Energy Management Systems". IEEE Transactions on Power Systems, Vol. 11(3), pp. 1393-1399, August 1996.

[32] Smith R., "Transformer Tap Estimation at Florida Power Corporation". IEEE Transactions on Power Apparatus and Systems, Vol. PAS-104, No. 12, pp. 3442-3445, December 1985.

[33] Stuart T., Herget C., "A Sensitivity Analysis of Weighted Least Squares State Estimation for Power Systems". IEEE Transactions on Power Apparatus and Systems, Vol. PAS-92, pp. 1696-1701, Sept./Oct. 1973.

[34] Teixeira P., Brammer S., Rutz W., Merritt W., Salmonsen J., "State Estimation of Voltage and Phase-Shift Transformer Tap Settings". IEEE Transactions on Power Systems, Vol. 7(3), pp. 1386-1393, August 1992.

[35] Van Cutsem T., Quintana V., "Network Parameter Estimation Using Online Data with Application to Transformer Tap Position Estimation". IEE Proceedings, Vol. 135, Pt C, No. 1, pp. 31-40, January 1988.

[36] Zarco P., Gómez A., "Off-Line Determination of Network Parameters in State Estimation". Proceedings 12th. Power System Computation Conference, pp. 1207-1213, Dresden, Germany, August 1996.

[37] Zarco P., "Network Parameter Estimation Using Recorded Measurement Data". Ph.D. Dissertation (in Spanish), University of Sevilla, 1997.

[38] Zarco P., Gómez, A."Power System Parameter Estimation: A Survey". IEEE Transactions on Power Systems, Vol. 15(1), pp. 216-222, February 2000.

Chapter 8

Topology Error Processing

8.1 Introduction

As explained in Chapter 1, the SE works with an electrical model provided by the *Topology Processor* (TP). This routine analyzes the status of all circuit breakers (CB) and switching devices in order to determine:

- The way physical nodes (bus sections) are interconnected to give rise to a reduced set of electrical nodes.
- The electrical node(s) to which every transmission element (line, transformer or shunt device) is connected.
- The energized or non trivial electrical islands.

In other words, the TP converts a bus section/switch detailed model into a compact and more useful bus/branch model. It is worth noting that, in this process, some measurements must be discarded (e.g., power flows through CBs) while others are merged into a single measurement point (e.g., injections measurements of several bus sections put together at a single electrical bus).

A physical CB, along with several isolating switches, constitute one or several 'logical' CBs. For instance, Figure 8.1 shows the two logical CBs necessary to define the connectivity of a line in the presence of two electrical nodes. In this chapter, the term CB will refer to such logical devices.

The correct statuses of all CBs in the system are known almost all the time. However, in some rare cases, the assumed status of certain CBs may

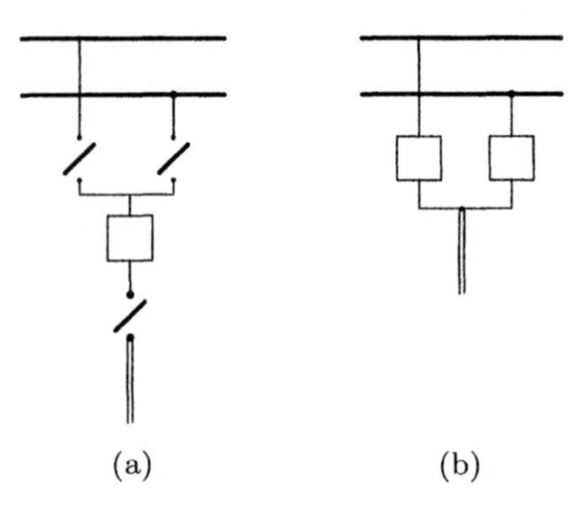

Figure 8.1. Physical (a) and logical (b) switching devices

be wrong. This will happen when some of the involved isolation switches, the majority of which are neither telemetered nor remotely operated, simply malfunction. Other reasons might be unreported breaker manipulation by maintenance teams, mechanical failure of signaling devices, etc. A more common situation is when the TP encounters a CB whose status is unknown. In cases like this, the TP must decide on the most likely CB status, for which it uses the status history for the same breaker and/or the values of related measurements as a guide. Hence, the risk of assuming the wrong status for the CB still will not be completely avoided.

When this happens, the bus/branch model generated by the TP is locally incorrect, leading to a *topological error*. Unlike the parameter errors discussed in the previous chapter, most of which remain undetected until a threshold is exceeded, topology errors usually cause the state estimate to be significantly biased. As a result, the bad data detection & identification routine may erroneously eliminate several analog measurements which appear as interacting bad data, finally yielding an unacceptable state. It is also possible for the SE process to diverge, or have serious convergence problems, in the presence of topology errors.

Therefore, there is a need to develop effective mechanisms intended to detect and identify this kind of gross errors. The aim of this chapter is to present classical and recent approaches to deal with topological errors and related matters.

It should be noted that the word 'substation', as used in this chapter, refers to a 'bus section group', i.e. the set of bus sections interconnected by switching devices. Real substations, usually containing power transformers, will be composed of as many groups as voltage levels.

8.2 Types of topology errors

Topology errors can be broadly classified in two categories:

- *Branch status errors:* Errors affecting the status of regular network branches (lines or transformers). An *exclusion error* takes place when an energized element is excluded from the model. The opposite, i.e., *inclusion error*, occurs when a disconnected element is assumed to be in service. The branches involved in this category of errors, which may affect one or more CBs, will always have non-zero impedances.

- *Substation configuration errors:* Errors affecting CBs whose purpose is to link bus sections within the substation. A *split error* arises when a single electrical bus is modeled as two buses. The opposite is called a *merging error*. Since all of them are CBs, no impedance can be associated with the branches involved in this category of errors.

As will be explained below, the techniques proposed to deal with the first type of errors are closely related with those described in the previous chapter for parameter estimation. The second type of errors, however, requires specific procedures in which the affected CBs appear explicitly in the SE model.

8.3 Detection of topology errors

Topology errors have, in general, a more dramatic influence on the measurement residuals than the parameter errors. This can be easily illustrated by considering for instance an exclusion error affecting a branch of admittance Y. Excluding this branch is equivalent to considering a branch of null admittance or, stated in other words, a branch whose admittance error is 100%. This value is an order of magnitude larger than the parameter errors found in practice (including those related with tap settings). The reader is referred to the figures shown in the preceding chapter, where the influence of branch susceptance errors as large as 25% are presented.

Conditions upon which topology errors can be detected are analyzed in detail in [6] and [31] (see section 8.6.1). A single branch error is detectable if the following two conditions are satisfied: (1) it is not an irrelevant branch (branch with no incident measurements), and (2) removal of any one of the measurements incident to this branch does not make it an unobservable branch.

The following two examples illustrate numerically what happens typically in the presence of topology errors.

Example 8.1:

Consider the 4-bus network shown in Figure 8.2-(a), where only active power measurements are relevant to solve the linear DC state estimation problem. The given measurements contain Gaussian noise ($\sigma^2 = 0.001$) and the same reactance is adopted for all lines ($x = 0.01$). Figure 8.2-(b) presents the resulting normalized residuals, which are well below the usual bad data threshold.

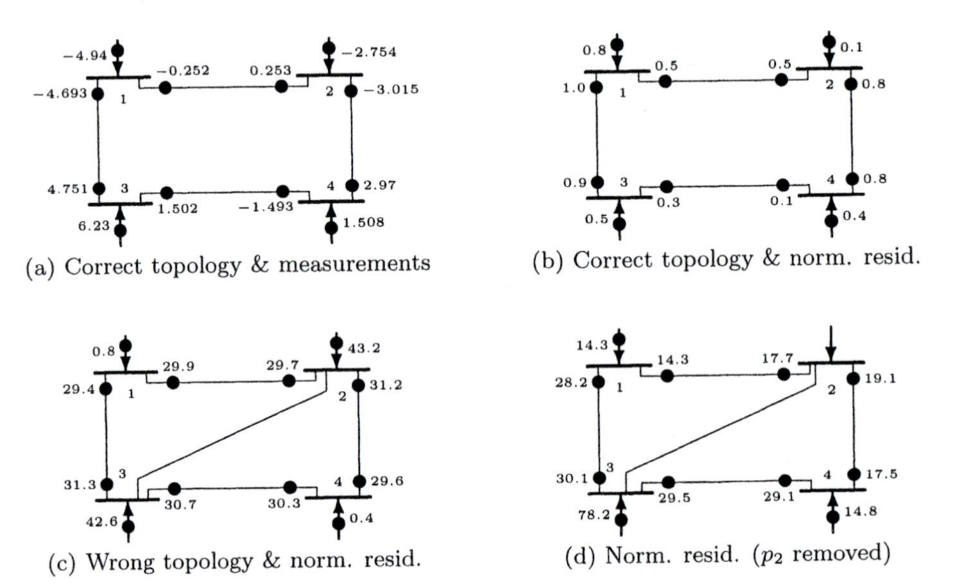

(a) Correct topology & measurements

(b) Correct topology & norm. resid.

(c) Wrong topology & norm. resid.

(d) Norm. resid. (p_2 removed)

Figure 8.2. Influence of branch inclusion error on normalized residuals

Assume now that a line, actually disconnected, is in service between buses 2 and 3. For the sake of coherence, it is also assumed that the power flows through this line are not telemetered, as their null values would be a clear indication of the topology error. Figure 8.2-(c) shows that, in this situation, the normalized residuals corresponding to injections 2 and 3, as well as those associated with power flows of incident branches, are very high. A conventional bad data analysis would conclude that p_2, the measurement with largest normalized residual, should be eliminated. When this is done, the residuals are still very high, as shown in Figure 8.2-(d). Although the residual of p_3 is now the largest, a significant increase can be noticed in the residuals of the power injections at buses 1 and 4, which are first neighbors of bus 2. If measurement p_3 is finally removed, the normalized residual values corresponding to the remaining measurements are very similar to those of Figure 8.2-(b), which is logical because the line 2-3 becomes an irrelevant branch when the two injections are removed.

Consequently, the topology error remains undetected and two correct measurements are discarded by the conventional SE.

An analysis of this example provides certain guidelines for the detection and identification of single branch status errors. A non-telemetered branch, regardless of its assumed status, will be suspected if the normalized residuals of the measurements which are incident to its terminal buses are among the largest. If the injection at one of those buses is missing, then large normalized residuals will also be expected at the first neighbors of such a bus.

A more systematic analysis of this kind of topology errors can be found in [16], where it is also concluded that the injections at the terminal buses of the wrong branch, when measured, will most likely be identified and suppressed as bad data.

Example 8.2:

Consider this time the 5-bus network and associated measurements shown in Figure 8.3-(a). As in Example 8.1, the measurements contain Gaussian noise ($\sigma^2 = 0.001$) and all line reactances are $x = 0.01$.

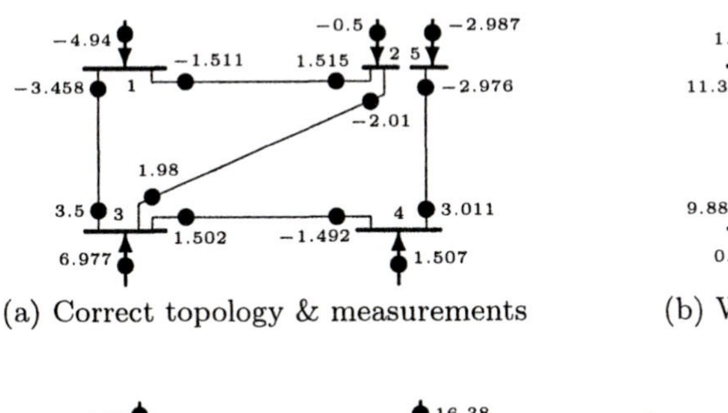

(a) Correct topology & measurements

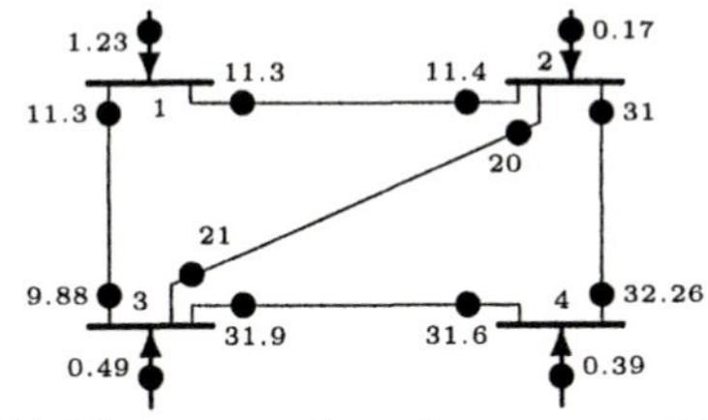

(b) Wrong topology & norm. resid.

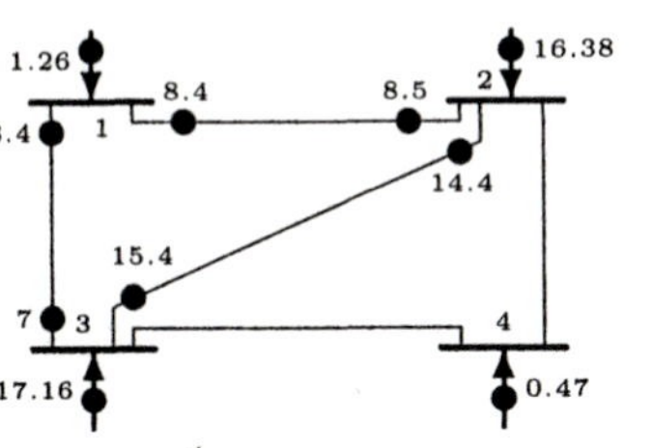

(c) Norm. res. ($p_{42}, p_{24}, p_{43}, p_{34}$ removed)

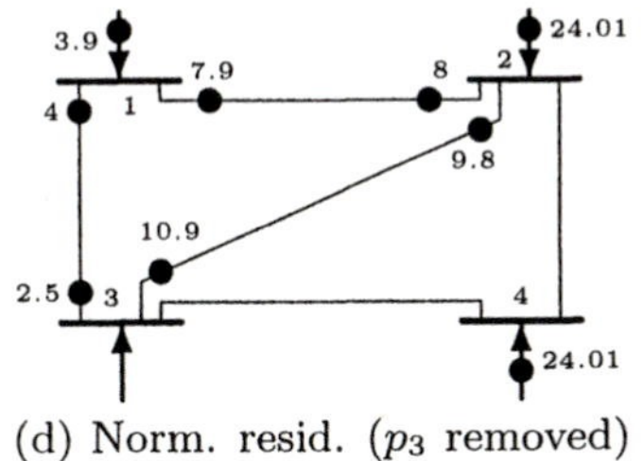

(d) Norm. resid. (p_3 removed)

Figure 8.3. Influence of bus merging error on normalized residuals

In the situation shown in Figure 8.3-(a), the largest normalized residual, corresponding to p_1, is 1.19. Now, the CB that couples buses 2 and 5, actually open, is assumed closed. This implies that p_2 and p_5 are collapsed into a single injection measurement by the topology processor, as shown in Figure 8.3-(b). Then, the bad data processor performs the following procedure:

Table 8.1. Estimated phase angles (radians)

	$\hat{\theta}_2$	$\hat{\theta}_3$	$\hat{\theta}_4$	$\hat{\theta}_5$
1. correct topology	0.0149	0.0348	0.0198	−0.0101
2. wrong topol. (initial)	0.0118	0.0379	0.0323	−
3. wrong topol. (final)	0.0149	0.0348	0.0449	−

- According to the normalized residuals presented in Figure 8.3-(b), the power flow p_{42} is considered bad data and removed. It should be observed that the residuals corresponding to all power flows, particularly those of the loop 2-3-4, are significantly affected by the topological error, whereas the power injections remain almost unaffected. Somewhat the opposite happens in Example 8.1, which can be explained by the fact that branch errors are essentially in conflict with Kirchhoff's current law (nodal equations), while bus configuration errors mainly contradict Kirchhoff's voltage law (loop equations).
- As expected, when p_{42} is removed, the largest normalized residual corresponds to p_{24}. Repeating the estimation cycle, p_{43} and p_{34} will be subsequently removed. The resulting residuals are given in Figure 8.3-(c).
- In this situation, p_3 leads to the largest residual and should be removed. As shown in Figure 8.3-(d), removal of p_3 causes both p_2 and p_4 to have identical residuals, since they now form a critical pair (see Chapter 5). If, for instance, p_4 is discarded (this is not shown in Figure 8.3), then p_2 will become critical and its residual will consequently be zero. As a result, all the remaining measurements will be consistent with the wrong topology (the largest normalized residual will be 1.19 and will again correspond to p_1).

In summary, the bus configuration error leads to the six measurements p_{42}, p_{24}, p_{43}, p_{34}, p_3 and p_4 (or p_2) being erroneously eliminated. The consequence is that the topology error remains hidden and that the value of $\hat{\theta}_4$ is totally wrong. Table 8.1 provides the estimated phase angles under three different conditions: (1) correct topology; (2) wrong topology, full set of measurements (initial); (3) wrong topology, six measurements discarded (final). Note that, while θ_2 and θ_3 are accurately estimated, the final value obtained for θ_4 is even worse than the one estimated before the bad data processing. This translates into very poor estimated values for the six discarded measurements.

Based on the results obtained in this example, bus configuration errors could be detected from a systematic analysis of the normalized residuals [6], provided the redundancy is large enough. The presence of large residuals at several branches forming a loop, coexisting with reduced residuals in the involved injections, should be considered a clear symptom of this kind of errors. However, unlike in Example 8.1, it is not so easy to develop *ad hoc* rules for the identification of a single suspected substation, once the topology error is detected. In the last example, the topology error may be located at any of the buses 2, 3 or 4.

In [13, 14] and [15], a method based on the number of measurements labelled as bad data is proposed. A counter which is initialized as zero, is assigned to each bus. A bad injection increases the counter of the respective bus by one unit, whereas a bad power flow increases the counters of its two terminal buses. Suspect buses are then identified as those at the top of the bus list, when sorted in decreasing order of their counters.

Given an estimate, and a list of measurements previously discarded as bad data, hypothesis testing is applied in [16] to determine alternative network topologies, that are consistent with the existing measurements.

In practice, there are cases in which the influence of topology errors on measurement residuals is not so significant. For exclusion and split errors, this happens when the power flow through the respective branch or CB is small, whereas inclusion and merging errors are less noticeable when the voltage drop between the two involved buses is small. In extreme cases, i.e., when the power flow or voltage drop is negligible, there is no way to detect the topology error, but then there is no need to worry either, as the state estimate will be acceptable.

By far, the most important factor affecting the capability to detect such errors is the presence of bad data on adjacent measurements. If redundancy is high enough, it is very unlikely that a single bad data can mislead the detection procedures described above, because they are based on abnormally high values of several selected residuals. Nevertheless, low redundancy combined with several interacting bad data constitute a risky scenario.

8.4 Classification of methods for topology error analysis

Methods for topology error analysis can be classified according to different criteria. The following two broad categories can be defined based on the adopted model:

- Bus-branch model: The relevant information for topology error analysis is obtained from the conventional bus-branch model generated by the TP.

- Bus section-switch model: The portion of the network presumably affected by the topology error, is modeled in detail including all the individual CBs present in that part of the network.

Until recently, use of such a detailed model for the entire network was assumed not to be viable, although this assumption may no longer hold true as will be discussed later in this chapter. Therefore, the application of the second approach currently requires proper identification of the area whose model is to be expanded. This leads to a two-stage procedure. First, based on the residuals provided by the conventional model, suspected substations or branches are identified. Second, the CBs associated with the candidate components are added to the model as explained later in this chapter, and the SE and bad data analysis is performed again. This computationally more expensive procedure is replacing the older techniques based solely on the bus-branch model, owing to its reliability and increased flexibility to analyze complex situations. An added benefit of the detailed model is that certain measurements discarded by the compact model can be taken into account.

Techniques for topology error analysis can also be categorized based on the moment at which the analysis is performed:

- *A priori* processing: Much in the same way as analog measurements are pre-filtered before entering the SE, the assumed status of CBs can be validated in advance by means of local consistency checks, rule-based techniques combined with recorded information, etc.

- *A posteriori* processing: The bad data analysis stage can be made more sophisticated so as to consider the possibility of topology errors being responsible for biased estimates. This type of processing is very similar to that of processing parameter errors (see Chapter 7). If the topology error is not clearly identified, i.e., there are several candidates, then the second SE run mentioned above, containing detailed models for the suspected areas, can be resorted to.

A priori processing methods by nature are based on fast, approximate techniques, usually applied at the substation level, and hence they may fail to properly identify the topological problem. On the other hand, post-processing methods rely on the results of a converged SE, which may not always exist in the presence of certain topology errors.

In the next section, preprocessing methods for topology validation are briefly addressed. The rest of the chapter will discuss the second class of methods, based on residual and Lagrange multiplier analysis for identifying both types of topology errors (branch and bus errors). Section 8.9 presents a recent development, allowing the information for topology error processing to be obtained, at a moderate cost, from a single execution of an implicitly constrained SE model. This new approach allows the detailed modelling of the entire network at little computational overhead and hence may soon replace the existing two-stage procedure which can incorporate explicit models of only a limited set of certain switching devices at the second step.

8.5 Preliminary topology validation

The following is a chronological account of the most outstanding contributions in this category of methods.

An original and systematic procedure, intended to locally validate switch indicators and analog measurements within a substation, is presented in [12]. The substation structure is modeled in terms of open and closed links between potential nodes, whose active power flows are estimated by solving the following linear programming (LP) problem:

$$
\begin{aligned}
\text{Min} \quad & \sum_{\text{all } k}(e_k + e_k') + \sum_{\text{all } j}(e_j + e_j') \\
\text{subject to} \quad & \sum_{i \in \ell} P_i = 0 \quad \text{for all } \ell \\
& P_k = P_{mk} + e_k - e_k' \quad \text{for all } k \\
& P_j = 0 + e_j - e_j' \quad \text{for all } j \\
& P_i \leq P_{i\max} \quad \text{for certain } i \\
& P_i \geq P_{i\min} \quad \text{for certain } i \\
& 0 \leq e_k, e_k', e_j, e_j'
\end{aligned}
$$

where:
P_i is the real power flow through link i
e_k, e_k', e_j, e_j' are the error terms
k is the power flow measurement number
j is the open switch number
ℓ is the node number, and $i \in \ell$ refers to the set of links incident to node ℓ
P_{imax}, P_{imin} upper and lower limits on the power flow through link i.

Given enough redundancy, the solution will provide the best estimates for all substation link flows. Measurements and switch indicators that are grossly in error will be automatically rejected by the LP solution. The estimated error variables for these erroneous measurements will be relatively large, indicating the misconfigured links. This method is developed as a local data validation tool, and therefore uses a lossless node-link based network model and linear network transport equations in terms of the real power flows through the links.

The approach proposed in [4] incrementally obtains an approximate network state. Starting from a measured bus, and giving preference to the best measurements, the state of adjacent buses is computed by means of available power flows. Eventually, by means of a breadth-first graph search, the most likely state for the entire network is obtained. Then, in those buses or loops where redundancy permits, consistency checks based on Kirchhoff's laws are performed to detect potential gross errors. Finally, a systematic analysis of the results provided by these checks should allow status errors to be identified.

A rule-based system, taking into account the temporal evolution of measurements and switch positions, is proposed in [26]. The basic idea is that, while normal evolution of load gives rise to smooth or incremental changes in the measurement values, switching operations are usually accompanied by more abrupt deviations. Trying to emulate the way an engineer would locally infer the coherency of present and past information around a given switch, including alarm and event lists, a knowledge base containing over 120 rules is developed. Practical experience gained in the integration of this expert system into a Swiss control center is reported in [27].

The application of artificial neural networks (ANN) to this problem is explored in [30]. The so-called 'normalized innovations', obtained in tracking mode for each measurement from an associated dynamic model, are used as inputs to self-organizing anomaly classifiers. The normalized innovations are, in the pre-filtering stage, the counterpart of normalized residuals in SE post-processing, but they are free from the 'smearing' effect that complicates identification of topology errors based solely on normalized residuals (as happened in Example 8.2).

More recently [17], a robust Huber estimator (see Chapter 6) based on an approximate decoupled model, is proposed as a means of pre-checking the assumed system topology. Under this approach, which ignores voltage measurements, active and reactive power flows through CBs and ordinary branches are adopted as state variables. This requires that branch losses be approximately expressed in terms of this alternative set of variables so that a single power flow can be used per branch. Once the iteratively reweighted estimator converges, branch statuses are determined on the basis of statistical tests applied to the 'normalized' estimated flows (see the next section). Since the adopted model is nearly linear, this procedure is less sensitive to the convergence problems suffered sometimes by conventional estimators in the presence of topology errors.

8.6 Branch status errors

As non-zero impedance branches explicitly appear in conventional estimators, there is no need to include detailed physical models if this is the only type of topology error of interest.

Similar to the case of parameter errors (see Chapter 7), branch status errors can be identified and corrected by means of either normalized residuals or state vector augmentation. Both solutions will be discussed separately.

8.6.1 Residual analysis

This approach makes use of the results of a converged state estimation in order to detect branch errors.

The effect of these types of errors on the measurement equations can be modeled in the following manner [31]:

$$H = H_e + E \tag{8.1}$$

where, H is the true Jacobian, H_e is the incorrect Jacobian due to topology errors, and E is the Jacobian error matrix. Substituting (8.1) in the linearized measurement model

$$z = Hx + e \tag{8.2}$$

yields:

$$z = H_e x + Ex + e \tag{8.3}$$

The statistical characteristics of the new residual vector can also be derived as below (see Chapter 4):

$$\begin{aligned} r &= z - H_e\hat{x} = (I - K_e)(Ex + e) \\ E(r) &= (I - K_e)Ex \\ \text{cov}(r) &= (I - K_e)R \end{aligned}$$

where

$$K_e = H_e(H_e^T R^{-1} H_e)^{-1} H_e^T R^{-1}$$

Using (8.2), each entry of the bias vector Ex can be written as a linear combination of errors in network branch flows. Let f be a vector of branch flow errors, and M be the measurement-to-branch incidence matrix. Then, the measurement bias Ex can be written as:

$$Ex = Mf \tag{8.4}$$

and the residual vector will be given by:

$$r = (I - K_e)Mf \tag{8.5}$$

Now, the expected value of the normalized residuals can be rewritten in terms of the branch flow errors:

$$E(r^N) = \Omega^{-\frac{1}{2}}(I - K_e)Mf = Sf \tag{8.6}$$

where $\Omega = \text{diag}\{\text{cov}(r)\}$, and $S = \Omega^{-\frac{1}{2}}(I - K_e)M$ is the sensitivity matrix for r^N with respect to branch flow errors, f.

Hence, topology error detection can proceed based on the normalized residual test, assuming that bad data in measurements have already been identified and eliminated.

Equation (8.6) states that $E(r^N)$ is a linear combination of the columns of S corresponding to branches whose status is wrong (as the remaining elements of f are null). However, finding a linear combination of columns that equals a given vector is not a trivial problem in the general case, as the solution is not unique. Therefore, even though the vector f may contain the power flow through any CB, using the normalized residuals for topology error identification is not advisable in complex cases, like those involving bus configuration errors.

A geometric interpretation of the measurement residuals is also possible [6]. Consider the expression for the residuals given in terms of the branch flows:

$$r = Tf \tag{8.7}$$

where $T = (I - K_e)M$. If branch j has an admittance error, then $f_j = \alpha$ and $f_k = 0$ for $k \neq j$, α being a scalar. Hence, the j-th column of T, T_j, will be colinear with r. Their dot product can thus be used to test colinearity by calculating:

$$\cos\theta_j = \frac{T_j^T r}{\| T_j \| \| r \|} \tag{8.8}$$

If $\cos\theta_j \cong 1.0$, and $\cos\theta_i < 1$ for all $i \neq j$, this will imply a single branch topology error in branch j. Single topology error detectability and identifiability conditions also follow from this colinearity condition:

1. Branch j is single topology error detectable if $T_j \neq 0$.

2. Branch j is single topology error identifiable if T_j is not colinear with any of the other columns of T.

Some results on detectability and identifiability of topology errors are derived based on the above formulation. Before stating these results, a few definitions are in order:

- A *critical branch* of a measured, observable network is one whose deletion leads to an unobservable network.

- A *critical pair of branches* are those whose simultaneous removal leads to an unobservable network, but neither one is a critical branch.

The proofs for the following results can be found in [6]:

1. A branch is not single topology error detectable if it is critical or it is incident only to critical measurements.

2. Single topology errors in either one of a critical pair of branches are not identifiable.

Note again that, in general, a unique linear combination of the columns of T that yields a vector colinear with r, can not be found.

Example 8.3:

Consider the 4-bus system shown in Figure 8.4. All branches have j1.0 p.u. impedance and all measurement weights are chosen as 1.0 for simplicity. Branch [a], shown by a dotted line, is actually disconnected, but will be erroneously assumed to be in service.

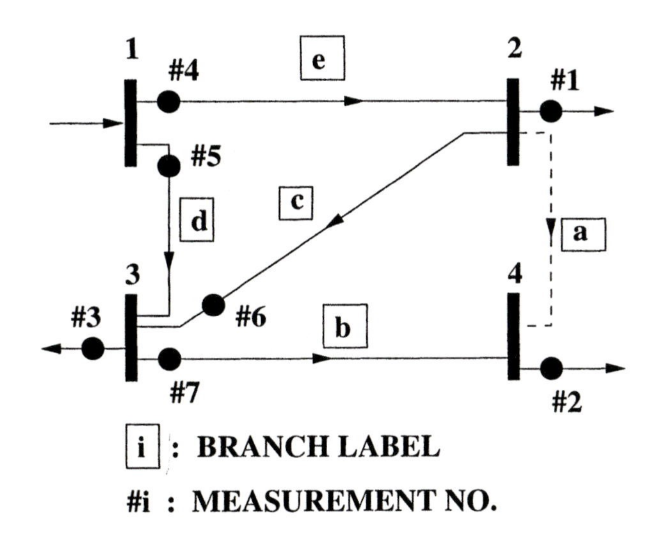

Figure 8.4. 4-Bus Test System for Example 8.3

Using the $P - \theta$ part of the decoupled model, the following matrices can be defined:

$$H_e = \begin{array}{c} \\ 1 \\ 2 \\ 3 \\ 4 \\ 5 \\ 6 \\ 7 \end{array} \begin{array}{c} \begin{array}{ccc} \theta_2 & \theta_3 & \theta_4 \end{array} \\ \left[\begin{array}{rrr} 3 & -1 & -1 \\ -1 & -1 & 2 \\ -1 & 3 & -1 \\ -1 & 0 & 0 \\ 0 & -1 & 0 \\ -1 & 1 & 0 \\ 0 & 1 & -1 \end{array} \right] \end{array}$$

$$H = \begin{array}{c} \begin{matrix} \theta_2 & \theta_3 & \theta_4 \end{matrix} \\ \begin{bmatrix} 2 & -1 & 0 \\ 0 & -1 & 1 \\ -1 & 3 & -1 \\ -1 & 0 & 0 \\ 0 & -1 & 0 \\ -1 & 1 & 0 \\ 0 & 1 & -1 \end{bmatrix} \end{array} \quad E = \begin{array}{c} \begin{matrix} \theta_2 & \theta_3 & \theta_4 \end{matrix} \\ \begin{bmatrix} -1 & 0 & 1 \\ 1 & 0 & -1 \\ 0 & 0 & 0 \\ 0 & 0 & 0 \\ 0 & 0 & 0 \\ 0 & 0 & 0 \\ 0 & 0 & 0 \end{bmatrix} \end{array}$$

$$M = \begin{array}{c} \\ 1 \\ 2 \\ 3 \\ 4 \\ 5 \\ 6 \\ 7 \end{array} \begin{array}{c} \begin{matrix} a & b & c & d & e \end{matrix} \\ \begin{bmatrix} 1 & 0 & 1 & 0 & -1 \\ -1 & -1 & 0 & 0 & 0 \\ 0 & 1 & -1 & -1 & 0 \\ 0 & 0 & 0 & 0 & 1 \\ 0 & 0 & 0 & 1 & 0 \\ 0 & 0 & -1 & 0 & 0 \\ 0 & 1 & 0 & 0 & 0 \end{bmatrix} \end{array}$$

Then, the sensitivity matrix S can be calculated as defined in (8.6).

The WLS estimator is run with the measurement set of Figure 8.4, but having branch [a] in service. The residuals are then used to evaluate (8.8) and the results are listed in Table 8.2. Note that, branches [a] and [b] have similar $\cos\theta$ values since they form a pair of critical branches.

Table 8.2. Results of $\cos\theta$ test

Branch	$\cos\theta$
a	0.984
b	−0.984
c	−0.425
d	−0.284
e	0.284

In [25], the so-called *correlation index* is used to identify topology errors. For a branch between buses i and j, this index is defined as follows:

$$e_{ij} = \frac{\mid \text{STM} \cap \text{SST}_{ij} \mid}{\mid \text{SST}_{ij} \mid} \tag{8.9}$$

where the vertical bar denotes the cardinal of a set, STM is the set of measurements identified as bad data and SST_{ij} refers to the set of measurements sensitive to the misconfiguration of branch i-j. This last set is obtained from the respective column of the sensitivity matrix, S, defined in (8.6), by discarding those entries which are not large enough. The scalar e_{ij} is a measure of how likely a topology error involving branch i-j is the real cause of the anomaly detected by the normalized residual test.

All of the above methods assume that the state estimator successfully converges and that the residual vector, r, is available for post processing. Under this assumption, they can effectively identify single branch topology errors. However, as explained before, they cannot identify multiple interacting branch flow errors, which may occur when the statuses of several CBs in a substation are wrong.

8.6.2 State vector augmentation

The status of a non-zero impedance branch can be represented by means of a single integer variable, k, that multiplies all branch admittances. When this variable is considered, the π model (see Chapter 2) is composed of the following admittances:

$$\begin{aligned} \text{series:} \quad & (g_{ij} + jb_{ij})k \\ \text{parallel:} \quad & jb^{p}_{ij}k \end{aligned} \tag{8.10}$$

Clearly, $k = 0$ represents a disconnected branch whereas $k = 1$ should be used if the branch is in service. The idea is to include the variable k into the state vector for any suspected branch. Note, however, the subtle difference between (8.10) and the equation adopted in the former chapter to estimate the relative line length.

As explained in [28], in a WLAV state estimator this extra variable is compensated by adding two conflicting pseudo-measurements, $k = 0$ and $k = 1$. If both pseudo-measurements are given the same weight, the estimator will automatically satisfy the one which is most coherent with the existing information, and will discard the other (see Chapter 6 for the properties of WLAV estimators).

When the WLS estimator is used, it makes no sense to enforce simultaneously the two contradictory constraints. If they are considered as very accurate pseudo-measurements and are given the same weight, then $k \approx 0.5$ will be obtained, which is useless. On the other hand, if both constraints are ignored, the estimated value of k will be dictated by the analog measurements. A value of k close to 1 is an indication of the branch being in service, while a value approaching 0 implies a disconnected element. In practice, however, owing to measurement errors, the estimated value of k may significantly differ from 0 or 1. Even worse, in the presence of nearby bad data, the estimated value of k may approach 0.5, in which case the status of the element is ambiguous. Additionally, the estimated values of the remaining state vector components would be less accurate than if k had the right integer value.

In order to elude these potential problems, the quadratic constraint

$$k(1-k) = 0 \tag{8.11}$$

is used to enforce the estimator to converge to either of the two feasible statuses [11]. If the adjacent measurements render k observable, then the above constraint constitutes a redundant information whose only purpose is to refine the estimated value of k. Otherwise, when k is unobservable, the equality constraint is useless because either of the two possible solutions can be reached depending on the starting point. This happens, for instance, when neither the power flows of the branch in hand nor its adjacent injections are measured (irrelevant branch). In such cases, even if the complex voltages of the terminal buses are observable, both the respective injections and the branch status remain undetermined.

This idea can be applied to any of the WLS state estimation algorithms described in Chapter 3. Although it is ideally suited to those methods in which equality constraints are explicitly modeled, (8.11) can also be handled as a very accurate measurement, provided orthogonal factorization of the Jacobian is employed in order to prevent the intrinsic ill-conditioning of the normal equations when large weighting factors are adopted.

Consider the case where the status of the branch connecting nodes i and j, represented by the variable k, is to be estimated. Then, the only rows of the Jacobian related to the new state variable are those corresponding to the respective power flows, adjacent power injections and the virtual measurement given by (8.11).

The power flows through the branch leaving bus i can be expressed as

$$P_{ij}(k) = kP_{ij} \tag{8.12}$$

$$Q_{ij}(k) = kQ_{ij} \tag{8.13}$$

where P_{ij} and Q_{ij} are the conventional power flows, computed as if $k = 1$. The power injections at bus i can be expressed as a function of k in a similar manner,

$$P_i(k) = \sum_{m \epsilon i, m \neq j} P_{im} + kP_{ij} \tag{8.14}$$

$$Q_i(k) = \sum_{m \epsilon i, m \neq j} Q_{im} + kQ_{ij} \tag{8.15}$$

Exchanging i for j in (8.12)-(8.15) the power injections at bus j and the opposite power flows through the branch can be obtained.

In addition to the only non-zero element due to (8.11),

$$\frac{\partial[k(1-k)]}{\partial k} = 1 - 2k \tag{8.16}$$

the following new terms will appear in the extra Jacobian column,

$$\frac{\partial P_{ij}(k)}{\partial k} = \frac{\partial P_i(k)}{\partial k} = P_{ij} \tag{8.17}$$

$$\frac{\partial Q_{ij}(k)}{\partial k} = \frac{\partial Q_i(k)}{\partial k} = Q_{ij} \tag{8.18}$$

along with their counterparts for node j. It is also clear from (8.10) that, in the remaining columns of the Jacobian, the series and shunt admittance parameters of the suspected branch should be multiplied by k.

Another important practical issue is the initial choice, k_0, for k. Experimental results show that, when (8.11) is treated as a very accurate measurement, the estimation process systematically converges to $k = 1$ if $k_0 > 1/2$ or $k = 0$ if $k_0 < 1/2$. Hence, the 'neutral' value $k_0 = 1/2$ should be adopted, which means that the Jacobian term given by (8.16) is null. Furthermore, since P_{ij} and Q_{ij} are also nearly zero at flat start, the Jacobian becomes ill-conditioned or even singular during the first iteration. As discussed in the previous chapter regarding the estimation of the line length, a way of circumventing this risk is by adding the new state variable, k, only after the first iteration.

Therefore, the WLS state augmentation technique can be summarized as follows:

- From flat start, perform one iteration and correct the state vector. At this step, the only difference with respect to a conventional state estimator is the factor $k_0 = 1/2$ which affects the parameters of the branch whose status is to be estimated.

- From the second iteration on, add the constraint (8.11) to the model and include k in the state vector. Note that the effect of the constraint will only be noticed at the end of the third iteration, since the value of k is still 1/2 during the second iteration.

Example 8.4:

Consider the 3-bus system of Figure 8.5, where all branch impedances are equal to 0.05j and all measurement weights are 1. Estimate the status of branch 1-2, with and without the constraint (8.11) added, for the following measurement values:

p_{13}	q_{13}	p_{23}	q_{23}	p_2	q_2	V_1	V_2
1.305	0.646	0.198	0.086	−0.807	−0.351	1.018	0.996

When the constraint $k(1-k) = 0$ is not added, the WLS estimator converges, after 14 iterations, to the following state vector:

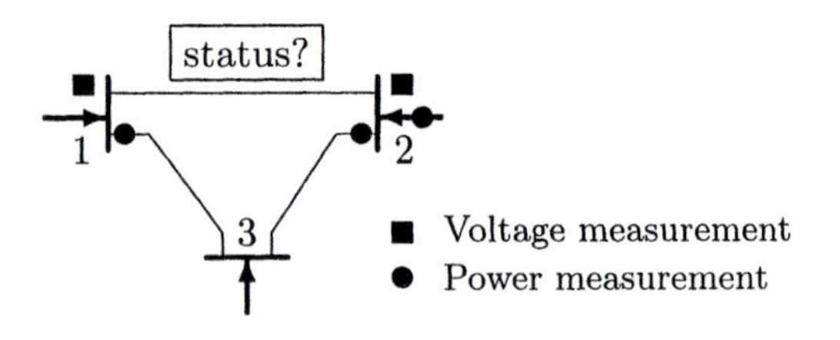

Figure 8.5. 3-bus system to illustrate branch status estimation

Magnitude	V_1	V_2	V_3	θ_2	θ_3	k
Est. value	1.0160	0.9904	0.9863	$-3.1485°$	$-3.7320°$	0.9105

which means that line 1-2 is 91% 'closed'. This physically meaningless state corresponds with the following estimated measurements:

Magnitude	p_{13}	q_{13}	p_{23}	q_{23}	p_2	q_2
Est. value	1.3045	0.647	0.1989	0.0837	-0.8075	-0.3499

and can be reached for any nonzero initial value k_0, the only difference being the number of iterations.

When the quadratic constraint is added, the estimator converges in 7 iterations to the following feasible state for $k_0 = 0.5$:

Magnitude	V_1	V_2	V_3	θ_2	θ_3	k
Est. value	1.0187	0.9942	0.9893	$-2.9893°$	$-3.6657°$	1.0000

which leads to the estimated measurements:

Magnitude	p_{13}	q_{13}	p_{23}	q_{23}	p_2	q_2
Est. value	1.2886	0.6392	0.2322	0.0996	-0.8241	-0.3583

However, the wrong status is reached when $k_0 < 0.5$. For instance, when $k_0 = 0.49$ the estimator converges, after 66 iterations, to the state:

Magnitude	V_1	V_2	V_3	θ_2	θ_3	k
Est. value	1.0021	0.0183	0.9720	$-69.2885°$	$-3.8410°$	0.0000

and measurements:

Magnitude	p_{13}	q_{13}	p_{23}	q_{23}	p_2	q_2
Est. value	1.3049	0.6468	-0.3238	-0.1412	-0.3238	-0.1412

The objective function value and normalized residuals suggest in this case the presence of bad data, originated by the incorrect line status. It is important, therefore, to start with the neutral value $k_0 = 0.5$ or to add the constraint only when the correct solution is sufficiently close.

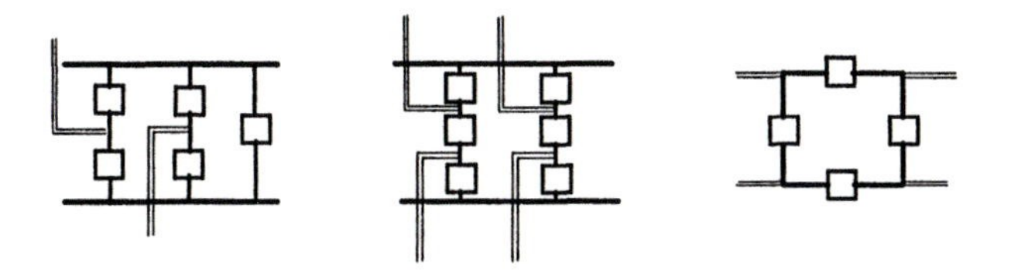

Figure 8.6. Usual substation configurations

8.7 Substation configuration errors

Circuit breakers and bus sections can be arranged in many different configurations within a substation, depending on the voltage level, role and importance of the substation, number of incident lines, etc. Figure 8.6 shows, from left to right, three of the most frequent configurations, namely: two buses with coupling, breaker-and-a-half and ring bus.

The common feature of all these cases is that CBs are interconnected through zero-impedance bus sections, unlike external connections which always involve a finite admittance. An arbitrarily small impedance might be inserted in series with every CB, so that the procedures presented in the former section could be applied. However, as explained in Chapter 3, this approach would lead to severe numerical problems and, most probably, to convergence difficulties and/or unacceptable results.

Furthermore, the number of electrical nodes determined by a particular substation depends on its topology and CB statuses. Frequently, two or more nodes exist within the same voltage level of a substation. In some cases, it is known for example that a branch is in service, but its terminal buses are not clearly identified. When this happens, the above procedures can not be applied because the extra variable k is intended to represent the on/off status of a particular branch located between a pair of well defined buses. Therefore, there is a need to discuss more flexible and general methodologies, intended to model the diverse situations arising in substation configuration errors and certain types of branch errors.

The techniques that are described in the remaining parts of this chapter can be considered the somewhat natural evolution of the LP-based, local pre-filtering procedure proposed in [12] (see section 8.5). Major differences and improvements are:

- A full CB model, including both active and reactive problems, is employed.
- Constraints for closed CBs are considered, in addition to zero power flow constraints.
- The entire power system, rather than a single substation, is analyzed,

either through the two-stage procedure discussed in section 8.4 or the implicit model presented in section 8.9.

8.7.1 Inclusion of circuit breakers in the network model

Effects of topology errors can be taken into account explicitly by representing the circuit breakers in terms of their real and reactive power flows, rather than as closed (zero impedance) or open (zero admittance) branches (see [22, 18, 19] for more details). The simple 4-bus test system shown in Figure 8.4 will be used to illustrate the modified formulation. Assume that bus 2 represents a substation which is composed of two separate buses, 2a and 2b, connected by a circuit breaker (see the expanded network in Figure 8.7). If the breaker status is unknown, then the breaker flows P_f, Q_f can be used as extra unknown variables in the equations of measurements incident to this breaker. For instance, the power injection measurements at bus 2a, will be written as:

$$\begin{aligned} P_{2a} &= P_f + P_{23} + P_{21} \\ Q_{2a} &= Q_f + Q_{23} + Q_{21} \end{aligned}$$

where P_{23}, P_{21}, Q_{23}, and Q_{21} are the real/reactive power flows along lines connecting bus 2a to its neighboring buses.

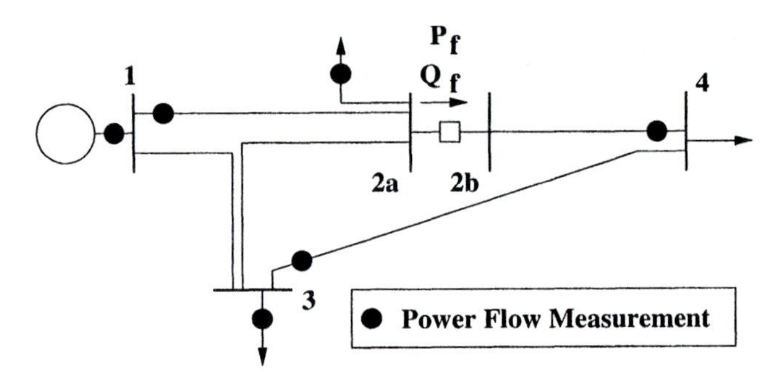

Figure 8.7. Circuit breaker modeling using power flow variables

Let us now consider a general or augmented set of measurement equations written in terms of the state variables and the breaker flows:

$$z_a = h(x) + Mf + e \tag{8.19}$$

where:
z_a is the generalized measurement vector,

h is the nonlinear vector function relating ordinary measurements to the states assuming all breakers are open,
x is the $n \times 1$ state vector containing bus voltages and phase angles,
M is the *measurement to circuit breaker* incidence matrix,
f is the vector of real/reactive power flows through the circuit breakers,
e is the measurement error vector.

Conventional state variables and CB flows can be combined to form an augmented state vector x_a as given below:

$$x_a^T = [x^T f^T]$$

In terms of this enlarged vector, (8.19) is written in compact form as:

$$z_a = \Phi(x_a) + e \tag{8.20}$$

The following three different kinds of 'measurements' should be considered when building (8.20) [7]:

1. Regular analog measurements, given by:

$$z = h(x_a) + e \tag{8.21}$$

2. Operational constraints imposed by the open or closed status of the switching branches. For a branch $k - m$ these will be $p_{km} = 0$, $q_{km} = 0$ if open, or $V_k - V_m = 0$, $\theta_k - \theta_m = 0$ if closed. These linear constraints can be compactly expressed by:

$$A_0 x_a + e_0 = 0 \tag{8.22}$$

Each operational constraint of (8.22) can be strictly enforced by setting $e_0 = 0$, or can be used as a measurement with finite uncertainty, i.e. $\text{cov}(e_0) \neq 0$.

3. Structural constraints imposed by the network connectivity, such as the zero injection constraints at certain nodes:

$$c(x_a) = 0 \tag{8.23}$$

Dropping the subscript a for simplicity, the state estimation problem can thus be written as follows:

$$\begin{aligned} \text{Minimize} \quad & J(r, r_0) \\ \text{Subject to} \quad h(\hat{x}) + r &= z \\ A_0\hat{x} + r_0 &= 0 \\ c(\hat{x}) &= 0 \end{aligned} \tag{8.24}$$

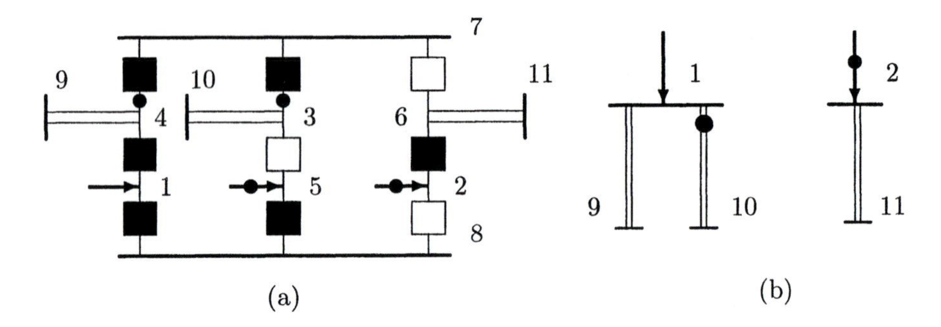

Figure 8.8. (a) 8-bus substation (closed CBs are shown in black); (b) bus-branch equivalent model.

where J is the objective function that depends upon r and r_0, the residuals for the analog measurements and operational constraints respectively, and $\hat{x}$ is the estimated state vector containing the estimated bus voltages, angles and switching branch real and reactive power flows.

The objective function, and the way the results are interpreted, is determined by the type of estimator adopted.

Example 8.5:

For the substation of Figure 8.8, where 9, 10 and 11 refer to external buses, develop the DC linearized estimator, both for the conventional bus-branch model and the augmented model described above.

Bus-branch model

As shown in Figure 8.8-(b), the bus-branch substation model contributes two unknowns to the state vector, i.e.,

$$x_{bb} = [\theta_1 \ \theta_2 \ x_e]^T \tag{8.25}$$

where x_e includes the external variables θ_9, θ_{10} and θ_{11}. The two resulting measurements are related to the state variables as follows:

$$\begin{aligned} p_2^m &= x_{2-11}^{-1}(\theta_2 - \theta_{11}) + \varepsilon_2 \\ p_{1-10}^m &= x_{1-10}^{-1}(\theta_1 - \theta_{10}) + \varepsilon_{1-10} \end{aligned} \tag{8.26}$$

where x_{1-10} and x_{2-11} denote the series reactance of external elements. Note that the aggregated measurement p_1^m is missing in this case, because one of its component injections, p_1, is not measured. It has been assumed that the Topology Processor is able to shift the power flow measurement p_{37}^m from the CB to the external branch 3-10, but sometimes this is not possible and these internal measurements can not be used by a conventional SE.

Full detailed model

The augmented state vector is, in this case:

$$x_a = [\theta_1\,\theta_2\,\theta_3\,\theta_4\,\theta_5\,\theta_6\,\theta_7\,\theta_8\,p_{47}\,p_{37}\,p_{67}\,p_{14}\,p_{35}\,p_{26}\,p_{18}\,p_{58}\,p_{28}\,x_e]^T \tag{8.27}$$

In terms of these variables, the available analog measurements are expressed as:

$$\begin{array}{lll} p_5^m = & -p_{35} + p_{58} & +\varepsilon_5 \\ p_2^m = & p_{26} + p_{28} & +\varepsilon_2 \\ p_{37}^m = & p_{37} & +\varepsilon_{37} \\ p_{47}^m = & p_{47} & +\varepsilon_{47} \end{array} \tag{8.28}$$

and the topological constraints ($e_0 = 0$):

$$\begin{array}{lll} \theta_4 - \theta_7 = 0; & \theta_3 - \theta_7 = 0; & \theta_1 - \theta_4 = 0 \\ \theta_4 - \theta_8 = 0; & \theta_5 - \theta_8 = 0; & \theta_2 - \theta_6 = 0 \\ p_{35} = 0; & p_{67} = 0; & p_{28} = 0 \end{array} \tag{8.29}$$

Finally, the structural constraints associated with zero-injection buses are:

$$\begin{array}{ll} p_4 = 0; & -p_{14} + p_{47} + x_{4-9}^{-1}(\theta_4 - \theta_9) = 0 \\ p_6 = 0; & -p_{26} + p_{67} + x_{6-11}^{-1}(\theta_6 - \theta_{11}) = 0 \\ p_3 = 0; & p_{35} + p_{37} + x_{3-10}^{-1}(\theta_3 - \theta_{10}) = 0 \\ p_7 = 0; & p_{47} + p_{67} + p_{37} = 0 \\ p_8 = 0; & p_{18} + p_{58} + p_{28} = 0 \end{array} \tag{8.30}$$

Table 8.3 compares the size of both models. From the figures shown, the following conclusions are obtained:

- The full model requires a very large number of variables (17 ignoring x_e), compared to the 2 variables of the 'electrical' model.
- The detailed model allows 4, rather than 2, measurements to be separately included and, hence, tested.
- As many topological constraints as CBs must be enforced, unless their status is not safely known.
- The number of structural constraints is also high (5 in this case). Typically, most bus sections where one or several CBs join a line or tranformer constitute zero-injection buses.

Table 8.3. Comparison between the bus-branch and the fully augmented model

Model	Analog meas.	Topol. const.	Struct. const.	State var.
Bus–branch	2	0	0	2
Full	4	9	5	17

Of course, observability can not be assessed based solely on the data presented in Table 8.3, as measurements located at nearby substations play an important role in this regard.

In spite of the substation size being moderate, this example clearly shows the impossibility of modeling on-line the whole set of substations in full detail, which calls for the two-stage procedure described above. At this point, the reader may wonder why so many state variables are added to the model if, at the same time, their value is forced to be null. The reason is that only when CBs are individually modeled, can their status be separately checked. However, as will be explained in sections 8.8 and 8.9, the number of state variables can be significantly reduced if, instead of this 'brute force' approach, a systematic analysis based on topological properties is performed.

8.7.2 WLAV estimator

The WLAV estimation method, described in Chapter 6, can be applied to solve the augmented model (8.24). This estimator is robust under bad data and can easily handle additional constraints. It solves the following optimization problem:

$$\text{Minimize } J(x) = \sum_{i=1}^{m} w_i(u_i + v_i) \tag{8.31}$$

$$\text{subject to } z = \Phi(x) + u - v \tag{8.32}$$

where u and v are nonnegative slack variables such that $(u - v)$ represents the measurement residual vector, w_i is the weight assigned to measurement i, and m is the number of measurements.

Additionally, the *operational* constraints referred to above can be appended to (8.32) to indicate a closed or open CB. Since the status information of the breakers may not be correct, such operational constraints are made soft by treating them as measurements with some uncertainty. For a breaker connected between $j - k$, one of the following constraints will be appended depending on its assumed status:

$$x_j - x_k + u_{m+i} - v_{m+i} = 0 \text{ [closed]} \tag{8.33}$$

$$f_{jk} + u_{m+i} - v_{m+i} = 0 \text{ [open]} \tag{8.34}$$

$$i = 1, 2, \ldots, b_s$$

where b_s is the number of CBs.

If any of the flow variables in x are unobservable in the absence of the operational constraints, then the status error for that breaker will not be detectable. Observability of breaker flows and cases of undetectable breaker status errors are identified by the WLAV estimator during the Phase I solution stage [1].

Example 8.6:

Consider the IEEE 14-bus system shown in Figure 8.9, whose data can be downloaded from [34]. The system is fully measured with injections and voltages at each bus, and power flows at both ends of each branch. The detailed bus-breaker configuration at substation 3 is also shown. Note that, as only the breaker 2 is closed, this substation is actually composed of three electrical nodes, one of them isolated (bus split). Measurements are simulated using the correct three-node model, while the WLAV estimator is run assuming breaker 1 is closed (two nodes only, merging error). Initial WLAV estimation results, omitted here for brevity, indicate bus 3 as a suspect substation due to the large percentage of incident measurements to bus 3 having large normalized residuals.

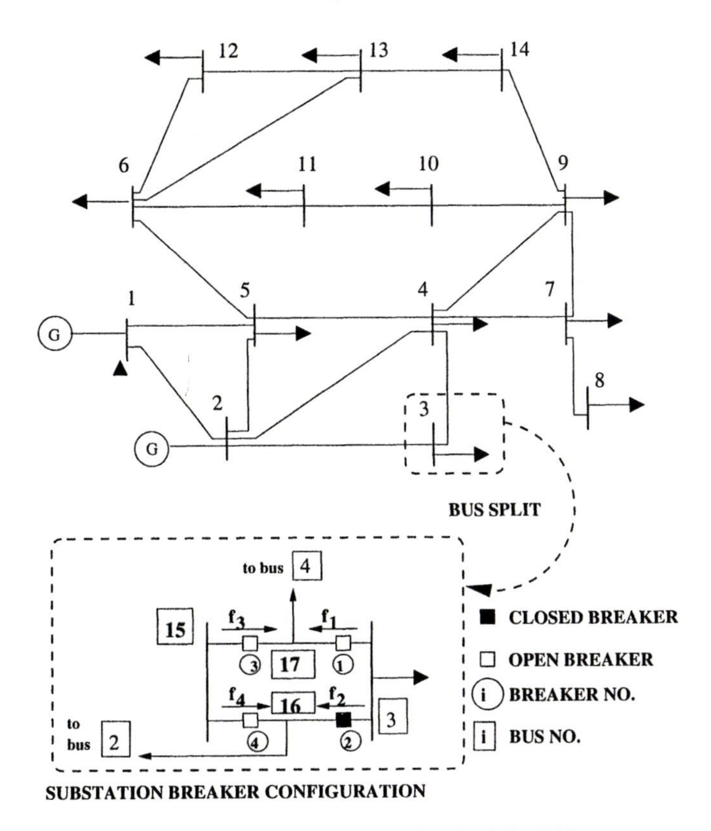

Figure 8.9. Detailed substation model at bus 3 of the 14-bus test system

Then substation 3 is modeled in detail. In addition to the existing measurements, zero injection measurements are added at newly created buses 15, 16 and 17. Closed breaker constraints for breakers 1 and 2, which are thought to be closed, and open constraints for breakers 3 and 4, are also included.

Running the WLAV estimator yields the estimated bus voltages listed in Table 8.4. Large power flows through breaker 2 (3-16) are compatible with this state, indicating that its status should be closed. On the other hand, bus 3 voltage is estimated correctly due to the constraint equations used between buses 3, 16 and 17. One of these constraints (3-17) is rejected by the WLAV estimator, since breaker 1 is actually open. The other one (3-16), however, is found consistent, allowing the estimation of the state of bus 3. Note that bus 15 voltage can not be estimated based on the available measurements, since it is isolated from the rest of the network by circuit breakers. So the WLAV estimator assigns bus 15 the flat start voltage at the solution. Note also that bus 4 and bus 17 have similar voltages, since 4-17 is an open ended line carrying a small line charging current only. As can be seen, the results of the state estimator using breaker models, should be analyzed with care due to these peculiarities arising from a nonconventional breaker model used.

Table 8.4. True versus Estimated States

Bus	True State		Estimated State	
No.	V	θ	V	θ
1	1.061	0.00	1.059	0.00
2	1.046	−5.15	1.044	−5.17
3	1.012	−15.64	1.010	−15.70
4	1.024	−9.36	1.022	−9.41
5	1.025	−8.12	1.023	−8.17
6	1.072	−13.43	1.069	−13.49
7	1.065	−12.43	1.063	−12.52
8	1.091	−12.43	1.089	−12.53
9	1.060	−14.04	1.058	−14.12
10	1.054	−14.21	1.052	−14.29
11	1.060	−13.95	1.057	−14.02
12	1.057	−14.29	1.055	−14.35
13	1.053	−14.36	1.050	−14.41
14	1.039	−15.16	1.037	−15.24
15	–	–	1.000	0.00
16	–	–	1.010	−15.70
17	1.027	−9.44	1.025	−9.48

8.7.3 WLS estimator

Alternatively, the objective function $J(r) = r^T R^{-1} r$ can be used in (8.24), leading to an equality-constrained WLS estimator [3, 9, 7].

Two possibilities arise at this point. One is to treat CB operational constraints as strict equality constraints, i.e. $r_0 = 0$. The resulting Lagrangian is,

$$\begin{aligned} \mathcal{L} &= \frac{1}{2} r^T R^{-1} r - \lambda^T (r - z + h(\hat{x})) \\ &- \lambda_0^T A_0 \hat{x} - \lambda_s^T c(\hat{x}) \end{aligned} \tag{8.35}$$

The other possibility consists of handling operational constraints as very accurate measurements, for which a very high weighting factor, ρ, will be assigned in the objective function,

$$\begin{aligned} \mathcal{L} &= \frac{1}{2} r^T R^{-1} r - \lambda^T (r - z + h(\hat{x})) \\ &+ \rho \hat{x}^T A_0^T A_0 \hat{x} - \lambda_s^T c(\hat{x}) \end{aligned} \tag{8.36}$$

As ρ grows, r_0 and its covariance tend to zero. Therefore, the normalized residual test can not be safely applied to this measurement type because of the numerical instability associated with the undefined 0/0 operation. However, as discussed in Chapter 3 (section 5), the following relationship holds:

$$\lambda_0 = \rho r_0 \tag{8.37}$$

and, consequently,

$$\lambda_i^N = \frac{\lambda_i}{\sqrt{\text{cov}(\lambda_i)}} = \frac{\rho r_i}{\sqrt{\text{cov}(\rho r_i)}} = \frac{\rho r_i}{\rho \sqrt{\text{cov}(r_i)}} = r_i^N \tag{8.38}$$

where $\text{cov}(\lambda_i)$ is computed as explained in [9]. This means that the normalized multipliers can be compared with, and used in the same manner as, the normalized residuals of ordinary measurements during the bad data detection & identification process. Statistically significant λ_i^N's imply inconsistent constraints, and the corresponding switch status is reversed. This procedure is repeated until all λ_i^N's decrease below a chosen threshold. Fine tuning of this procedure to avoid cycling may be needed (see [7]).

The above discussion does not apply to CBs whose status is unknown, because no operational constraints are added in this situation. However, based on the results provided by the WLS estimator, hypothesis testing can still be used to check for the most likely CB status [20]. Let $Ax = 0$ be the p constraints comprising the null hypothesis. Then, it can be shown that the performance index,

$$J(A\hat{x}) = (A\hat{x})^T [A(H^T W H)^{-1} A^T]^{-1} (A\hat{x}) \tag{8.39}$$

follows a χ^2 distribution with p degrees of freedom. This provides a threshold to validate hypotheses about open/closed statuses.

Another problem related to unknown CBs is the fact that, due to measurement noise, the estimated power flow through open devices is not completely null, nor is the voltage drop across closed switches. Accuracy of the estimation can be easily improved if the following constraints [32],

$$\begin{aligned} c_p &= P_{ij}(\theta_i - \theta_j) = 0 \\ c_q &= Q_{ij}(V_i - V_j) = 0 \end{aligned} \tag{8.40}$$

are added to the model for a breaker between buses i and j. When building the Jacobian, these extra rows lead to the following new terms:

$$\begin{aligned} &\frac{\partial c_p}{\partial P_{ij}} = \theta_i - \theta_j \\ &\frac{\partial c_p}{\partial \theta_i} = -\frac{\partial c_p}{\partial \theta_j} = P_{ij} \\ &\frac{\partial c_q}{\partial Q_{ij}} = V_i - V_j \\ &\frac{\partial c_q}{\partial V_i} = -\frac{\partial c_q}{\partial V_j} = Q_{ij} \end{aligned} \tag{8.41}$$

Since these terms are either null or very small at flat start, the above constraints are useful only after the first iteration.

Note that the status of a breaker forming a loop with closed CBs is irrelevant. The same happens with a CB belonging to a cut-set where the remaining CBs are open. In such cases, there is no need to add extra variables and constraints.

Example 8.7:

Consider again the 3-bus system of Figure 8.5, slightly modified as shown in Figure 8.10. In the modified system, bus 2 is actually composed of two bus-bar sections which are linked by a CB whose status is uncertain. For the same impedance values and weights adopted for Example 8.4, estimate the status of the unknown CB, with and without the pair of constraints (8.40) added. Available measurements comprise those provided in Example 8.4, plus a voltage measurement at bus section '2b'. It is assumed that 75% of all power injected at bus 2 is incident to section '2a', while the remaining 25% lies at section '2b'. The following two tables provide the measurement values:

p_{13}	q_{13}	p_{2b3}	q_{2b3}	p_{2a}	q_{2a}	p_{2b}	q_{2b}
1.305	0.646	0.198	0.086	−0.6053	−0.2632	−0.2018	−0.0877

V_1	V_{2a}	V_{2b}
1.018	0.996	1.002

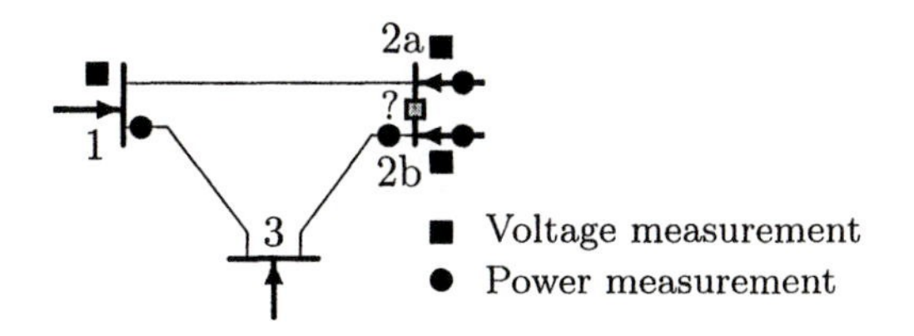

Figure 8.10. 3-bus system with unknown CB

When the constraints in (8.40) are not added, the WLS estimator converges after 4 iterations to the following state vector:

Bus	Magnitude	Angle (°)
1	1.0215	0.0000
2a	0.9983	−2.8245
2b	0.9962	−3.1182
3	0.9919	−3.6922

This state is compatible with the following power flows through the CB:

$$p_{2a,2b} = 0.3998 \quad ; \quad q_{2a,2b} = 0.1745$$

which suggests that the CB is closed, even though the voltages at its terminal buses are not identical. This physically meaningless state can be avoided by adding the constraints (8.40). In this case, the estimator converges in 6 iterations to the state:

Bus	Magnitude	Angle (°)
1	1.0214	0.0000
2a	0.9969	−2.9939
2b	0.9969	−2.9939
3	0.9921	−3.6524

where the complex voltages of bus sections '2a' and '2b' are identical. The resulting CB power flows are:

$$p_{2a,2b} = 0.4437 \quad ; \quad q_{2a,2b} = 0.1913$$

These values are about 10% larger than those obtained formerly, showing the importance of adding the constraints (8.40) as a means of improving the accuracy of the estimate.

Before ending this section it is interesting to point out another important difference between the models (8.35) and (8.36), related to the potential inclusion of linearly dependent operational constraints. Such constraints arise, for instance, in the presence of loops (cut-sets) exclusively composed of closed (open) CBs. If these constraints are modeled as pseudo-measurements, like in (8.36), there is no need to worry about their number and linear independence, so long as the weighting factor ρ is not so high that numerical problems arise. Assume now that the model given by (8.35), rewritten for simplicity as

$$\mathcal{L} = J(x) - \lambda_0^T A_0 x \tag{8.42}$$

is adopted. The first-order optimality conditions, when applied to (8.42), allow both x and λ_0 to be determined, provided A_0 is of full rank (linearly independent constraints). But there are cases, where linear independence of the rows of A_0 can not be assured. Let A_I and A_D denote the set of linearly independent and redundant rows of A_0 respectively. The above constrained model becomes,

$$\mathcal{L} = J(x) - \lambda_I^T A_I x - \lambda_D^T A_D x \tag{8.43}$$

which, taking into account the linear relationship between the rows of A_D and A_I,

$$A_D = K A_I \tag{8.44}$$

can be also written as,

$$\mathcal{L} = J(x) - (\lambda_I + K^T \lambda_D) A_I x \tag{8.45}$$

This means that only an equivalent multiplier vector

$$\lambda = \lambda_I + K^T \lambda_D \tag{8.46}$$

can be computed, but not its separate components λ_I and λ_D, unless further assumptions are made. For instance, it is easy to show that assigning the same weighting factor to all constraints,

$$\mathcal{L} = J(x) + \rho x^T A_I^T A_I x + \rho x^T A_D^T A_D x \tag{8.47}$$

translates into the following additional relationship,

$$\lambda_D = K \lambda_I \tag{8.48}$$

In matrix form, (8.46) and (8.48) yield,

$$\begin{bmatrix} I & K^t \\ K & -I \end{bmatrix} \begin{bmatrix} \lambda_I \\ \lambda_D \end{bmatrix} = \begin{bmatrix} \lambda \\ 0 \end{bmatrix} \tag{8.49}$$

Therefore, when the objective function (8.35), rather than (8.36), is minimized, only linearly independent constraints should be used. The Lagrange multipliers of arbitrary sets of constraints could be obtained, if needed, by means of (8.49).

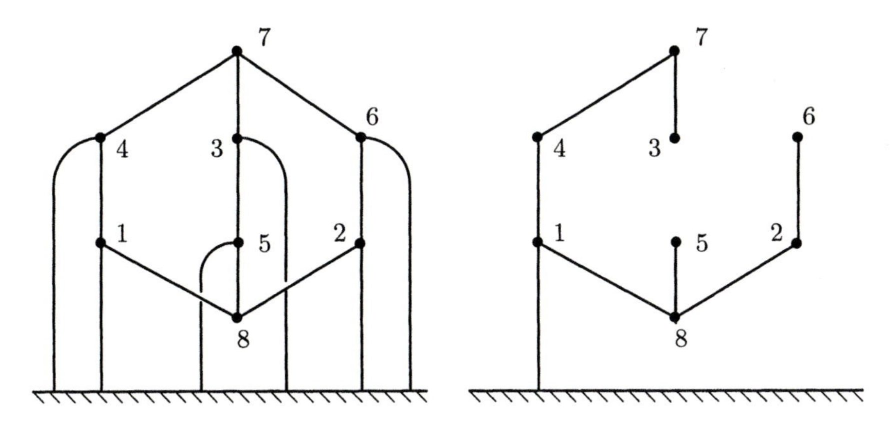

Figure 8.11. Graph corresponding to the 9-CB substation of Figure 8.8 and a possible tree

8.8 Substation graph and reduced model

This section will show that, by properly exploiting topological properties of circuits, i.e., Kirchhoff laws, only a subset of power flows through switching devices must be added to the state vector [10]. This requires that the substation graph (SG) be previously introduced, from which linearly independent operational constraints can be easily identified as well. For the sake of clarity, only the active subproblem of the SE will be analyzed in detail in the following development, although any noteworthy difference with the reactive problem will be indicated. It is assumed that the reader is familiar with basic linear circuit theory (see [5] and Chapter 4).

The SG is composed of as many internal nodes as bus-bar sections plus a virtual "ground" node intended to represent the external system. The n_i internal nodes of this graph are interconnected by b_s internal branches representing switching elements. Furthermore, two types of external branches can be distinguished in the general case: a) b_z non-zero impedance branches corresponding to lines, transformers or shunt devices; b) b_p branches connecting non-zero injection nodes to ground (external sources or sinks of power). The number of zero-injection nodes is therefore $n_i - b_p$.

Example 8.8:

Figure 8.11 shows the graph corresponding to the substation of Figure 8.8.

In this case, there are $n_i = 8$ buses, $b_s = 9$ internal branches (switches), $b_z = 3$ regular branches and $b_p = 3$ external injections. Note that there are 5 zero-injection nodes.

Each branch of the substation graph contributes a pair of unknowns to the active problem, namely its active power flow and its voltage phase drop (and another pair to the reactive problem). Let us analyze separately both sets of unknowns.

On the one hand, by means of Kirchhoff's voltage law, any branch phase angle (or voltage magnitude drop in the reactive problem) can be expressed in terms of a base of n_i branch phase angles. This set of basic phase angles can be composed of the branches of an arbitrary tree, but in conventional nodal-based formulations the nodal phase angles, which serve for the same purpose, are preferred (note that this particular base corresponds to a star-shaped tree composed of real or virtual branches connecting all internal nodes to ground). This explains why n_i phase angles are retained in the state vector, as in the full model described in the former section.

Similarly, Kirchhoff's current law allows any power flow to be expressed solely in terms of a base of power flows determined by the links of a tree. For a connected graph, with the notation adopted above, the number of such links is in principle $(b_s + b_p + b_z) - n_i$. However, the b_z power flows of regular branches can be expressed in terms of the nodal phase angles of the substation in hand and adjacent substations, already contained in the state vector. Therefore, the number of truly independent power flow variables that should be included in the state vector is $(b_s + b_p) - n_i$, instead of the b_s power flows adopted by the fully augmented model, which means a reduction of $n_i - b_p$ power flows.

The following remarks are in order:

- $n_i - b_p$ is also the number of null-injection nodes, whose respective constraints are no longer enforced since they are implicitly taken into account by the SG definition. Hence, the same amount of state variables and constraints are removed from the model.

- The power flows retained in the state vector should be selected in such a way that they are links of a certain tree in which regular non-zero impedance branches are also links by default.

- Consequently, it is not strictly necessary that the power flows included in the state vector correspond to those of CBs. As will be shown in the next section, it is sometimes more convenient to consider certain power injections as state variables, and hence links of the tree. However, for the tree to be connected to ground, at least one of the power injections should be a tree branch.

Example 8.9:

The full model of Example 8.5, corresponding to the substation of Figure 8.8, will be reconsidered in the light of the above development. In this case, the state vector must be augmented with just $(b_s + b_p) - n_i = 4$ power flows, selected as explained earlier. Among the several possible choices, the following state vector is adopted:

$$x_r = [\theta_1\, \theta_2\, \theta_3\, \theta_4\, \theta_5\, \theta_6\, \theta_7\, \theta_8\, p_5\, p_2\, p_{67}\, p_{35}\, x_e]^T \tag{8.50}$$

where x_e contains θ_9, θ_{10} and θ_{11}. This selection of power flows corresponds to a tree composed of CBs 1-8, 1-4, 4-7, 3-7, 2-6, 2-8 and 5-8, plus the injection at bus 1 (refer to Figure 8.11).

In terms of these state variables, the available analog measurements are expressed as:

$$\begin{aligned} p_5^m &= p_5 & &+\varepsilon_5 \\ p_2^m &= p_2 & &+\varepsilon_2 \\ p_{37}^m &= -p_{35} - x_{3-10}^{-1}(\theta_3 - \theta_{10}) & &+\varepsilon_{37} \\ p_{47}^m &= p_{35} + x_{3-10}^{-1}(\theta_3 - \theta_{10}) - p_{67} & &+\varepsilon_{47} \end{aligned} \tag{8.51}$$

and the topological constraints:

$$\begin{array}{lll} \theta_4 - \theta_7 = 0; & \theta_3 - \theta_7 = 0; & \theta_1 - \theta_4 = 0 \\ \theta_4 - \theta_8 = 0; & \theta_5 - \theta_8 = 0; & \theta_2 - \theta_6 = 0 \\ p_{35} = 0; & p_{67} = 0; & \\ p_{28} = p_2 - p_{67} - x_{6-11}^{-1}(\theta_6 - \theta_{11}) = 0 & & \end{array} \tag{8.52}$$

Comparing (8.50) with (8.27) it is apparent that the reduced model saves 5 state variables with respect to the full model. Note that the 5 structural constraints associated with zero-injection buses are not explicitly enforced, which compensates for the 5 missing power flow variables.

A substation without external power injections constitutes a special case, as there is no way to connect the internal tree branches to ground if all non-zero impedance branches must be links. This particular case is handled by allowing a zero-injection virtual branch to be retained at any of the substation buses, leading to the non-linear structural constraint appearing also in the bus-branch model. Hence, the reduced model discussed in this section involves the same number of structural constraints as those of the bus-branch model.

Further model reductions are many times possible by previously identifying critically observable state variables and their associated critical constraints. For instance, the operational constraint of an open CB is critical, and hence can be removed from the model, if none of the available power measurements is formulated in terms of the respective power flow variable, which can be also dropped from the state vector. This possibility is systematically explored in [10], where carefully selected trees are employed both for the voltage and power flow submodels. The reduction achieved by this technique depends anyway on the measurement distribution and redundancy.

Before ending this section it is worth noting that the SG provides also a simple means of finding which operational constraints are linearly independent. For this purpose, a particular tree is selected so that it includes as many closed CBs and excludes as many open CBs as possible. When such a tree contains all closed but no open CB, all operational constraints are linearly independent. As explained in the former section, redundant constraints, corresponding to loops of closed CBs and cut-sets of open CBs, should be removed from the model if they are handled as equality constraints, in order to avoid rank-deficient matrices.

8.9 Implicit substation model: state and status estimation

Up to now, at least three trees of the SG have been considered, namely:

- A branch voltage tree intended to apply Kirchhoff's voltage law. The branch voltages of this tree should be included in the state vector. For convenience, however, this tree is seldom built, as the nodal voltages constitute a simpler alternative to get an equivalent number of variables.
- A branch power flow tree intended to apply Kirchhoff's current law. The power flows of the links of this tree should be included in the state vector, except for those power flows which are non-linear functions of the bus voltages.
- A tree intended to identify the largest set of linearly independent operational constraints.

In this section, a single tree, called the **Proper Tree**, useful for the three goals discussed above, is first introduced. This tree requires that some node voltages be replaced by branch voltages but, as shown later, the structure of the resulting equations leads naturally to a model in which none of the operational constraints needs to be explicitly handled. In turn, this model provides a philosophically different approach to perform state and status estimation [33].

The Proper Tree is selected keeping the following goals in mind:

1. The conventional bus-branch model should be embedded in the resulting state vector. This requires that electrical nodes be previously identified and their voltages be included in the state vector. An

electrical node is any bus-bar section, or group of bus-bar sections interconnected by closed CBs, directly connected to at least one external branch. In Figure 8.12, bus sections 1-3-4-5-7-8 constitute the electrical node 1, while sections 2-6 lead to node 2.

2. The power flow through non-zero impedance branches is a non-linear function of electrical bus voltages, which are already included in the state vector. Therefore, it is convenient for these regular branches to be links of the tree so that their power flows can be directly removed from the state vector.

3. As many closed CBs as possible should belong to the tree, and the opposite can be stated for open CBs. This way, independent loops of closed CBs and cut-sets of open CBs, leading to linearly dependent constraints, are automatically identified.

Consequently, the Proper Tree is defined as follows:

- Exclude all non-zero impedance branches.
- Include as many closed CBs as possible.
- Select, for every electrical bus, a non-zero injection node that will be termed the **base node** (this is the node retained in the bus-branch model). Include the respective injection branch in the tree (note that any other injection of the same electrical bus becomes a link). An electrical bus exclusively connected to non-zero impedance branches constitutes a special case that is handled by adding a virtual null-injection branch to any of its elemental buses. This extra branch, also included in the tree, is responsible for the null-injection constraints conventionally used by state estimators.
- If the case arises, complete the tree with CBs whose status is unknown.
- Exclude as many open CBs as possible.

Example 8.10:

Figure 8.12 shows the Proper Tree for the substation of Figure 8.8. The state vector is composed of the voltages corresponding to tree branches,

$$x_\theta = [\theta_1, \theta_2, \theta_{14}, \theta_{18}, \theta_{58}, \theta_{47}, \theta_{37}, \theta_{26}, \theta_e]^t \tag{8.53}$$

and the power flows/injections corresponding to links, excluding those of actual lines or transformers,

$$x_p = [p_{28}, p_{67}, p_{35}, p_5]^t \tag{8.54}$$

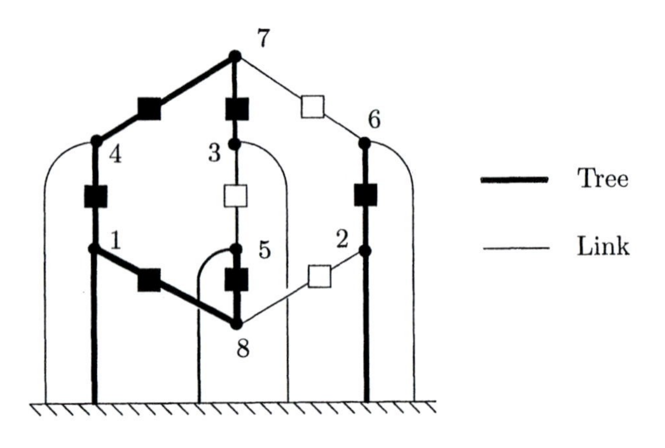

Figure 8.12. Proper Tree

As the only difference with Example 8.9 lies in the chosen tree, the total number of state variables in (8.53) and (8.54) is the same as those of (8.50), i.e., five less than the full model. The reader should check that any existing or potential measurement can be formulated solely in terms of the state variables contained in (8.53) and (8.54). For instance, if the voltage magnitude at bus-bar 5 is measured, then the following expression results for the reactive problem,

$$V_5^m = V_{58} - V_{18} + V_1 + \varepsilon$$

where V_5^m is the measured value and ε the associated noise.

An important additional reason for selecting the proper tree as suggested above is the fact that linearly independent CB constraints involve identity matrices only. In this case, (8.29) becomes,

$$\left[\begin{array}{cc|cccccc} 0 & 0 & 1 & 0 & 0 & 0 & 0 & 0 \\ 0 & 0 & 0 & 1 & 0 & 0 & 0 & 0 \\ 0 & 0 & 0 & 0 & 1 & 0 & 0 & 0 \\ 0 & 0 & 0 & 0 & 0 & 1 & 0 & 0 \\ 0 & 0 & 0 & 0 & 0 & 0 & 1 & 0 \\ 0 & 0 & 0 & 0 & 0 & 0 & 0 & 1 \end{array}\right] \left[\begin{array}{c} \theta_1 \\ \theta_2 \\ \hline \theta_{14} \\ \theta_{18} \\ \theta_{58} \\ \theta_{47} \\ \theta_{37} \\ \theta_{26} \end{array}\right] = 0 \tag{8.55}$$

$$\left[\begin{array}{ccc|c} 1 & 0 & 0 & 0 \\ 0 & 1 & 0 & 0 \\ 0 & 0 & 1 & 0 \end{array}\right] \left[\begin{array}{c} p_{28} \\ p_{67} \\ p_{35} \\ \hline p_5 \end{array}\right] = 0 \tag{8.56}$$

Note that, as the tree contains all closed CBs but none open CB, all topological constraints are linearly independent in this example.

As a consequence of the way the proper tree is selected, the reduced model can be written in compact form as follows (refer to the above example):

$$z = h(x_I, x_{CB}) + \varepsilon \tag{8.57}$$

$$c(x_I, x_{CB}) = 0 \tag{8.58}$$

$$x_{CB} = 0 \tag{8.59}$$

$$K \cdot x_{CB} = 0 \tag{8.60}$$

where the notation adopted is:

- x_I: Component of the state vector containing,
  - Voltage magnitude and phase angle of all base nodes (each one representing an electrical node).
  - Voltage magnitude and phase angle across unknown and open CBs belonging to the proper tree.
  - Power flows through links of the proper tree corresponding to injections, as well as unknown and closed CBs.
- x_{CB}: Component of the state vector comprising,
  - Voltage magnitude and phase angle across closed CBs included in the proper tree.
  - Power flows through open CBs excluded from the proper tree.

The vector z contains all available measurements, including those discarded by conventional estimators, $h(\cdot)$ is the vector of linear and non-linear functions relating z with x, and $c(\cdot)$ comprises the non-linear constraints contributed by null-injection base nodes.

In the above model, (8.59) represents the linearly independent CB constraints, i.e., those corresponding to closed CBs in the tree and open CBs in the co-tree. As explained in the former example, one of the reasons for selecting the proper tree as suggested in this section is to keep (8.59) as simple as possible (identity matrix involved). Equation (8.60), on the other hand, refers to linearly dependent CB constraints, i.e., those corresponding to loops of closed CBs and cut-sets of open CBs. As discussed in section 8.7.3, both types of constraints can be included in the model provided they are handled as very accurate measurements. However, when the Lagrange multiplier method is adopted, only (8.59) can be included and its Lagrange multiplier vector, λ, be computed. If needed, this vector can be split into two components λ_I, λ_D, corresponding to (8.59) and (8.60) respectively, by solving (8.49). For simplicity, (8.60) will be discarded in the sequel.

With these premises, the Lagrangian of the proper model becomes,

$$\mathcal{L} = \frac{1}{2} r^t W r - \mu^t c(x_I, x_{CB}) - \lambda^t x_{CB} \tag{8.61}$$

and the following first-order optimality conditions can be written,

$$H_I^t W r + C_I^t \mu = 0 \tag{8.62}$$

$$H_{CB}^t W r + C_{CB}^t \mu + \lambda = 0 \tag{8.63}$$

$$c(x_I, x_{CB}) = 0 \tag{8.64}$$

$$x_{CB} = 0 \tag{8.65}$$

where $r = z - h(x_I, x_{CB})$ is the residual vector, and H_I, H_{CB}, C_I and C_{CB} are the respective Jacobian matrices.

The following remarks are in order:

1. Equation (8.63) allows the multiplier vector of topological constraints, λ, to be computed from

$$\lambda = T \begin{bmatrix} r \\ \mu \end{bmatrix} \tag{8.66}$$

where,

$$T = -\begin{bmatrix} WH_{CB} \\ C_{CB} \end{bmatrix}^t \tag{8.67}$$

is properly called the **Topological Sensitivity** matrix. In order to normalize the multiplier vector, λ, its covariance can be also obtained from,

$$\text{cov}(\lambda) = T \text{cov} \left(\begin{bmatrix} r \\ \mu \end{bmatrix} \right) T^t \tag{8.68}$$

2. Equation (8.65) can be used to eliminate x_{CB} from the proper model. Let $h_0(x_I)$ and $c_0(x_I)$ represent $h(x_I, 0)$ and $c(x_I, 0)$ respectively. Then, it can be shown that the so-called **Implicit Model**,

$$z = h_0(x_I) + \varepsilon \tag{8.69}$$

$$c_0(x_I) = 0 \tag{8.70}$$

provides exactly the same results as the original model without explicitly handling topological constraints [33].

Example 8.11:

The implicit model corresponding to the active subproblem of the substation of Figure 8.8, when the proper tree is that of Figure 8.12, gives rise to the following state vector,

$$x_I = [\theta_1, \theta_2 | p_5]^t \tag{8.71}$$

Compared to the conventional bus-branch model, a single extra variable, p_5, is added. However, this makes it possible to include three extra measurements in the state estimation (p_5^m, p_{47}^m, p_{37}^m) and, what is more important, to detect/identify topological errors.

In the general case, the implicit model requires the following additional state variables, with respect to conventional estimators:

1. Injections which are links of the tree. When there are several injections at the same electrical node, only one of them is a tree branch, which forces the remaining to be links and, hence, state variables.

2. Voltages across unknown CBs in the tree and power flows through unknown CBs in the co-tree. These variables remain in the model because the respective topological constraints are missing.

3. Voltages across open CBs in the tree and power flows through closed CBs in the co-tree. These CBs are, so to speak, "misplaced" in the tree, leading to redundant constraints which are useless to decrease the size of the state vector (when the variable is a voltage drop, the constraint refers to a power flow, and viceversa).

Some of the additional variables, particularly those of items 2 and 3, are clearly unobservable in many cases and could be removed.

Typically, as in the case analyzed later, the number of additional variables in the implicit model is a very small percentage of the total. Therefore, the cost of solving the implicit model is comparable to that of the bus-branch model. However, the implicit model allows topological error analysis to be performed at the end of the state estimation process by means of (8.66)-(8.68), much in the same way as bad data is processed in conventional estimators. This way, there is no need to carry out the two-stage estimation procedure in which those substations declared suspicious after the first run are modeled in detail during the second.

The overall process based on the implicit model can be summarized as follows:

1. Topology Processing. Among others, the following tasks are performed at every substation: Identification of electrical nodes, selection of proper tree and state vector configuration, the structure of T is built and many of its elements, which are constant, are determined.

2. **State Estimation.** Except for the size of the state and measurement vectors, this stage is essentially identical to existing formulations. As a byproduct, the residual and Lagrange multiplier vectors, r, μ, as well as associated covariances are obtained. Non-constant elements of T are computed.

3. **Bad Data and Topology Error Analysis.** This task is carried out at the substation level. First, normalized residuals are computed. If they are small enough, then stop. Otherwise, obtain normalized Lagrange multipliers corresponding to CB constraints by means of (8.66) and (8.68). If needed, individual CB multipliers can be computed from (8.49). Select the largest normalized value. If this corresponds to a measurement, remove it or make it dormant and go to step 2. Otherwise, change the status of the respective CB and go to step 1 (topological information must be updated in this case).

Note that bad data and topology error analysis is performed simultaneously. Compared to conventional estimators, where only the diagonal elements of covariance matrices are sought, further computations are needed to obtain the covariances of CB constraints. However, as matrix T is composed of as many decoupled blocks as substations, these computations involve small submatrices which are dealt with sequentially for every suspected substation.

Example 8.12:

Exact measurements are generated for the substation of former examples (DC state estimation). Assigning same weights to all measurements ($w_i = 1000$) and identical line reactances ($X = 0.02$ p.u.), two cases, whose relevant normalized residuals/multipliers are collected in Table 8.5, are run:

1. Assuming the given topology is correct, a gross error is added to p_{47}^m. The largest normalized residual is 10.3 and, as expected, it is associated with p_{47}^m. As there are several large residuals, it is decided to obtain normalized multipliers. The largest value, corresponding to $p_{67} = 0$, is $7.8 < 10.3$. Hence, it is concluded that the topology is correct and the bad data is properly identified.

2. The assumed open status for the CB 6-7 is wrong (i.e., exact measurements are compatible with CB 6-7 being closed). The largest normalized residual is now 15.5 and is associated again with p_{47}^m. However, when normalized multipliers are computed, the largest one is $20.5 > 15.5$ and corresponds to $p_{67} = 0$. Therefore, it is concluded this time that the status of CB 6-7 is wrong.

Table 8.5. Normalized values for the 9-CB substation

Measurement/ Constraint	Normalized residual/multiplier	
	Case 1	Case 2
p_{73}^m	5.2	7.7
p_{47}^m	10.3	15.5
p_2^m	0	13.4
$p_{28} = 0$	0	13.4
$p_{67} = 0$	7.8	20.5
$p_{35} = 0$	5.1	7.7

In both cases, the same matrix T is employed:

$$
\begin{array}{cc}
 & \begin{array}{ccccccc} p_{73}^m & p_{47}^m & p_5^m & p_2^m & p_{9-4}^m & p_{11-6}^m & p_{10-3}^m \end{array} \\
T = 10^3 & \left[\begin{array}{ccccccc}
50 & 50 & 0 & 0 & -50 & 0 & -50 \\
0 & 0 & 0 & 0 & 0 & 0 & 0 \\
0 & 0 & 0 & 0 & 0 & 0 & 0 \\
50 & 50 & 0 & 0 & 0 & 0 & -50 \\
50 & 50 & 0 & 0 & 0 & 0 & -50 \\
0 & 0 & 0 & 50 & 0 & -50 & 0 \\
0 & 0 & 0 & -1 & 0 & 0 & 0 \\
0 & 1 & 0 & -1 & 0 & 0 & 0 \\
-1 & -1 & 0 & 0 & 0 & 0 & 0
\end{array}\right]
\begin{array}{l}
\theta_{14} = 0 \\
\theta_{18} = 0 \\
\theta_{58} = 0 \\
\theta_{47} = 0 \\
\theta_{37} = 0 \\
\theta_{26} = 0 \\
p_{28} = 0 \\
p_{67} = 0 \\
p_{35} = 0
\end{array}
\end{array}
$$

A null row in this matrix indicates that the respective constraint is critical ($\theta_{18} = 0$ and $\theta_{58} = 0$ in this example), but other constraints may be critical as well (e.g., $\theta_{14} = 0$, $\theta_{47} = 0$ and $\theta_{26} = 0$).

Example 8.13:

Every substation of the IEEE 14-bus test system has been modeled in detail, as in [10], resorting to typical bus-bar arrangements (see the one-line diagram in Figure 8.13).

Table 8.6 provides, for each substation, the number of state variables required by the substation models discussed in this chapter (bus-branch, full, reduced and implicit). Both the active and reactive problems are considered. Note that the structure of the measurement set is irrelevant to build this table, except for the super-reduced model of [10], based on the notion of equivalent constraints, which is not included. Of course, if the redundancy is not enough, some of those variables will be unobservable.

The number of additional variables required by the implicit model has essentially to do with the number of unknown CBs and multiple injection branches at the same electrical bus. Note that a cut-set of open CBs exists at substation 5.

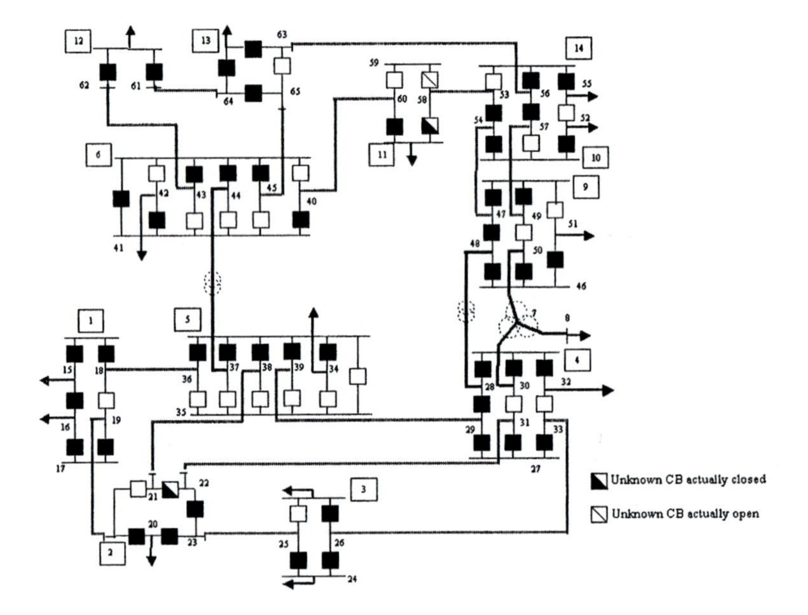

Figure 8.13. IEEE 14-bus system with substations modeled in detail

Table 8.6. Number of state variables required by different models

Subst.	B/B	Full	Reduced	Implicit
1	2	24	16	4
2	4	20	14	6
3	2	16	12	4
4	2	34	20	2
5	2	36	24	4
6	2	36	24	2
9	2	30	18	2
10/14	4	34	22	4
11	4	16	12	6
12	2	10	6	2
13	2	16	10	2
Total	28	272	178	38

Some buses, like 35 (substation 5) and 59 (substation 11) are clearly unobservable, since their phase angles do not appear in any measurement/constraint. If the respective variables are removed from the state vector, then the full and implicit models comprise 268 and 34 variables respectively. This means that the implicit model requires only 6 extra variables compared to the conventional bus-branch model.

8.10 Observability analysis revisited

Conventional observability analysis, as presented in Chapter 4, assumes that the state vector is composed exclusively of bus voltage magnitudes and phase angles, which is not the case when any of the augmented models discussed above are resorted to. Therefore, observability analysis should be reconsidered in the light of the generalized estimation concepts introduced in this chapter. For the topological observability analysis, this is done in [8], while numerical observability determination is generalized in [18]. The reader is referred to those references for further details.

Irrespective of the adopted model being full or implicit, the presence of certain power flows (or injections) in the state vector gives rise to the following major differences with respect to the conventional formulation:

- The notion of topological island somewhat vanishes and that of observable island is generalized. Conventionally, a topological island, separated from other islands by open CBs, may lead to one or several observable islands. For each observable island, one of its phase angles can be arbitrarily chosen without any influence on the power flows of observable branches. Many trivial islands, composed of bus sections and switches, or isolated nodes, are identified and discarded by the topology processor at the very beginning. However, according to the basic observability definition, a branch is said to be observable when its power flow can be obtained from the available information. Therefore, open CBs should be considered as observable branches in the new paradigm, which means that two or more topological islands can lead to a single generalized observable island, i.e., a set of interconnected branches and switches whose power flows are observable. Network configuration is hence translated into a set of additional constraints which, along with existing measurements, determine observable branches. A single generalized observable island, anyway, requires as many phase angles to be specified as conventional observable islands it comprises.

- Numerically based observability must be able to cope with zero pivots arising in columns corresponding to power flow variables. When this happens, the null pivot is replaced by 1, the power flow is assigned an arbitrary value and the respective branch is flagged as unobservable. Back substitution will determine subsequently the remaining unobservable branches. Sometimes, because of the new state variables, conventionally observable islands become partly unobservable.

These philosophical differences are illustrated in the following examples.

Example 8.14:

Analyze whether the network shown in Figure 8.14-(a) is observable or not. Repeat the analysis when p_2 is measured.

Initial situation

To begin with, it is very easy to conclude that, when the bus/branch model is adopted, the resulting 3-bus network is observable by means of the two power flow measurements, for which a single reference angle is required. The status of CBs cannot be checked, however, with this model.

The full topological model would require 6 power flow variables plus 8 nodal phase angles. Instead, the implicit model based on the proper tree of Figure 8.14-(b) will be adopted. Note that the only additional variable retained in the model is the injection at bus 2 (tree link). The Jacobian corresponding to the DC state estimator, assuming unity reactances, is given in this case by:

$$\begin{array}{c} \\ p_{75}^m \\ p_{86}^m \end{array} \begin{array}{c} \begin{array}{cccc} \theta_1 & \theta_7 & \theta_8 & p_2 \end{array} \\ \left[\begin{array}{cccc} -1 & 1 & 0 & 0 \\ -1 & 0 & 1 & 0 \end{array} \right] \end{array}$$

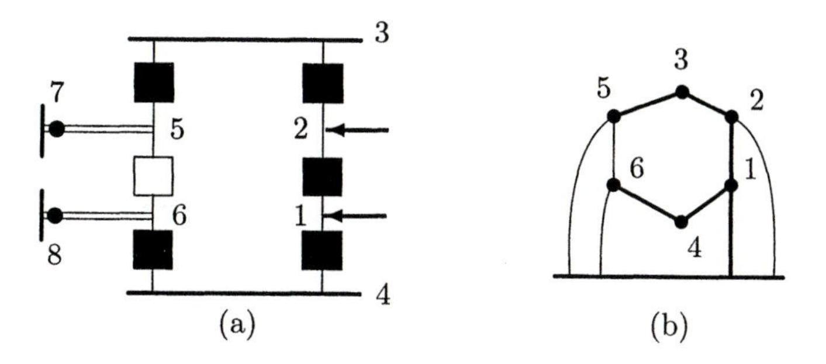

Figure 8.14. 6-CB substation (a) and associated proper tree (b) to illustrate generalized observability concepts

Clearly, p_2 is not observable (zero pivot), which means that p_1 and p_{12}, expressed as:

$$-p_1 = p_2 + p_{86} + p_{75} \quad ; \quad -p_{12} = p_2 + p_{75}$$

are not observable either. Hence, in spite of a single reference angle being required, there are two observable islands, given by branches {5-7, 3-5, 2-3} and {6-8, 4-6, 1-4} respectively.

Injection measurement added

When p_2^m is added, the Jacobian corresponding to the implicit model becomes:

$$\begin{array}{c} \\ p_{75}^m \\ p_{86}^m \\ p_2^m \end{array} \begin{array}{c} \begin{array}{cccc} \theta_1 & \theta_7 & \theta_8 & p_2 \end{array} \\ \left[\begin{array}{cccc} -1 & 1 & 0 & 0 \\ -1 & 0 & 1 & 0 \\ 0 & 0 & 0 & 1 \end{array} \right] \end{array}$$

which is of full rank so long as an angle pseudomeasurement is added. Therefore, the whole network is observable.

The same conclusions would have been reached if the full substation model had been adopted.

Example 8.15:

Perform generalized observability analysis for the network of Figure 8.15-(a), where the status of CB 1-2 is unknown.

From the perspective of a conventional state estimator, a decision must be first taken about the status of CB 1-2. Assume that, based for instance on measurement values, it is decided that CB 1-2 is open. Then, two observable islands exist, each one matching its corresponding topological island.

When the implicit model developed in the former section is employed, the

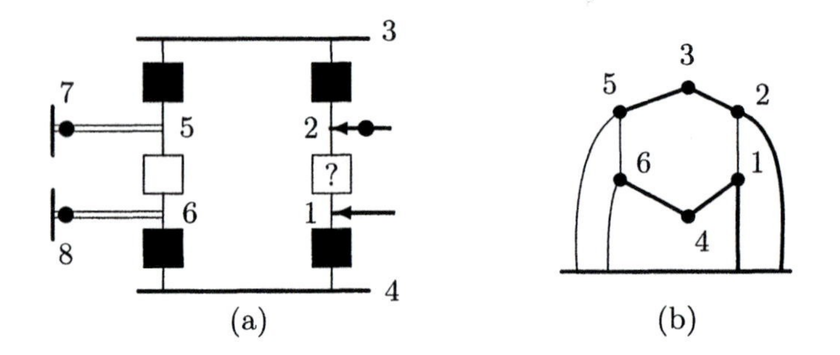

Figure 8.15. 6-CB substation (a) and associated proper tree (b) to illustrate an observable network requiring two reference angles

proper tree shown in Figure 8.15-(b) results, which leads to the following Jacobian:

$$\begin{array}{c} \\ p_{75}^m \\ p_{86}^m \\ p_2^m \end{array} \begin{array}{c} \begin{array}{ccccc} \theta_1 & \theta_2 & \theta_7 & \theta_8 & p_{12} \end{array} \\ \left[\begin{array}{ccccc} 0 & -1 & 1 & 0 & 0 \\ -1 & 0 & 0 & 1 & 0 \\ 0 & 1 & -1 & 0 & -1 \end{array} \right] \end{array}$$

This time, p_{12} is observable, and hence the entire network, for which two reference angles must be chosen. Note that there is no need to decide in advance whether one or two topological islands exist.

8.11 Problems

1. Consider the 4-bus system of Example 8.1, Figure 8.2-(a). Assume that line 1-2, actually closed, is erroneously assumed open (exclusion error). Ignoring the power flow measurements corresponding to this branch, carry out the WLS estimation followed by a conventional bad data analysis. Are the conclusions reached for the inclusion error studied in Example 8.1 valid in this case?

2. Repeat the previous analysis, taking into account this time the power flow measurements at branch 1-2. Compare the results obtained in both cases, paying special attention to the measurements with largest residual.

3. In the network corresponding to Example 8.2, Figure 8.3-(a), the topology processor is misled by wrong field information, concluding that bus 3 is composed of two bus sections (bus split error). Branch 1-3, along with 50% of the net power injection is incident to one

of the sections, while the other half of the injected power plus the remaining branches give rise to the second section. Is this topology error detectable? Why?

4. Consider again the 4-bus system of Example 8.1. Build the sensitivity matrix that relates normalized residuals with power flows, as defined by (8.6). Then, obtain the correlation index given by (8.9) for each branch. Is the inclusion error properly identified?

5. In the 3-bus network corresponding to Example 8.4, Figure 8.5, the active and reactive power injection measurements at bus 2 are successively decreased from their original values in steps of 10%. This leads eventually to a pair of bad data along with the unknown status of branch 1-2. Compare the results of running the WLS estimator with and without the quadratic constraint (8.11) added, in terms of number of iterations, accuracy of estimated values, particularly the branch status, and capability to detect the bad data. Note that, when the constraint is missing, the branch status becomes ambiguous (value of k close to 0.5) in the presence of the bad injections, and also that bad data take longer to be flagged.

6. Consider the system of Figure 8.10 analyzed in Example 8.7. The following sets of Gaussian measurements are generated which are compatible with the CB '2a-2b' being open:

p_{13}	q_{13}	p_{2b3}	q_{2b3}	p_{2a}	q_{2a}	p_{2b}	q_{2b}
1.305	0.646	0.198	0.0862	−0.807	−0.351	0.234	0.0756

V_1	V_{2a}	V_{2b}
1.0144	1.0023	0.9754

For the same impedance values and weights adopted in Example 8.4, perform the WLS estimation with and without the pair of constraints (8.40) added, and compare the power flow through the CB in both cases.

7. Repeat the analysis of Examples 8.10 and 8.11 when the CB 1-4 in the substation of Figure 8.8 is open. Compare the size and structure of the full, bus/branch and implicit models.

8. Repeat the analysis of Examples 8.10 and 8.11 when the status of the CB 2-6 in the substation of Figure 8.8 is unknown. Compare again the different models.

9. Confirm that the same conclusions of Examples 8.14 and 8.15 are reached by using the full rather than the implicit substation model.

10. Assume that, in the substation of Example 8.14, the CB 1-2 is open and p_2^m is not available. Are all power flows observable? How many phase angle references are needed?

References

[1] Abur A., Kim H., Çelik M., "Identifying the Unknown Circuit Breaker Statuses in Power Networks". *IEEE Trans. on Power Systems*, Vol. 10(4), pp. 2029-2037, November 1995.

[2] Alsac O., Vempati N., Stott B., Monticelli A., "Generalized State Estimation". *IEEE Transactions on Power Systems*, Vol. 13, No. 3, pp. 1069-1075, August 1998.

[3] Amerongen R.A.M., "On the exact incorporation of virtual measurements in orthogonal transformation based state-estimation procedures". *International Journal of Electrical Power and Energy Systems*, Vol. 13, No. 3, pp. 167-174, June 1991.

[4] Bonanomi P., Gramberg G., "Power System Data Validation and State Calculation by Network Search Techniques", *IEEE Transactions on Power Apparatus and Systems*, Vol. PAS-102,No. 1, pp.238-249, January 1983.

[5] Chua L.O., Desoer C.A., Kuh E.S., *Linear and Nonlinear Circuits*, McGraw-Hill, New York, 1987.

[6] Clements K.A., Davis P. W., "Detection and Identification of Topology Errors in Electric Power Systems", *IEEE Transactions on Power Systems*, Vol. 3, No. 4, pp. 1748-1753, November 1988.

[7] Clements K.A., Simões-Costa A., "Topology Error Identification Using Normalized Lagrange Multipliers", *IEEE Transactions on Power Systems*, Vol. 13(2), pp. 347-353, May 1998.

[8] Costa, A.S.; Lourenco, E.M.; Clements, K.A., "Power System Topological Observability Analysis Including Switching Branches", *IEEE Transactions on Power Systems*, Vol. 17(2), pp. 250-256, May 2002.

[9] Gjelsvik A., Aam S., "The significance of the Lagrange Multipliers in WLS State Estimation with Equality Constraints". *Proceedings of the 11th Power Systems Computation Conference*, pp. 619-625, Avignon, Aug-Sep 1993.

[10] Gómez Expósito A., Villa A., "Reduced Substation Models for Generalized State Estimation", *IEEE Transactions on Power Systems*, Vol. 16(4), pp. 839-846, Nov. 2001.

[11] Gómez Expósito A., Zarco P., Quintana V. H., "Accurate estimation of circuit breaker positions". *Large Engineering Systems Conference on Power Engineering*, Halifax, June 1998.

[12] Irving M.R., Sterling M.J., "Substation Data Validation", *Proceedings of IEE*, Vol. 129, Part C, No. 3, pp.119-122, May 1982.

[13] Koglin H. J., Oeding D., Schmitt K. D., "Identification of Topological errors in State Estimation", *IEE International Conference on Power System Monitoring and Control*. Durham, pp. 140-144, 1986.

[14] Koglin H. J., Neisius H. T., "Treatment of Topological Errors in Substations", *Proc. 10th Power Systems Computation Conference*. Butterworth, Graz, pp. 1045-1053, 1990.

[15] Koglin H. J., Neisius H. T., "A Topology Processor Based on State Estimation". *Proceedings of the 11th Power Systems Computation Conference*, pp. 633-638, Avignon, Aug-Sep 1993.

[16] Lugtu R. L., Hackett D. F., Liu K. C., Might D. D., "Power System State Estimation: Detection of Topological Errors". *IEEE Trans. on Power Apparatus and Systems*, Vol. PAS-99, pp. 2406-2412, Nov/Dec 1980.

[17] Mili L., Steeno G., Dobraca F., French D. "A Robust Estimation Method for Topology Error Identification". *IEEE Trans. on Power Systems*, Vol. 14, No. 4, pp. 1469-1476, Nov 1999.

[18] Monticelli A., "Modeling Circuit Breakers in Weighted Least Squares State Estimation", *IEEE Transactions on Power Systems*, Vol. 8(3), pp. 1143-1149, August 1993.

[19] Monticelli A., "The impact of modelling short circuit branches in State Estimation", *IEEE Trans. on Power System*, Vol. 8, No.1, pp. 364-370, Feb 1993.

[20] Monticelli A., "Testing Equality Constraint Hypotheses in Weighted Least Squares State Estimators", *PICA 99*, Santa Clara, 1999.

[21] Monticelli A., "State Estimation in Electric Power Systems. A generalized Approach". Kluwer Academic Publishers, 1999.

[22] Monticelli A., Garcia A., "Modeling Zero Impedance Branches in Power Systems State Estimation", *IEEE Transactions on Power Systems*, Vol. 6(4), pp. 1561-1570, November 1991.

[23] Prais M., Bose A., "A Topology Processor that tracks Network modifications over time". *IEEE Trans. on Power System*, Vol. 3, No.3 pp. 992-998, Aug. 1988.

[24] Sasson A.M., Ehrmann S.T., Lynch P., Van Slyck L.S., "Automatic Power System Network Topology Determination". *IEEE Trans. on Power Apparatus and System*, Vol. PAS-92, pp. 610-618, March/April 1973.

[25] Simões-Costa A., Leao J. A., "Identification of Topology Errors in Power System State Estimation". *IEEE Transactions on Power Systems*, Vol. 8, pp. 1531-1538, Nov. 1993.

[26] Singh N., Glavitsch H., "Detection and Identification of Topological Errors in Online Power System Analysis", *IEEE Transactions on Power Systems*, Vol. 6(1), pp. 324-331, February 1991.

[27] Singh N., Oesch F., "Practical Experience with Rule-Based On-Line Topology Error Detection", *IEEE Transactions on Power Systems*, Vol. 9(2), pp. 841-847, May 1994.

[28] Singh H., Alvarado F.L., "Network Topology Determination using Least Absolute Value State Estimation", *IEEE Transactions on Power Systems*, Vol. 10, No. 3 , pp. 1159-1165, Aug. 1995.

[29] Slutsker I., Mokhtari S., "Comprehensive Estimation in Power Systems: State, Topology, and Parameter Estimation". *Proceedings of the American Power Conference*, Vol. 57-I, pp. 149-155, Chicago, Illinois, April 1995.

[30] Souza J.C.S., Leite da Silva A.M., Alves da Silva A.P., "On line topology determination and bad data suppression in power system operation using artificial neural networks". *IEEE Transactions on Power Systems*, Vol. 13, No. 3, pp. 796-803, Aug. 1998.

[31] Wu F.F., Liu W.H.E., "Detection of of topology errors by State Estimation". *IEEE Transactions on Power Systems*, Vol. 4, pp. 176-183, Aug. 1989.

[32] Villa A., Gómez Expósito A., "Modeling Unknown Circuit Breakers In Generalized State Estimators". *Power Tech 2001*, Oporto, Portugal, September 2001, paper 221.

[33] Villa A., Gómez Expósito A., "Implicitly Constrained Substation Model for State Estimation", *IEEE Transactions on Power Systems*, Vol. 17(3), pp. 850-856, Aug. 2002.

[34] http://www.ee.washington.edu/research/pstca/

Chapter 9

State Estimation Using Ampere Measurements

9.1 Introduction

In the preceding chapters, line current magnitude (ampere) measurements are intentionally left out of the presented analysis. Despite their widespread availability in substations and their utilization for protective relaying functions, ampere measurements are seldom telemetered to the bulk transmission system's control center. Instead, utilities prefer to invest in a large number of transducers that supply real and reactive power measurements to the substation RTUs. This is certainly justifiable given the importance of reliability in monitoring the operation of the high voltage system.

Recent developments in the power industry as a result of the electric business deregulation initiated the separation of the formerly integrated generation, transmission and distribution activities. This has led to the creation of several electric utilities that essentially plan and operate regional distribution networks, while the so-called Independent System Operator (ISO) manages the operation of the transmission network. Efficient management of such distribution networks calls for improved monitoring capabilities, where ampere measurements can play a vital role as will be discussed in this chapter.

In this context, the notion of "distribution network" comprises not only the radially operated low and medium voltage feeders, but also those subtransmission systems with voltage levels below 138-kV or 220-kV, which are typically not managed by the ISO. Higher quality-of-service standards, as well as legal requirements for open access to the wires by any market partner, are challenging the way such distribution networks have been traditionally operated. In many cases, simple SCADA systems that report directly to the main EMS, are installed by these utilities. Their main responsibilities have to do with component loading and bus voltage monitoring, organization of maintenance tasks, prompt service restoration after a blackout, etc. Taking advantage of advances in computers and communications, such local and modest systems are nowadays being converted into true distribution management systems (DMS), in much the same way the advanced energy management systems emerged in the late sixties and seventies. Needless to say, this upgrading relies strongly on the development of a state estimator, which should be designed taking the following characteristics of distribution networks into consideration:

1. High R/X ratios, sometimes well above unity.
2. Very limited measurement set. Power measurements are usually restricted to 132-kV or 138-kV and above. On the other hand, many ampere measurements are available at the lower voltage levels. These are cheaper to install and facilitate the checking for line overloads by the system operator.
3. Small number of loops compared to transmission networks. In order to reduce short-circuit currents, sub-transmission systems (50-132-kV, 34.5-138-kV) are usually operated through separated ring structures. Lower voltage distribution systems are radially operated for the sake of simplicity and economy.
4. No major generators, if any, connected to networks under 132-kV or 138-kV (eolic farms and co-generation systems are becoming an important exception to this rule). Each sub-transmission loop is typically supplied power from two transmission buses through appropriate transformers, in order to increase system reliability.
5. Line susceptances are almost negligible.

The first item precludes the application of the decoupling principle. Furthermore, as will be shown below, ampere measurements cannot be clearly coupled to any of the resulting subproblems. Therefore, decoupled state estimators will not be addressed in this chapter.

The second item implies a reduced redundancy and hence filtering capability, as many pairs of power measurements are replaced by single current measurements. This redundancy deterioration is aggravated by the third feature, as the branch-to-node ratio approaches the unity for nearly radial networks.

The third and fourth features are sometimes helpful in the process of determining power flow directions in the presence of ampere measurements. This is the case, for instance, of active power flows in radial networks fed from a single point (more caution should be exercised in general regarding reactive power flow directions).

The last item is not so relevant, but in the presence of current measurements it tends to deteriorate the Jacobian condition number.

This chapter describes the main difficulties associated with the use of ampere measurements in state estimators. Then, it presents various techniques that have been developed in the last decade in order to overcome these problems. The notion of observable network will be reconsidered and extended to cope with new situations arising in the presence of current measurements.

9.2 Modeling of Ampere Measurements

For a branch connecting nodes i and j, the following equation relates its current magnitude to the state variables [19]:

$$I_{ij} = \left[AV_i^2 + BV_j^2 - 2V_iV_j(C\cos\theta_{ij} - D\sin\theta_{ij})\right]^{1/2} \tag{9.1}$$

where the following coefficients have been defined:

$$\begin{aligned} A &= g_{ij}^2 + (b_{ij} + b_{sh})^2 \\ B &= g_{ij}^2 + b_{ij}^2 \\ C &= g_{ij}^2 + b_{ij}(b_{ij} + b_{sh}) \\ D &= g_{ij}b_{sh} \\ g_{ij},\, -b_{ij} &: \text{ series conductance \& susceptance} \\ b_{sh} &: \text{ 1/2 charging susceptance} \end{aligned}$$

When the line charging susceptance b_{sh} is neglected, the simplified expression provided in Chapter 2 is obtained:

$$I_{ij} = \left[(g_{ij}^2 + b_{ij}^2)(V_i^2 + V_j^2 - 2V_iV_j\cos\theta_{ij})\right]^{1/2} \tag{9.2}$$

Note that the "sine" term in (9.1) vanishes, which is one of the main sources of trouble when using ampere measurements.

Figure 9.1 represents both I_{ij} and I_{ij}^2 as a function of V_i and θ_i for $V_j = 1$ and $\theta_j = 0$ (assuming unity admittance).

From the above expression it is easy to obtain the following Jacobian entries for the state variables of bus i (those of bus j are obtained by simply

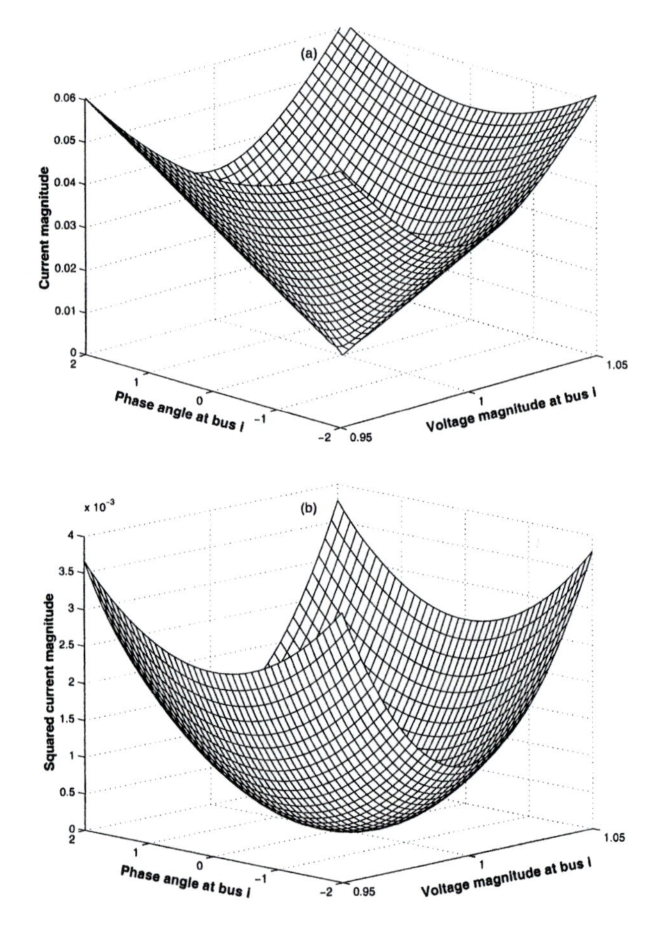

Figure 9.1. Current magnitude (a) and squared current magnitude (b)

exchanging subscripts i and j below):

$$\begin{aligned}
\frac{\partial I_{ij}}{\partial V_i} &= \frac{g_{ij}^2 + b_{ij}^2}{I_{ij}}(V_i - V_j \cos\theta_{ij}) \\
\frac{\partial I_{ij}}{\partial \theta_i} &= \frac{g_{ij}^2 + b_{ij}^2}{I_{ij}} V_i V_j \sin\theta_{ij}
\end{aligned} \tag{9.3}$$

Figure 9.2 shows the derivatives of I_{ij} as a function of V_i and θ_i for $V_j = 1$ and $\theta_j = 0$ (assuming unity admittance). Note that the smaller the difference $V_i - V_j$ the steeper the resulting sigmoid-like and impulse-like functions. In the limit ($V_i \approx V_j$, $\theta_{ij} \approx 0$), both derivatives will be undefined, as evident also from Figure 9.1.

It is also possible for the squared current magnitude measurement, I_{ij}^2, to be included in the SE model, in which case the following terms are of interest:

$$\begin{aligned}
\frac{\partial I_{ij}^2}{\partial V_i} &= 2(g_{ij}^2 + b_{ij}^2)(V_i - V_j \cos\theta_{ij}) \\
\frac{\partial I_{ij}^2}{\partial \theta_i} &= 2(g_{ij}^2 + b_{ij}^2) V_i V_j \sin\theta_{ij}
\end{aligned} \tag{9.4}$$

Note that if I_{ij}^2 is adopted instead of I_{ij}, the measurement covariance will double.

Figure 9.3 is the counterpart of Figure 9.2 for I_{ij}^2. The resulting functions are much smoother in this case and can be approximated, around flat start, as follows:

$$\begin{aligned}
\frac{\partial I_{ij}^2}{\partial V_i} &\approx 2(g_{ij}^2 + b_{ij}^2)(V_i - V_j) \\
\frac{\partial I_{ij}^2}{\partial \theta_i} &\approx 2(g_{ij}^2 + b_{ij}^2)\theta_{ij}
\end{aligned}$$

Therefore, both derivatives tend to zero when $V_i \approx V_j$ and $\theta_{ij} \approx 0$.

It is apparent from those figures that ampere measurements can not be coupled *a priori* with either the active or reactive subproblem. Current measurements will be coupled with the active (or reactive) component only if θ_{ij} (or $V_i - V_j$) is large enough. Hence, for heavily loaded lines, current measurements tend to couple both problems.

Note that the net current injection at a bus is frequently the sum of several external contributions but current magnitudes can not be algebraically added into a single measurement. Hence, the possibility of having current injection measurements will be neglected throughout this chapter. However, should such a measurement exist, it could be also included in the estimation process, much in the same way and with the same difficulties as for the branch current magnitude measurements.

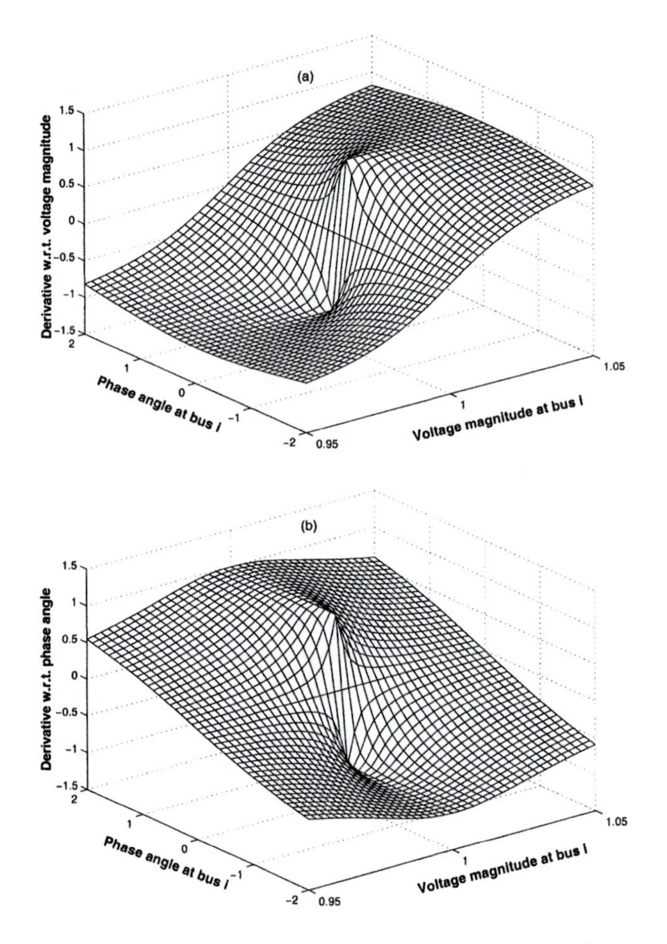

Figure 9.2. Derivatives of current magnitude with respect to voltage magnitude (a) and phase angle (b)

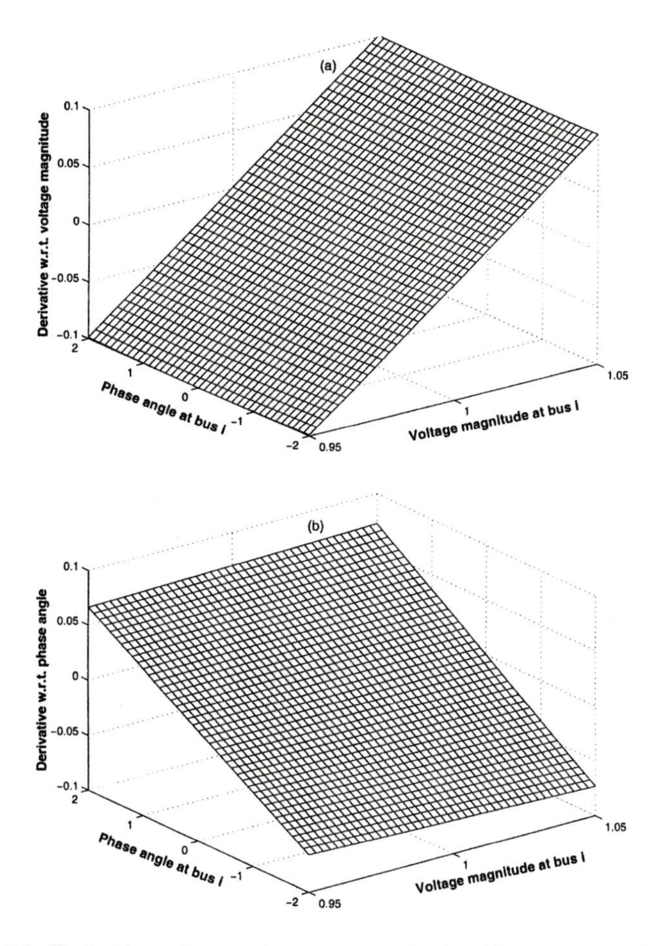

Figure 9.3. Derivatives of squared current magnitude with respect to voltage magnitude (a) and phase angle (b)

9.3 Difficulties in Using Ampere Measurements

The use of ampere measurements will lead to various problems, which in turn may seriously deteriorate the performance of the state estimators. We will first present the problems and then develop methods to alleviate them. The following problems related to numerical and/or observability issues are likely to arise:

- For flat start, Jacobian elements are undefined if the raw value I_{ij} is used (Figure 9.2), or null when I_{ij}^2 is adopted (Figure 9.3), which means that current measurements are useless in this situation. Consequently, any observability analysis in the presence of such measurements should be based on a Jacobian computed at a different point. The following solutions have been suggested to circumvent this problem:

  - Add artificial shunt elements which are removed after the first iteration.
  - Initialize state variables with a random small perturbation.

  Both schemes perform reasonably well provided the current actually flowing through the line exceeds a certain threshold. Otherwise, i.e., when the line loading is negligible, the estimated value $\hat{\theta}_{ij}$ approaches zero and the Jacobian becomes ill-conditioned if the ampere measurement is needed to assure observability. It is better in such cases to substitute the measurement $I_{ij} \approx 0$ by $P_{ij} \approx 0$.

- Abrupt changes around the origin of these Jacobian terms when I_{ij} is used (Figure 9.2), due to the strong non-linearity inherent to (9.1). This dependence of Jacobian terms upon the state vector may cause convergence difficulties for lightly loaded lines, unless the step-length is carefully chosen during the iterative process. Note that this problem will be alleviated if I_{ij}^2 is used instead (Figure 9.3), at the expense of its deteriorated covariance.

- In the absence of power measurements, the only information about phase angles that can be obtained, according to (9.2), is:

$$\cos\theta_{ij} = \frac{V_i^2 + V_j^2 - I_{ij}^2(r_{ij}^2 + x_{ij}^2)}{2V_iV_j} \tag{9.5}$$

  Therefore, two opposite values $\pm\theta_{ij}$ are compatible with a given set of voltage and ampere magnitude measurements. This translates into two different power flow solutions, because of the presence of the "sine" term when writing the power flow equations as follows (line susceptance ignored):

$$\begin{aligned} P_{ij} &= b_{ij}V_iV_j\sin\theta_{ij} + \frac{1}{2}\left[g_{ij}(V_i^2 - V_j^2) - I_{ij}^2 r_{ij}\right] \\ Q_{ij} &= \frac{1}{2}\left[b_{ij}(V_i{}^2 - V_j{}^2) - I_{ij}{}^2 x_{ij}\right] - g_{ij}V_iV_j\sin\theta_{ij} \end{aligned} \tag{9.6}$$

For a lossless line ($r_{ij} = g_{ij} = 0$) a unique Q_{ij} value but two P_{ij} values with opposite signs are obtained. In subtransmission systems, however, where $r \approx x$, multiple and different solutions are expected both for the active and reactive power flows.

The multiplicity of solutions issue is not exclusively due to ampere measurements, but may arise also in the presence of power measurements. Assuming the values of V_i and θ_i are given, Table 9.1 shows the number of possible solutions for V_j and θ_j when different measurement pairs are available. Conventional estimators ignore the possibility of having multiple solutions by assuming that power measurements come in pairs.

Table 9.1. Number of possible solutions for different measurement combinations

Measured magnitudes	No. of solutions	
	θ_j	V_j
P_{ij}-Q_{ij}	1	1
P_{ij}-V_j	1	1
Q_{ij}-V_j	2	1
I_{ij}-V_j	2	1
P_{ij}-I_{ij}	2	2
Q_{ij}-I_{ij}	2	2

The following example illustrates the multiplicity of solutions issue discussed above. In order to simplify the analysis it will be assumed throughout this chapter, unless otherwise noticed, that all measurements are exact.

Example 9.1:

Consider the three-bus system of Figure 9.4 comprising six measurements. The corresponding phasor diagram is shown in Figure 9.5. Arbitrarily assuming the phase angle of bus 1 voltage as zero, the complex voltage at bus 2 can be determined either by the pair P_{12}-Q_{12} or P_{12}-V_2. On the other hand, the voltage phasor at bus 3 must lie at the intersection of two circles, one centered at "P"

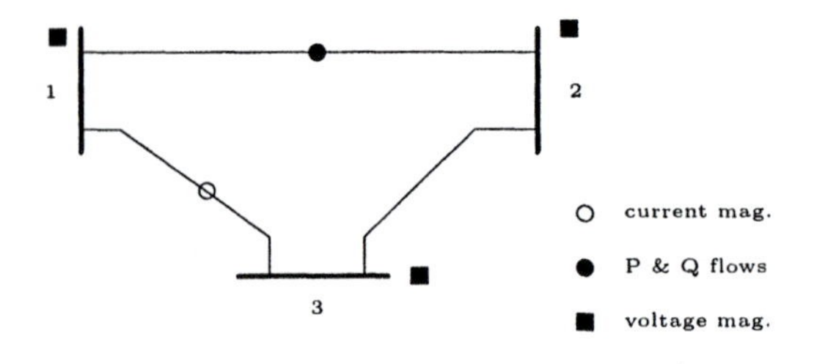

Figure 9.4. 3-bus system with six measurements

with radius V_3, the other centered at "Q" with radius $I_{13}z_{13}$. Consequently, there are two possible solutions for V_3, marked by $V_3^{(1)}$ and $V_3^{(2)}$. Provided that the linearization is not performed at flat start, it is easy to show that the column rank of the measurement Jacobian for this example is 5, which implies observability in the conventional sense. However, as apparent from the diagram, the solution is not unique.

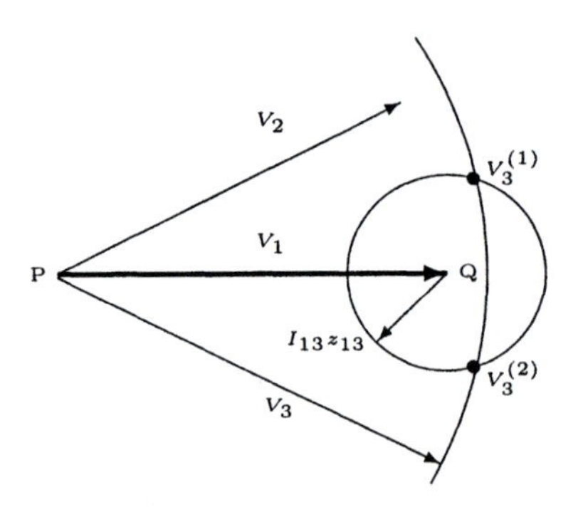

Figure 9.5. Phasor diagram illustrating the two possible solutions for the system of Figure 9.4

Of all the inconveniences cited above, the possibility of multiple solutions is by far the most important, as it affects several well-established notions in conventional state estimators, like that of observability.

Observability is defined as the ability to uniquely estimate the state of the system using the given measurements. The commonly used observability algorithms [18, 15] are based on the assumption that if a solution can be found to the state estimation problem, then it will be unique. This is a valid assumption as long as the measurements come in real and reactive pairs and the observability tests can be done using decoupled models. As shown above, the natural decoupling between the real and reactive measurements is lost when current magnitude measurements are included in the measurement set. But the coupled model no longer guarantees uniqueness of the solution, even if a solution can be reached numerically. That is, the column rank of the measurement Jacobian H, may remain full throughout the iterative solution steps, indicating a *conventionally* observable or solvable system, even though it may have multiple possible solutions. Hence, when using current magnitude measurements, a full rank Jacobian does not necessarily imply a *uniquely* observable system.

A system is said to be non-uniquely observable if more than one state can be found based on the given set of measurements. Accordingly, those branches whose power flows can assume more than one value that satisfies all the system measurements will be labeled as non-uniquely observable branches. On the other hand, uniquely observable systems will have a unique state and a unique set of branch power flows through every branch. State estimation will yield the same solution, irrespective of the starting point, for uniquely observable systems. However, it may converge arbitrarily to any one of the possible solutions, based on the starting point, for systems with multiple solutions.

Besides the notion of unique solvability, there is also a need to reconsider the way existing measurements are classified as critical or redundant, as well as the capability of a state estimator to detect bad data in the presence of current measurements. The rest of the chapter will be focused on these issues.

9.4 Inequality-Constrained State Estimation

Previous work on the use of ampere measurements in WLS state estimators is restricted to cases where the system is already observable without these measurements. So, the main purpose of including them is to improve accuracy rather than extending the observable network. The results of simulations where different types of measurements are added to a critical set of measurements are presented in [20]. It is demonstrated that for the chosen performance index, line currents are comparable to active power flows and better than reactive power flows. In [12] a comparison is made between the inclusion of voltage magnitudes and line currents. It is concluded that improved estimation accuracy is provided by the voltage set. Reference [14] mentions the risk of overflow for flat start and the lack of directional information as the two major problems posed by the inclusion of line-current measurements. The conclusion is that such measurements should not be considered to extend the observable network. Reference [16] is the first work specifically devoted to the topic of using line currents in state estimation.

Assuming the system is already observable by other types of measurements, inclusion of line-current measurements to improve redundancy (i.e., accuracy), especially for the $P-\theta$ subproblem, is recommended. The use of the squared value of current magnitudes in order to avoid the overflow problem is suggested in [9]. Furthermore, for those lines in which the direction of power flow is known *a priori*, it is suggested that current measurements can be safely used for extending observability. This is frequently true in radial networks, but seldom so in general meshed networks where power flow directions may change arbitrarily during daily operation.

The first attempt to systematically deal with the multiplicity of solutions issue in the presence of ampere measurements is presented in [19]. The basic idea is that many, if not all, of the potential solutions that may be reached when ampere measurements are included, can be discarded by checking the net power injected at those buses for which its sign is known *a priori*. In high-voltage distribution networks, active power is *always* delivered to low voltage buses via distribution transformers. On the other hand, active power taken from the transmission level or small generators is injected into a few buses of the distribution network. Consequently, it is almost always possible to assert whether the specified real power for each node is less, greater than or equal to zero (due to the presence of capacitor banks at lower levels, this is not so clear in certain cases for the reactive power). This knowledge can be incorporated into the estimation process in the form of *inequality constraints*. The estimation problem thus becomes that of minimizing a non-linear function subject to non-linear equality and inequality constraints. Mathematically stated:

$$\begin{array}{lrcl} \text{Minimize} & J(x) & = & [z-h(x)]^t W [z-h(x)] \\ \text{Subject to} & c_i(x) & \geq & 0, \quad i=1,2,\cdots,N \end{array} \tag{9.7}$$

This model is a natural extension of the equality-constrained state estimator discussed in Chapter 3, and can be used for other purposes as well in conventional estimators, like forcing certain magnitudes of the external equivalent to lie within an acceptable range, etc. Numerical solution of the above model is obviously more complex, and the use of interior-point techniques is strongly recommended [13].

A theoretical analysis is presented in [19] showing that the number of inequality constraints strictly required in practice to assure uniqueness of solution can be rather low, particularly for meshed networks, provided all voltage and line-current magnitude measurements are available. The worst case arises in radial networks, where all inequality constraints are needed.

Example 9.2:

Assume a single loop like the one shown in Figure 9.6, where all branch resistances are negligible and all reactances are identical. Besides, it is assumed that all voltage magnitudes are 1 pu and all branch current magnitudes are measured.

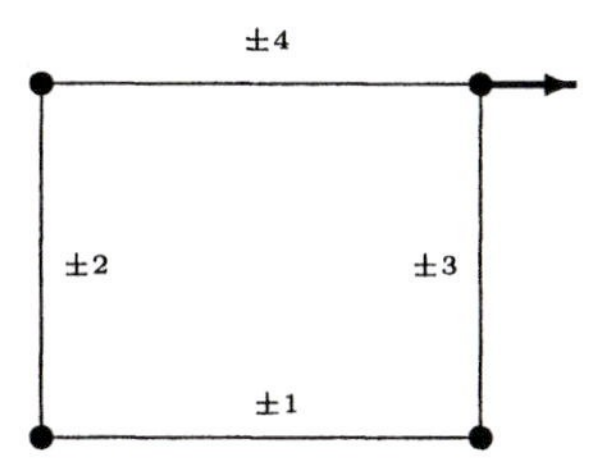

Figure 9.6. 4-bus ring with two possible solutions

From the available measurements, the absolute values of all branch phase angles can be obtained from (9.5):

$$|\theta_{ij}| = \arccos\left\{1 - \frac{1}{2} I_{ij}^2 x_{ij}^2\right\}$$

The resulting values are shown next to the corresponding branches in Figure 9.6. A simple analysis of those values indicates that only the following two combinations of signs are possible for Kirchhoff's voltage law to be satisfied:

$$1 - 2 - 3 + 4 = 0 \qquad ; \qquad -1 + 2 + 3 - 4 = 0$$

But, according to (9.6), it is also clear that those solutions lead to opposite active power flows and, consequently, opposite power injections. Therefore, a single inequality constraint (marked up with an arrow in Figure 9.6) would suffice in theory to force the estimator to converge to the right solution.

In certain cases, however, numerical coincidences may lead to a larger number of feasible combinations, particularly when measurement noise is taken into account.

Example 9.3:

Consider the 5-bus ring shown in Figure 9.7, where two phase angles are, by chance, very similar (the two respective branches might correspond, for instance, to identical transformers approximately sharing one half of the total load). Note that, this time, because of the redundancy and the presence of noise, the phase angles provided *a priori* by (9.5) do not exactly obey Kirchhoff's voltage law (of course, the estimated values will do!).

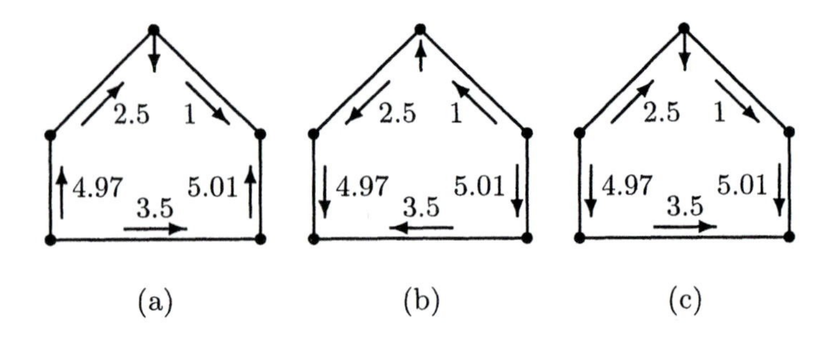

Figure 9.7. 5-bus ring with more than two solutions

In this case, a single inequality constraint at the top corner bus is able to discriminate between the two opposite solutions (a) and (b) but not between (a) and (c). Hence, at least two bus constraints are needed (the reader is referred to the end-of-chapter Problem 2 to continue the analysis).

Unfortunately, even if *all* inequality constraints are added to the model, there is no guarantee that a unique solution exists when the measurement set is exclusively composed of voltage and current magnitudes (i.e., unsigned measurements). This happens for instance in radial networks which are fed from several buses and in meshed networks with incomplete measurement sets, as illustrated by the following examples.

Example 9.4:

Figure 9.8(a) shows a 5-bus radial network obtained by unfolding the loop of Example 9.2. The arrows at each bus indicate the sign of the respective power injection.

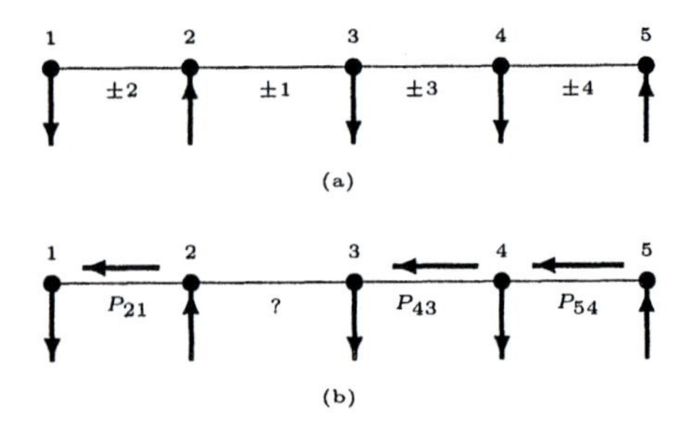

Figure 9.8. Radial network with two solutions

Because of the constraints at buses 1 and 5, the signs of power flows P_{21} and P_{54} can be determined. Also, considering the relative values of $|\theta_{23}|$ and $|\theta_{43}|$ and the constraint at bus 3, it can be also concluded that P_{43} is positive. However, as suggested by the question mark in Figure 9.8(b), the sign of P_{23} can not be safely ascertained, in spite of all constraints being available. Note that, unlike in Example 9.2, the measurement set is critical in this case, because of the extra bus. Therefore, not only the measurement set, but also the topology is relevant when analyzing the solution uniqueness.

Example 9.5:

Consider the system shown in Figure 9.9. Assume that all bus voltage magnitudes (all equal to 1.0) and the line currents I_1 and I_2 are measured and $I_1 > I_2$. Further assume that bus 2 is a load bus ($P_2 < 0$) and bus 3 is a generator bus ($P_3 > 0$). The possible solutions for θ_2 and θ_3 are shown in Figure 9.10. It can be shown that the choice of $V_2^{(1)}$ will imply $P_{12} > 0$ and $P_{23} < 0$, yielding $P_2 < 0$ which is consistent with the inequality constraint on the net power injection at this load bus. On the other hand, the choice of $V_2^{(2)}$ will imply $P_{12} < 0$ and $P_{23} > 0$, yielding $P_2 > 0$ which contradicts the inequality constraint. Thus, $V_2^{(1)}$ is chosen as the solution. However, for this chosen solution, one can choose either one of $V_3^{(1)}$ or $V_3^{(2)}$, since both voltages can yield solutions satisfying the inequality constraint $P_3 > 0$ at the generator bus 3. So, despite the use of inequality constraints on the bus injections, possibility of multiple (in this case two) solutions can not be completely avoided.

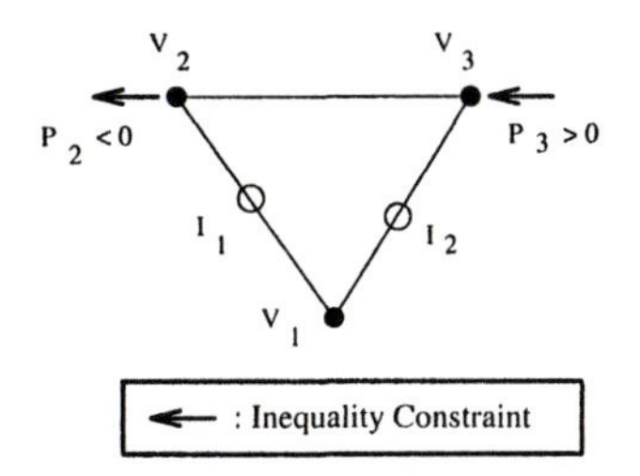

Figure 9.9. System with two possible solutions

However, contrary to what the pioneers in this research subject area initially believed, ampere measurements can be useful, not only to increase the redundancy, but also to extend the uniquely observable network when the measurement set contains certain strategically located power measurements, without having to enforce inequality constraints. This will be illustrated by the following example.

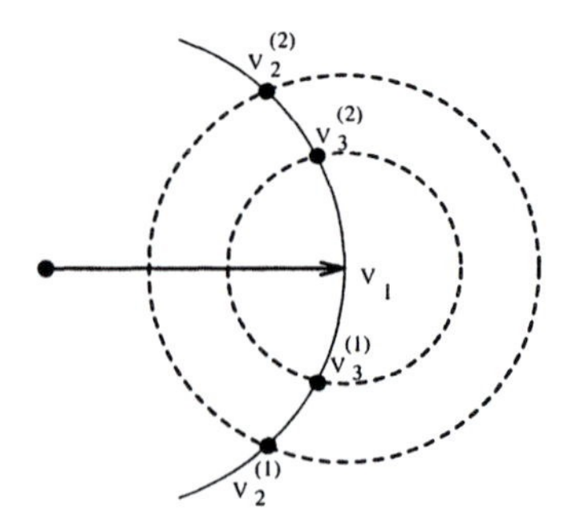

Figure 9.10. Phasor diagram for possible solutions

Example 9.6:

In the system of Figure 9.11, comprising 4 buses and 8 measurements, branches 1-4 and 4-3 are fully observable by means of the respective power flow measurements.

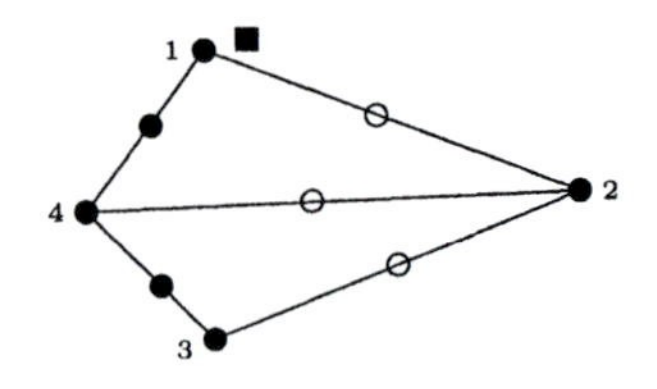

Figure 9.11. 4-bus system with a unique solution

Given the voltage phasors of buses 1, 3 and 4, the voltage of bus 2 is uniquely determined by the intersection of the three circles centered at the tips of the three phasors with radii $I_{2k}z_{2k}$, for $k = 1, 3, 4$. This point is labelled as P in Figure 9.12. It should be noted that any one of the line currents is redundant. Also, note that point P becomes ill-defined as the three phasors tend to be in phase. That is, the fact that θ_{13}, θ_{14} and θ_{34} are not null is crucial for the unique observability of the voltage at bus 2.

In order to take advantage of this possibility, several *ad hoc* procedures have been developed in the last decade aimed at determining, for a given topology and combination of measurements, which branches are uniquely observable, nonuniquely observable and fully unobservable. The following two sections will be devoted to summarize these developments.

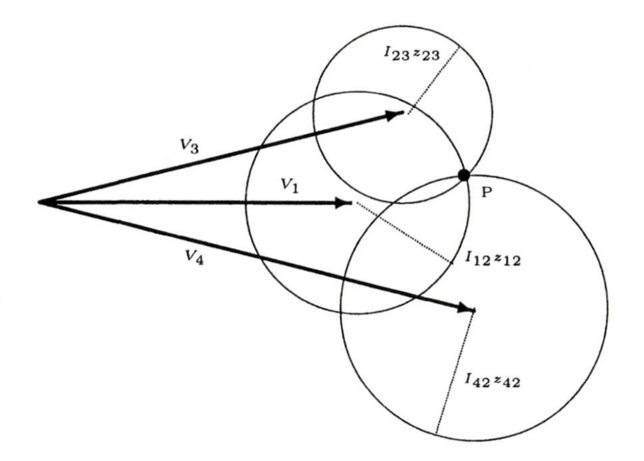

Figure 9.12. Phasor diagram for the system of figure 9.11

9.5 Heuristic Determination of P-θ Solution Uniqueness

As voltage magnitudes for nearly all buses are commonly included in most measurement sets, usually in a redundant manner, it makes sense to restrict the solution uniqueness analysis to just the P-θ subproblem by assuming that all voltage magnitudes are known.

At first glance, it seems that the massive presence of current magnitudes in the measurement set gives rise, when considered individually, to a large number of possible combinations in the signs of branch phase angles, each one leading to a different power flow solution. However, as apparent from the preceding results and discussions, Kirchhoff's laws establish additional restrictions among certain subsets of signs in such a way that only two combinations remain valid, provided that the possibilities of exact numerical cancellations are ignored. Furthermore, whenever power flow or injection measurements get involved in any of those equations, one of the two solutions will be automatically excluded. This can be formalized in the following observability rules:

Loop rule: In a given loop, if all the branches are measured (either a current magnitude or a power flow measurement), then all branches will be observable provided at least one of them has a power flow measurement (or equivalently, at least one of them is observable). If all branches of the loop carry only current measurements, then there will be two possible solutions, corresponding to the possible polarities of θ_{ij} in each branch.

Node rule: At a given injection measured node, if all the branches connected to that node are measured (either current magnitude or power flow), all connected branches will be observable. One exception for lossless branches is the zero-injection node (see the explanation in the next paragraph), in which case two solutions for the branch variables remain, unless at least one incident power flow is measured.

According to (9.6), when all branches are lossless, like in the above examples, the two opposite values of θ_{ij} lead also to identical but opposite active power flows, which means that zero-injection constraints are useless to discriminate the correct solution. However, in real distribution systems, the relatively large value of r_{ij} gives rise to probably contrary but different power flows, in which case virtual zero-injection measurements are also helpful to prevent multiplicity of solutions (there are no exceptions to the node rule).

A heuristic procedure to determine uniquely observable branches can be devised by sequentially applying the above two rules, starting with fully measured loops and nodes and progressing in a systematic manner [3]. As distribution networks are operated by splitting them into simpler subnetworks, such a procedure will almost always succeed in correctly finding the maximum observable subnetwork.

In many cases, one or several inequality constraints may be needed in order to rule out those physically infeasible solutions, in the absence of power measurements.

Example 9.7:

Consider the network and measurement set of Figure 9.13, where buses have been labeled with letters and branches with numbers.

The following independent loop equations can be written:

$$0 = \hat{\theta}_1 \mp \hat{\theta}_2 \mp \hat{\theta}_3 \mp \hat{\theta}_6 \mp \hat{\theta}_5 \mp \hat{\theta}_4 \tag{9.8}$$

$$0 = \mp\hat{\theta}_5 \mp \hat{\theta}_6 + \theta_7 + \theta_9 + \theta_8 \tag{9.9}$$

where the variables carrying the "^" symbol can be directly obtained from existing current or power flow measurements.

Also, assuming unity branch reactances, the DC power measurement model leads to:

$$P_c = \mp\hat{\theta}_3 \mp \hat{\theta}_6 + \theta_7 \tag{9.10}$$

$$P_e = \pm\hat{\theta}_4 \mp \hat{\theta}_5 - \theta_8 \tag{9.11}$$

Using the "Loop rule" in (9.8), the signs of the branch variables θ_2, θ_3, θ_4, θ_5 and θ_6 will be uniquely defined. Substituting these in (9.10) and (9.11), θ_7 and θ_8 can also be uniquely determined. Finally, (9.9) will yield the remaining unknown

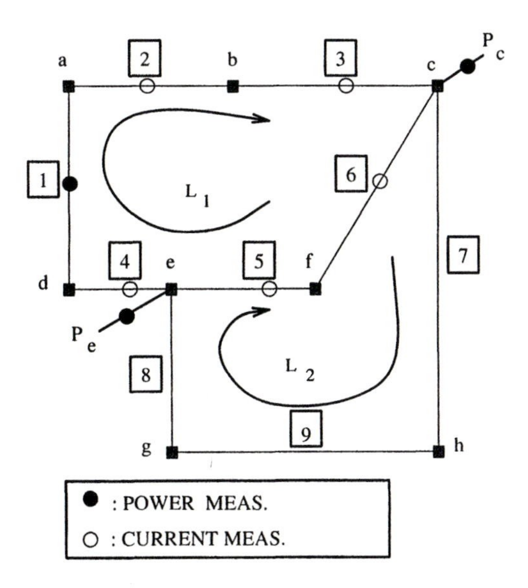

Figure 9.13. 8-bus system for Example 9.7

θ_9. Therefore, the network is uniquely observable with the given measurement configuration.

Example 9.8:

Consider the same system with a different measurement configuration, as shown in Figure 9.14.

In this case, the following 2 independent loop and 2 node equations can be written:

$$0 = \hat{\theta}_1 \pm \hat{\theta}_2 + \theta_3 + \theta_6 \pm \hat{\theta}_5 \pm \hat{\theta}_4 \tag{9.12}$$

$$0 = \mp\hat{\theta}_5 - \theta_6 \pm \hat{\theta}_7 + \theta_9 \pm \hat{\theta}_8 \tag{9.13}$$

$$P_c = -\theta_3 + \theta_6 \pm \hat{\theta}_7 \tag{9.14}$$

$$P_e = \pm\hat{\theta}_4 \mp \hat{\theta}_5 \mp \hat{\theta}_8 \tag{9.15}$$

Using the "Node rule" in (9.15), the signs of the branch variables θ_4, θ_5 and θ_8 can be uniquely determined. Substituting these in (9.12) and (9.13), multiple solutions for θ_3, θ_6 and θ_9 can be found corresponding to the possible choices of +/- signs for θ_2 and θ_7 in (9.12), (9.13) and (9.14). Thus, this network is not uniquely observable for the given measurement configuration.

Note that, in both examples above, the same number of current, power

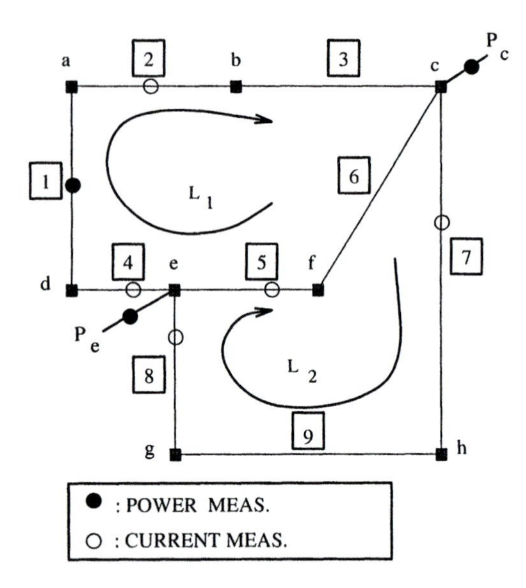

Figure 9.14. 8-bus system for Example 9.8

flow and injection measurements are available but in different configurations. While they make the network uniquely observable in the first case, they fail to uniquely define the state of the second network.

9.6 Algorithmic Determination of Solution Uniqueness

Even though the above heuristic procedures may suffice in simple, nearly radial network structures, like those usually found in systems under 132 kV, a fully automated technique, preferably based on existing numerical routines and matrices, would be most welcome for the general and more complex case.

Current magnitude measurements can be considered as measurements with possible structural errors (because of the unknown signs of the Jacobian entries and that these signs are arbitrarily set at the initial guess). If there were other measurements which were functions of the same states as these current magnitude measurements, they would force the states to move towards the unique solution even if the initial signs of the Jacobian entries for the current magnitudes were incorrect. Obviously, this would not be possible, if any of the current magnitude measurements were critical, since critical measurements would be linearly independent of the rest of the measurements. So, the existence of a critical current magnitude measurement implies that the system is not uniquely observable.

When a redundant (non critical) measurement is in error, the set of measurements that will be affected by this error, forms the *residual spread component* corresponding to these measurements [7]. Consider a current measurement belonging to a residual spread component, which contains only other current and voltage magnitude measurements but no other type of measurements. Then there is a possibility that more than one solution will satisfy all the current measurements in the residual spread component without affecting any of the remaining measurements outside the residual spread area. Such residual spread components will be referred to as *v-i residual spread components.* If the residual spread component contains at least one power flow or injection measurement, then the possibility of multiple solutions will be avoided. This is due to the fact that the power flow or injection measurements can only be satisfied by a single solution, forcing the current magnitude measurements to satisfy that single solution among the several possible ones. Note that a critical measurement is nothing but a residual spread component with a single member. Therefore, in general, we can conclude that if a v-i residual spread component can be found, then the system will not be uniquely observable.

The presence of v-i residual spread components and/or critical current measurements, leading to multiple solutions, can be detected in several ways, like the ones discussed in the two subsections that follow.

9.6.1 Procedure based on the residual covariance matrix

As derived in Chapter 5, the residual covariance matrix Ω can be obtained from

$$\Omega = S \cdot R = R - HG^{-1}H^T \tag{9.16}$$

where:

H: measurement Jacobian matrix
R: covariance matrix of measurement error vector
$G = H^T R^{-1} H$: gain matrix
S: residual sensitivity matrix.

If a measurement is critical, then the corresponding column of the matrix Ω will be zero. Furthermore, if a measurement z_k, belongs to a residual spread component, then the k'th column of Ω will be zero except for the entries corresponding to the measurements belonging to the same residual spread component. Therefore, by computing the corresponding column of Ω, one can find out whether such a measurement is critical or not and, in this case, which measurements belong to the same residual spread component.

Since the criticality and the residual spread components need to be determined only for the current magnitude measurements, only those columns of Ω corresponding to the current measurements are required. Any desired column, Ω_k, can be calculated according to the following procedure:

- Solve $G\, y_k = h_k^t$
  where h_k : k'th row of H.

- Compute Ω_k as: $\Omega_k = R_k - H y_k$
  where R_k is the k'th column of R (containing all zeros except for R_{kk} as the k'th entry).

Note that the sparse triangular factors of G are already available from the state estimation solution and h_k^t is a very sparse vector, hence y_k can be obtained by sparse forward/back substitutions very efficiently.

The steps of the uniqueness determination procedure can now be presented [5]. It is assumed that a conventional observability analysis program is available to determine the solvability of the system. The following procedure is intended for detecting the possibility of multiple solutions only. So, it starts out with the assumption that the given network is solvable, but has possibly several solutions corresponding to the given measurement set.

1. Compute the columns of Ω corresponding to the current magnitude measurements. Note that, as the numerical values of R are irrelevant, the residual sensitivity matrix S is as good as R for this purpose.

2. If any column, k, contains a nonzero entry corresponding to a power flow or an injection measurement, then skip that column. Otherwise, flag the current measurement together with all the other measurements with nonzero entries in that column, as a v-i residual spread component that has potential to yield multiple solutions. If the column is completely zero, then flag the current magnitude measurement as critical.

If no measurement is flagged, then the system will be uniquely observable. If one or more current magnitude measurements are flagged either as critical or as a member of a v-i residual spread component, then the system will not be uniquely observable.

During step 2 the entries of Ω are to be checked for possible zeros. The decision on a zero has to be made based on a numerical threshold which may not be the same for different systems and measurement configurations. In order to have a decision threshold independent of the systems tested, the entries of Ω should be normalized with respect to the absolute maximum diagonal element, Ω_{max}. Then, Ω_{jk} will be assumed to be zero if

$|\Omega_{jk}|/\Omega_{max} \leq \epsilon$, where ϵ is of the same order as the convergence threshold used for the state estimation algorithm (e.g. $\epsilon = 1.0\text{e-}4$).

It is important to realize the implications of the above threshold and the potential limitations of this approach when very few power measurements exist and the residual spread components embrace electrically distant measurements. In some exceptional cases, the presence of a single power measurement might not suffice to prevent the estimator from converging to an incorrect state.

Compared to a conventional state estimator, in which only the diagonal elements Ω_{kk} are needed for the bad data identification cycles, the above procedure requires that whole columns of Ω be computed. Note, however, that *only* those columns corresponding to the current measurements not belonging to the residual spread components that are already analyzed, need to be processed. This will reduce the computational burden, especially for systems containing few but large size residual spread components containing current magnitude measurements.

Example 9.9:

Consider the 4-bus system and measurement configuration of Figure 9.15. Assuming unity reactances and null resistances, for the arbitrary (but non flat

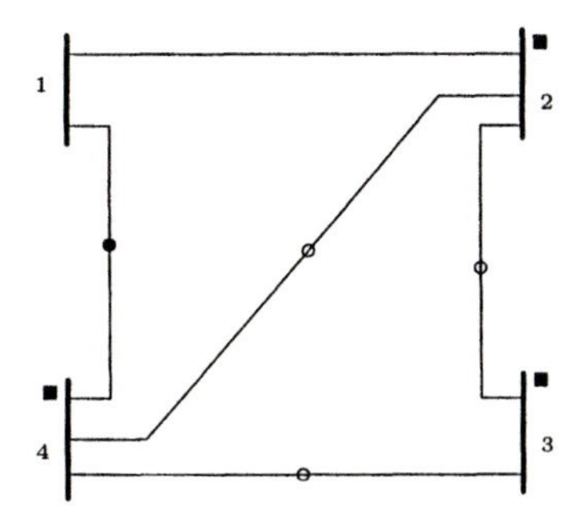

Figure 9.15. 4-bus system for Example 9.9

start) state given below:

Bus	V	θ (°)
1	1	0
2	1	−10
3	0.9	−15
4	1.1	−5

the resulting Jacobian matrix will be:

$$
H = \begin{array}{c} \\ P_{14} \\ Q_{14} \\ V_2 \\ V_3 \\ V_4 \\ I_{23}^2 \\ I_{24}^2 \\ I_{34}^2 \end{array}
\begin{array}{c} \theta_2 \quad \theta_3 \quad \theta_4 \quad V_1 \quad V_2 \quad V_3 \quad V_4 \\
\left[\begin{array}{rrrrrrr}
0 & 0 & 1.096 & 0.096 & 0 & 0 & 0.087 \\
0 & 0 & -0.096 & 0.904 & 0 & 0 & -0.996 \\
0 & 0 & 0 & 0 & 1.000 & 0 & 0 \\
0 & 0 & 0 & 0 & 0 & 1.000 & 0 \\
0 & 0 & 0 & 0 & 0 & 0 & 1.000 \\
0.157 & -0.157 & 0 & 0 & 0.207 & -0.192 & 0 \\
-0.192 & 0 & 0.192 & 0 & -0.192 & 0 & 0.208 \\
0 & -0.344 & 0.344 & 0 & 0 & -0.367 & 0.427
\end{array}\right]\end{array}
$$

If R is assumed to be the identity matrix, then the residual covariance matrix Ω will be:

$$
\begin{array}{c} \\ P_{14} \\ Q_{14} \\ V_2 \\ V_3 \\ V_4 \\ I_{23}^2 \\ I_{24}^2 \\ I_{34}^2 \end{array}
\begin{array}{c} P_{14} \quad Q_{14} \quad V_2 \quad V_3 \quad V_4 \quad I_{23}^2 \quad I_{24}^2 \quad I_{34}^2 \\
\left[\begin{array}{rrrrrrrr}
0.000 & 0.000 & 0.000 & 0.000 & 0.000 & 0.000 & 0.000 & 0.000 \\
0.000 & 0.000 & 0.000 & 0.000 & 0.000 & 0.000 & 0.000 & 0.000 \\
0.000 & 0.000 & 0.001 & -0.001 & -0.001 & -0.027 & -0.022 & 0.012 \\
0.000 & 0.000 & -0.001 & 0.000 & 0.000 & 0.013 & 0.011 & -0.006 \\
0.000 & 0.000 & -0.001 & 0.000 & 0.000 & 0.013 & 0.011 & -0.006 \\
0.000 & 0.000 & -0.027 & 0.013 & 0.013 & 0.532 & 0.435 & -0.243 \\
0.000 & 0.000 & -0.022 & 0.011 & 0.011 & 0.435 & 0.356 & -0.198 \\
0.000 & 0.000 & 0.012 & -0.006 & -0.006 & -0.243 & -0.198 & 0.111
\end{array}\right]\end{array}
$$

Inspection of the entrees in Ω reveals that the measurement subset $\{V_2, V_3, V_4, I_{23}, I_{24}, I_{34}\}$ forms a v-i residual spread component, while P_{14} and Q_{14} are critical, which means that there are two possible solutions for θ_2 and θ_3. Note that, despite the redundancy of the current magnitude measurement I_{23}, multiple solutions can not be avoided due to the symmetry of the resulting phasor diagram (this is left as an end-of-chapter exercise). The reader is encouraged to verify that by adding the measurement I_{12} the full system can be made uniquely observable.

9.6.2 Procedure based on the Jacobian matrix

The information concerning the presence of v-i residual spread components can also be obtained from the Jacobian matrix, without having to compute the residual covariance matrix.

For the sake of simplicity, it will be assumed again that unobservable branches have been previously removed from the model, by following any of the procedures discussed in Chapter 4. In [11] it is explained how the identification of fully unobservable and nonuniquely observable branches can be combined into a single generalized observability procedure.

The linearized and error-free measurement equations can be partitioned so that the first n will correspond to a linearly independent set, yielding:

$$\left[\begin{array}{c} H_1 \\ H_2 \end{array} \right] x = \left[\begin{array}{c} z_1 \\ z_2 \end{array} \right] \tag{9.17}$$

Such a partitioning is a byproduct of the Peters-Wilkinson decomposition of the Jacobian matrix:

$$H = \left[\begin{array}{c} H_1 \\ \cdots \\ H_2 \end{array} \right] = \left[\begin{array}{c} L \\ \cdots \\ M \end{array} \right] [U] \tag{9.18}$$

where:

$H_1 = L \cdot U$ is a $n \times n$ square matrix
L is a $n \times n$ lower triangular matrix
U is a $n \times n$ upper triangular matrix
H_2 and M are $(m - n) \times n$ rectangular matrices.

As unobservable branches have been removed, no zero pivots should be encountered during the Jacobian factorization, provided the linearization has not been carried out at flat start.

Based on the Jacobian factorization it is straightforward to express the redundant measurements, z_2, as a linear combination of the linearly independent set, z_1:

$$\begin{aligned} z_2 &= H_2 \cdot H_1^{-1} \cdot z_1 \\ &= M \cdot L^{-1} \cdot z_1 \\ &= S_z \cdot z_1 \end{aligned} \tag{9.19}$$

Let us call matrix S_z the *measurement sensitivity matrix*, since its elements carry the sensitivity information between z_1 and z_2. Each row of S_z corresponds to a redundant measurement. Nonzero elements in each row indicate those measurements in z_1 that belong to the same residual spread component as that redundant measurement. Note on the other hand that a null column in S_z implies that the corresponding measurement in z_1 is *critical.* As proved in [11], use of S_z matrix in (9.19) is equivalent to the use of the residual covariance matrix Ω for measurement classification (the residual sensitivity matrix S can be expressed in terms of the smaller and simpler to obtain S_z matrix).

The j-th row of S_z, denoted by s_j, can be efficiently computed by solving:

$$L^T s_j^T = M_j^T \tag{9.20}$$

where M_j is the j-th row of M. This is just a backsubstitution using the transpose of the triangular factor L and the respective row of M as the right hand side vector.

Then, from the nonzero pattern of S_z, the subset of current measurements in z_1 belonging to v-i residual spread components can be easily identified (critical measurements constitute a particular case).

An alternative but equivalent procedure, that traces sequentially in L and M the lower factorization paths corresponding to current measurements, instead of explicitly computing the matrix S_z, can be found in [4]. As factorization paths of currents belonging to previously traced paths need not be obtained, this implementation is probably the most efficient way of obtaining v-i residual spread components.

Example 9.10:

Consider again the 4-bus system of Example 9.9, whose Jacobian rows are already ordered as required by the partition suggested in (9.18), i.e.,

$$
\begin{bmatrix} H_1 \\ \cdots \\ H_2 \end{bmatrix} =
\begin{array}{c} \\ P_{14} \\ Q_{14} \\ V_2 \\ V_3 \\ V_4 \\ I_{23}^2 \\ I_{24}^2 \\ I_{34}^2 \end{array}
\begin{array}{c}
\begin{array}{ccccccc} \theta_2 & \theta_3 & \theta_4 & V_1 & V_2 & V_3 & V_4 \end{array} \\
\left[\begin{array}{ccccccc}
0 & 0 & 1.096 & 0.096 & 0 & 0 & 0.087 \\
0 & 0 & -0.096 & 0.904 & 0 & 0 & -0.996 \\
0 & 0 & 0 & 0 & 1.000 & 0 & 0 \\
0 & 0 & 0 & 0 & 0 & 1.000 & 0 \\
0 & 0 & 0 & 0 & 0 & 0 & 1.000 \\
0.157 & -0.157 & 0 & 0 & 0.207 & -0.192 & 0 \\
-0.192 & 0 & 0.192 & 0 & -0.192 & 0 & 0.208 \\
\hline
0 & -0.344 & 0.344 & 0 & 0 & -0.367 & 0.427
\end{array}\right]
\end{array}
$$

Consequently, the matrix S_z is:

$$
S_z = H_2 H_1^{-1} = [0,\, 0,\, -0.1096,\, 0.0537,\, 0.0543,\, 2.1911,\, 1.7917]
$$

As expected, it is concluded that both P_{14} and Q_{14} are critical measurements, the remaining ones constituting a v-i redundant set.

9.7 Identification of Nonuniquely Observable Branches

A majority of algorithms and procedures associated with state estimation are more easily implemented and more efficiently solved when the conventional nodal formulation is adopted. However, virtually all of them could be reformulated by resorting to branch variables. This is the case, for instance, of the procedures presented in the former section to detect v-i residual spread components. In fact, expressions (9.17) to (9.20) remain valid if x denotes the whole set of branch variables and z contains, in addition to regular measurements, a set of independent loop equations, which compensates for the increased number of state variables. This model was adopted in section 4.5 to develop a noniterative numerical observability procedure.

For the purpose of this section, as the solution uniqueness issue is a direct consequence of the signs of branch phase angles being undefined in the presence of ampere measurements, the use of branch variables offers added advantages. The material that follows constitutes a simple extension of the observability algorithm described in section 4.5.

Assume, therefore, that the branch-based Jacobian has been factorized like in (9.18), and that the subset of current measurements in z_1 belonging to v-i residual spread components has been identified. Then, for each member z_k of this measurement set, a list of nonuniquely observable branches is obtained by solving the triangular system:

$$U x_k = e_k \tag{9.21}$$

and flagging the nonzero elements of x_k, where e_k contains a single nonzero value at position k. The union of all branches flagged during these fast backward processes constitutes the nonuniquely observable network portion.

A convenient alternative to the full branch model is the reduced model where all the variables associated with the links and corresponding loop equations are eliminated. In obtaining this model, it is recommended that the chosen tree includes all power flow measurements. The resulting measurement vector of this reduced branch model is identical to that of the nodal approach (in fact, the same matrix S_z is obtained) and the tree branch variables remaining in the state vector are related to the nodal unknowns through a regular transformation matrix. The procedure described above remains valid but provides information only about tree branches. Solution uniqueness for a given link could be subsequently determined by applying for instance the node and loop rules of section 9.5.

Example 9.11:

Consider the 4-bus system with the given measurement set shown in Figure 9.16.

Let us assume that all voltage magnitudes are measured and that only phase angle observability is to be analyzed. The Jacobian matrix in the branch reference

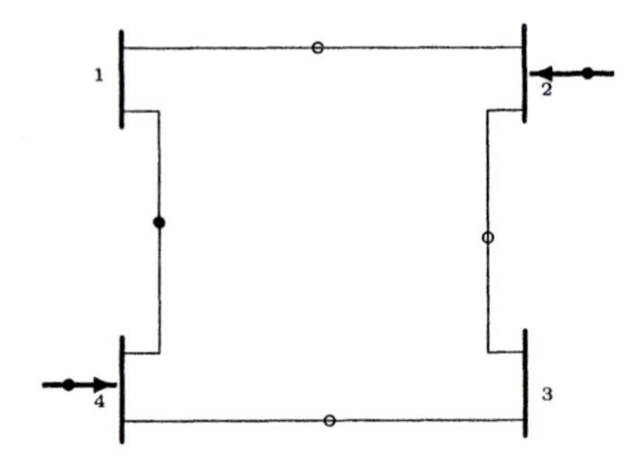

Figure 9.16. 4-bus system for Example 9.11

frame can be written as:

$$
\begin{array}{r} \\ \Delta I_{12}^2 \\ \Delta I_{23}^2 \\ \Delta I_{34}^2 \\ \theta - loop \\ \Delta P_4 \\ \Delta P_2 \end{array}
\left[\begin{array}{cccc}
\Delta\theta_{12} & \Delta\theta_{23} & \Delta\theta_{34} & \Delta\theta_{41} \\
x & & & \\
 & x & & \\
 & & x & \\
x & x & x & x \\ \hline
 & & x & x \\
x & x & &
\end{array} \right]
$$

where x's represent the nonzero entries. In this simple case, the Jacobian is already lower trapezoidal, which means that $H_1 = L$, $H_2 = M$ and U is the identity matrix. Without explicitly building S_z it is easy to conclude that no v-i residual spread components exist. Hence, the system is declared uniquely observable.

Now, let us remove the injection measurement at bus 4. Then the corresponding Jacobian will become:

$$
\begin{array}{r} \\ \Delta I_{12}^2 \\ \Delta I_{23}^2 \\ \Delta I_{34}^2 \\ \theta - loop \\ \Delta P_2 \end{array}
\left[\begin{array}{cccc}
\Delta\theta_{12} & \Delta\theta_{23} & \Delta\theta_{34} & \Delta\theta_{41} \\
x & & & \\
 & x & & \\
 & & x & \\
x & x & x & x \\ \hline
x & x & &
\end{array} \right] \tag{9.22}
$$

As the third and fourth columns of M are null, it is apparent that I_{34} and the loop equation are critical to estimate θ_{34} and θ_{41} respectively. Also, as U reduces to the identity matrix, no extra branches are flagged during the backsubstitution processes. Thus, branches 3-4 and 4-1 will be declared as not uniquely observable for this case.

Example 9.12:

The IEEE 14-bus test system with the measurement set shown in Figure 9.17 will be used to illustrate the uniqueness determination procedure. The exact measurement values adopted for this network are provided in Tables 9.2 and 9.3.

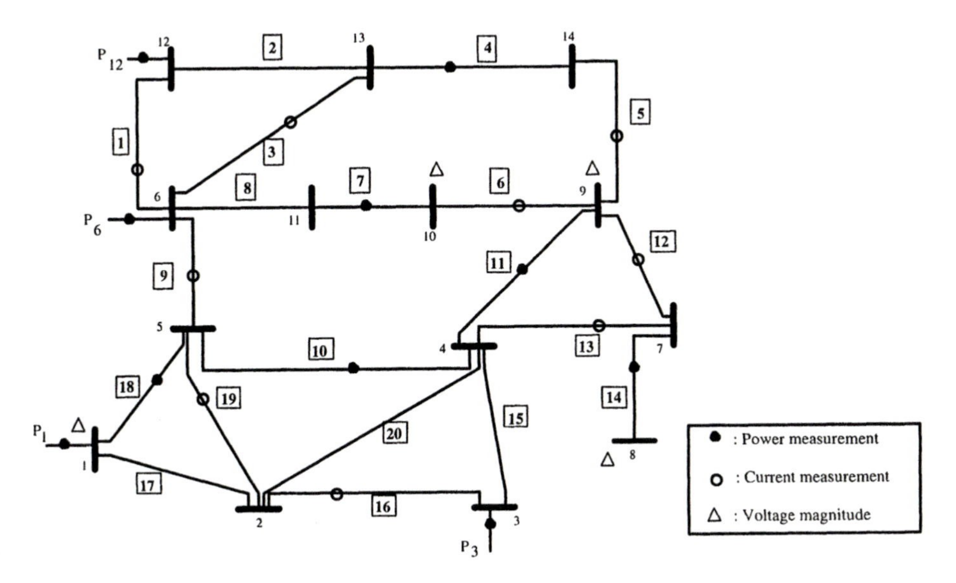

Figure 9.17. IEEE 14-bus system and available measurements for Example 9.12

Note that the system will become unobservable if the current measurements are excluded. Thus, the line current measurements are useful to extend observable islands in this example. Also, note that existing observability methods based on the rank of H (numerical [18] or topological [15]), can not be used to test network observability due to the existing current measurements.

The results of two cases will be presented. In the base case, the network equations are composed of 7 real/reactive loop, 6 real/reactive power flow, 4 real/reactive injection, 4 voltage magnitude and 9 current magnitude equations. The procedures described earlier declare the whole system as uniquely observable.

Then, the measurement configuration is modified by deleting the line current measurements in branches 5 and 9. The observability algorithm declares this system as unobservable and determines the branches 1,2,3,5,6,8 and 9 as unobservable. This corresponds to three observable islands formed by branches 10 through 20 (island 1), branch 4 (island 2) and branch 7 (island 3) respectively.

Two different solutions satisfying exactly all the available measurements are provided in Table 9.4. Note that the two solutions differ at buses 6 and 10 through 14. However, the voltage values at the pairs of buses 10-11 and 13-14 are such that a unique power flow results for branches 4 and 7.

Table 9.2. Bus-related measurements for Example 9.12

Bus	V	P_{inj}	Q_{inj}
1	1.060	2.324	−0.169
3	–	−0.942	0.044
6	–	−0.112	0.047
8	1.090	–	–
9	1.056	–	–
10	1.051	–	–
12	–	−0.061	−0.016

9.8 Measurement Classification and Bad Data Identification

When the measurement set does not contain current measurements, only n linearly independent measurements are strictly required to estimate a network whose state vector is composed of n variables. If there exists any current magnitude measurement then the system state will not be uniquely observable by only n linearly independent measurements, as shown in the preceding sections. Based on this observation, measurements can be reclassified as follows [6]:

Noncritical: when deleted, the system remains uniquely observable.

Critical: when removed, the system becomes unobservable.

Uniqueness-Critical: when eliminated the system is not uniquely observable, i.e., several solutions are possible.

Critical measurements containing bad data can not be detected because they will be exactly satisfied, yielding null residuals. On the other hand, bad data in a uniqueness-critical measurement can be detected. However, its identification depends upon the magnitude of the bad data, initial state chosen to iterate and proximity in the domain of convergence of other possible solutions. Even when it is correctly identified and removed, a unique solution can not be guaranteed. Therefore, only bad data affecting noncritical measurements can be safely identified and eliminated.

Table 9.3. Branch-related measurements for Example 9.12

Branch	$\mid I \mid$	P_{flow}	Q_{flow}
1	0.076366	–	–
3	0.178826	–	–
4	–	0.05632	0.01692
5	0.095849	–	–
6	0.064200	–	–
7	–	−0.03774	−0.01529
9	0.449792	–	–
10	–	−0.61217	0.15669
11	–	0.16090	−0.00321
12	0.270054	–	–
13	0.290826	–	–
14	–	0.00000	−0.16910
16	0.701174	–	–
18	–	0.75551	0.03504
19	0.397307	–	–

Unlike the uniqueness-related observability issues, the required modifications to the bad data identification procedures depend on the particular estimator adopted. The LS and LAV estimators will be separately discussed below.

9.8.1 LS Estimation

Bad data detection and identification is carried out as a post estimation procedure, which can be summarized as follows (see Chapter 5):

1. Use the estimated state to calculate the normalized measurement residuals.

2. Sort them in descending order.

3. If the largest residual exceeds a threshold, eliminate the respective measurement and repeat the process.

When using only conventional measurements, there is no risk of eliminating a critical measurement during the identification cycles, as it will be pushed always to the bottom of the sorted list. However, in the presence

Table 9.4. Two possible solutions for the second case of Example 9.12

	Solution 1		Solution 2	
Bus	Voltage	Angle (°)	Voltage	Angle (°)
1	1.060	0.00	1.060	0.00
2	1.045	−4.98	1.045	−4.98
3	1.010	−12.72	1.010	−12.72
4	1.018	−10.32	1.018	−10.32
5	1.020	−8.78	1.020	−8.78
6	1.070	−14.06	1.061	−13.61
7	1.062	−13.37	1.062	−13.37
8	1.090	−13.37	1.090	−13.37
9	1.056	−14.95	1.056	−14.95
10	1.051	−14.79	1.051	−15.10
11	1.057	−14.48	1.057	−14.80
12	1.055	−14.92	1.041	−14.15
13	1.050	−15.00	1.035	−13.81
14	1.035	−15.88	1.020	−14.72

of current magnitude measurements, there is a need to check whether or not the identified bad data corresponds to a noncritical measurement, in the sense defined above. According to the conditions for solution uniqueness found in the preceding sections, noncriticality of measurements can be checked by the following procedure:

1. If the measurement belongs to a residual spread component containing only power and voltage measurements, then declare it as noncritical. Else, continue.

2. If the measurement refers to a power flow or injection and the residual spread component does not contain any other power measurement, then declare it as uniqueness-critical. Else, continue.

3. Check if any of the remaining ampere measurements in the same residual spread component will become critical when this measurement is eliminated. If the answer is yes, then declare the measurement as uniqueness-critical, else declare it as noncritical.

Note that, in order to carry out step 3, the diagonals of the residual covariance matrix must be updated when the candidate measurement is removed. This process can be notably simplified if the techniques described in [17] are used.

9.8.2 LAV Estimation

As discussed in Chapter 6, the main advantage of the LAV estimation is its ability to reject bad data as part of the estimation process. Therefore, if leverage points are properly dealt with, the estimated state will be unbiased since bad data are discarded by the LAV estimator. This is true as long as the measurement set contains only conventional measurements. In the presence of current measurements, further postprocessing is required to assure uniqueness of the state reached.

The normalized residuals of rejected measurements can be obtained, and those exceeding a threshold can be declared as bad data [1]. The list of suspect measurements must then be tested for noncriticality per the three step procedure presented above. If any of the suspect measurements is found to be uniqueness critical, then the results of the LAV estimator can not be trusted due to the possibility of multiple solutions.

Example 9.13:

Consider the 4-bus system of Figure 9.18 whose data are given in Table 9.5.

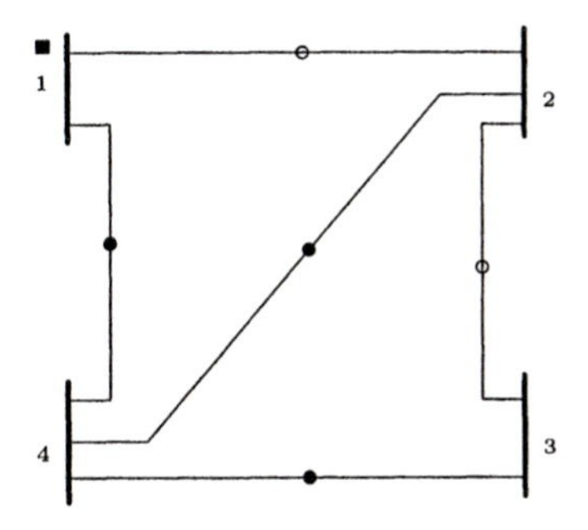

Figure 9.18. 4-bus system for Example 9.13

Table 9.5. Data for the system of Figure 9.18

From bus	To bus	R	X	Current magnitude	P/Q flows
1	4	.0001	0.2	–	1.083/0.626
2	4	.0001	0.2	–	0.340/0.204
3	4	.0001	0.5	–	−0.1234/0.09
1	2	.0001	0.1	1.5983	–
2	3	.0001	0.2	0.6820	–

The system is uniquely observable, even in the absence of current measurements. Power flow measurements of branch 2-4 are chosen to be inconsistent with the remaining measurements (bad data). The results provided by the two estimators will be separately analyzed.

LS Estimation

The sorted list of normalized residuals are given in Table 9.6. In this case the estimator converges to a point close to the true solution and the normalized residual test is able to identify the two bad data of branch 2-4.

Table 9.6. Normalized residuals provided by the LS estimator for the system of Example 9.13

Measurement	$\lvert r_N \rvert$
P_{24}	22.7
P_{14}	20.8
I_{12}	20.6
Q_{24}	19.7
Q_{14}	17.8
V_1	14.1
Q_{34}	8.0
P_{34}	8.0
I_{23}	8.0

However, if P_{2-4} and Q_{24} are eliminated, I_{12} and I_{23} will become critical and the system is no longer uniquely observable. The two possible solutions are shown in Table 9.7.

Table 9.7. Two possible solutions when power flows of branch 2-4 are eliminated

Bus	V	θ (°)	V	θ (°)
1	1.050	0.00	1.050	0.00
2	0.992	−8.37	1.034	−8.75
3	1.000	−16.21	1.000	−16.21
4	0.953	−12.50	0.953	−12.50

LAV Estimation

The LAV estimator converges to a state close to one of the two solutions given in Table 9.7, depending upon the starting point. Hence, even though the system is uniquely observable, due to the presence of bad data the state estimation solution will not be reliable.

9.9 Problems

1. Consider again the 3-bus system of Example 9.1. The voltage measurement V_3 is replaced by the current measurement I_{23}. Analyze the solution uniqueness based on the respective phasor diagram. Note in passing that the same situation would arise in the system of Example 9.6 if one of the current measurements were missing.

2. In the 5-bus ring of Example 9.3, analyze which buses are the right candidates for the extra inequality constraint to be added so that the correct solution (a) can be discriminated from (c).

3. Prove that, if P_e in the 8-bus system of Example 9.7 is substituted by P_g or P_h all branches will remain uniquely observable.

4. Assume that, in the 8-bus system of Example 9.8, buses "b" and "c" are generation buses, while the remaining ones are load buses. Prove that, thanks to this extra information, the system becomes uniquely observable. *Hint:* consider the cut-set composed of branches 2, 6 and 7, whose power flow directions are determined by the signs of bus injections.

5. Obtain the phasor diagram corresponding to the 4-bus system of Example 9.9, taking V_4 as phase origin, and check to see if there are two possible solutions for θ_2 and θ_3. Furthermore, by means of this diagram, prove that the addition of I_{14} renders the system uniquely observable. In this situation, analyze what happens if: a) I_{24} is removed; b) I_{23} is removed.

6. Utilize the topological concepts to manually solve the two cases of Example 9.12. *Hints:* First of all, replace those network portions which are observable through power measurements by a single super node. It is also useful to "associate" injections 6 and 12 to branches 2 and 8 respectively. Then, apply the loop and node rules of section 9.5 to the small equivalent system that results.

7. Using a Matlab script file build the Jacobians for the two cases analyzed in Example 9.12, detect v-i residual spread components by computing the respective matrices S_z and identify unobservable branches in the second case.

8. Repeat the analysis of Example 9.13 when the voltage magnitude at bus 2 is added to the measurement set.

References

[1] A. Abur, "A Bad Data Identification Method for Linear Programming State Estimation", *IEEE Transactions on PWRS*, vol. 5(3), August 1990, pp. 894-900.

[2] A. Abur and A. Gómez Expósito, "Observability and Bad Data Identification When Using Ampere Measurements in State Estimation", *IEEE International Conference on Circuits and Systems*, paper no. 958, May 3-6, 1993, Chicago, IL.

[3] A. Abur and A.G. Expósito, "Algorithm for Determining Phase-Angle Observability in the Presence of Line-Current-Magnitude Measurements", *IEE Proc. Gener. Transm. Distrib.*, vol 142(5), September 1995, pp. 453-458.

[4] A. Abur and A. Gómez Expósito, "Multiple Solutions and Unique Observability in State Estimation", *Power Systems Computation Conference*, pp. 1200-1206, August 1996, Dresden.

[5] A. Abur and A. Gómez Expósito, "Detecting Multiple Solutions in State Estimation in the Presence of Current Magnitude Measurements", *IEEE Transactions on PWRS*, vol. 12(1), Feb. 1997, pp. 370-375.

[6] A. Abur and A. Gómez Expósito, "Bad Data Identification When Using Ampere Measurements", *IEEE Transactions on PWRS*, vol. 12(2), May 1997, pp. 831-836.

[7] K.A. Clements, G.R. Krumpholz and P.W. Davis, "Power System State Estimation Residual Analysis: An Algorithm Using Network Topology ", *IEEE Trans. on Power Apparatus and Systems*, vol. PAS-100, No.4, April 1981, pp. 1779-1787.

[8] D.M. Falcão and M.A. Arias, "State Estimation and Observability Analysis Based on Echelon Forms of the Linearized Measurement Models", *IEEE Trans. on Power Systems*, vol.9, No.2, May 1994, pp. 979-987.

[9] K.I. Geisler, "Ampere Magnitude Line Measurements for Power System State Estimation ", *IEEE Trans. on Power Apparatus and Systems*, vol. PAS-103, No.8, Aug.1984, pp. 1962-1969.

[10] A. Gómez Expósito *et al.*, "Development of a State Estimator Based on Line-Current Measurements", *TOP*, Official Journal of Spanish Statistical and Operations Research Society, vol 2(1), June 1994, pp. 85-104.

[11] A. Gómez Expósito and A. Abur, "Generalized Observability Analysis and Measurement Classification", *IEEE Transactions on PWRS*, vol. 13(3), August 1998, pp. 1090-1095.

[12] H.L. Fuller and T.A. Hughes, "State Estimation for Power Systems with Mixed Measurements". IEEE/PES Summer Meeting Paper C74 363–8, Anaheim, 1974.

[13] E. Handschin, M. Langer and E. Kliokys, "An Interior Point Method for State Estimation with Current Magnitude Measurements and Inequality Constraints", *Proceedings of the PICA Conference*, Salt Lake City, Utah, May 7-12, 1995, pp. 385-391.

[14] J.S. Horton and R.D. Masiello, "On–line Decoupled Observability Processing". PICA Conf. Proc., 1977.

[15] G.R. Krumpholz, K.A. Clements and P.W. Davis, "Power System Observability: A Practical Algorithm Using Network Topology", *IEEE Trans. on Power Apparatus and Systems*, Vol. PAS-99, No.4, July/Aug. 1980, pp. 1534-1542.

[16] J.L. Marinho, P.A. Machado and C. Bongers, "On the Use of Line Current Measurement for Reliable State Estimation in Electric Power Systems", 1979 PICA Conference.

[17] L. Mili and T. Van Cutsem, "Implementation of HTI Method in Power System State Estimation", *IEEE Transactions on PWRS*, vol. 3(3), August 1988, pp. 887-893.

[18] A. Monticelli and F.F. Wu, "Network Observability: Theory", *IEEE Transactions on PAS*, Vol.PAS-104, No.5, May 1985, pp. 1042-1048.

[19] J.M. Ruiz Muñoz and A. Gómez Expósito, "A Line Current Measurement Based State Estimator", *IEEE Transactions on PWRS*, vol. 7(2), May 1992, pp. 513-519.

[20] E.J. Wolters, "Results of a Numerical and a Hybrid Simulation of a Power System Tracking State Estimator with an Emphasis on Line-Current Measurements", Proceedings of the PSCC, Grenoble, France, 1972.

Appendix A

Review of Basic Statistics

In this appendix, a brief review of the basic definitions and concepts in statistics will be given. The intent is to highlight some of the relevant concepts in statistics that are frequently referred to in various chapters of the book. The review starts with some definitions for the random variables and commonly used statistical functions.

A.1 Random Variables

A real valued function, that is defined on the sample space **S**, is called a random variable (r.v.). A simple example of a r.v. is the outcome of rolling a dice. Here, the sample space is the set of numbers 1 through 6, i.e. all possible outcomes of rolling a dice.

A.2 The Distribution Function (d.f.), F(x)

It is a real valued function of a real number x, defined as:

$$F(x) = Pr(X \leq x), \quad -\infty < x < \infty \tag{A.1}$$

$$0 \leq F(x) \leq 1 \tag{A.2}$$

where, Pr stands for "probability of".

The distribution function, d.f. has several useful properties some of which are listed below:

Property-1 F(x) is non-decreasing as x is increasing, i.e.

$$\text{if } x_1 < x_2, \text{ then } F(x_1) < F(x_2)$$

Property-2 Lower and upper limits of d.f are zero and one respectively, i.e.

$$\lim_{x \to -\infty} F(x) = 0, \text{ and, } \lim_{x \to \infty} F(x) = 1$$

Property-3 A d.f at a given point x is defined by its limit from the right, i.e.

$$F(x) = F(x^+) \quad \text{for all } x.$$

A.3 The Probability Density Function (p.d.f), f(x)

A non-negative function f, defined on the real line, is called a probability density function if for any interval T, the following is satisfied:

$$Pr(X \in T) = \int_T f(x)dx$$

The p.d.f has the following properties:

$$f(x) \geq 0 \quad (A.3)$$

$$\int_{-\infty}^{\infty} f(x)dx = 1 \quad (A.4)$$

Note that when a random variable X has a continuous distribution, then $Pr(X = x) = 0$. Also, the distribution and density functions are related as shown below:

$$Pr(X \leq x) = \int_{-\infty}^{x} f(u)du \quad (A.5)$$

$$= F(x) \quad (A.6)$$

or as long as the p.d.f is continuous, the d.f. will be differentiable:

$$\frac{dF(x)}{dx} = f(x)$$

A.4 Continuous Joint Distributions

Joint distributions are defined for two or more random variables. Consider the case of two random variables, X and Y. A non-negative function f(x,y) is called the joint probability density function of X and Y, if for a region T in the x-y plane, the following holds true:

$$Pr[(X, Y) \in T] = \int\int_T f(x, y)dxdy$$

Again, similar properties can be written for the joint p.d.f:

$$f(x,y) \geq 0 \tag{A.7}$$

$$\int_{-\infty}^{\infty}\int_{-\infty}^{\infty} f(x,y)dxdy = 1 \tag{A.8}$$

Similarly, joint distribution function of two random variables X,Y can be defined as:

$$F(x,y) = Pr(X \leq x \text{ and } Y \leq y) \tag{A.9}$$

$$f(x,y) = \frac{\partial^2 F(x,y)}{\partial x \partial y} \tag{A.10}$$

A.5 Independent Random Variables

X and Y are independent random variables if

$$f(x,y) = f_1(x)f_2(y) \text{ and } F(x,y) = F_1(x)F_2(y)$$

A.6 Conditional Distributions

The conditional probability density function g, of a random variable X when another random variable Y is already known to have assumed a value y, is given by:

$$g(x \mid y) = \frac{f(x,y)}{f_2(y)} \qquad -\infty < x < \infty$$

A.7 Expected Value

Expected (or mean) value of a random variable X, is denoted by $E(X)$ and defined as:

$$E(X) = \int_{-\infty}^{\infty} xf(x)dx$$

Expected value has the following properties:

1. Expected value of a random variable $Y = aX + b$, will be given by $E(Y) = aE(X) + b$.
2. If $Pr(X \geq a) = 1$, then $E(X) \geq a$. If $Pr(X \leq b) = 1$, then $E(X) \leq b$.
3. $E(X_1 + X_2 + \ldots + X_n) = E(X_1) + E(X_2) + \ldots + E(X_n)$

A.8 Variance

Variance of a random variable X is denoted by σ^2 and defined as:

$$\sigma^2 = Var(X) = E[(X - \mu)^2], \quad \text{where} \quad \mu = E(X)$$

Variance has the following properties:

1. $Var(aX + b) = a^2 Var(X)$
2. $Var(X) = E(X^2) + [E(X)]^2 - 2\mu E(X) = E(X^2) - [E(X)]^2$
3. $Var(\sum_{i=1}^{n} a_i X_i) = \sum_{i=1}^{n} a_i^2 Var(X_i)$, for independent X_i's.

A.9 Median

Median of a distribution of a random variable X, is defined as the value m along the real line, such that

$$Pr(X \leq m) \geq \frac{1}{2} \quad \text{and} \quad Pr(X \geq m) \geq \frac{1}{2}$$

Also, note that:

$$Pr(X > m) = 1 - Pr(X \leq m) \leq \frac{1}{2} \tag{A.11}$$

$$Pr(X > m) \leq \frac{1}{2} \leq Pr(X \leq m) \tag{A.12}$$

A.10 Mean Squared Error

It can be shown that the value of z that will minimize the expected value of the squared error, i.e. $[X - z]^2$, is the expected value of the random variable X. Consider the expression

$$E[(x - z)^2] = E(X^2) - 2zE(X) + z^2$$

Choosing $E(X) = z$ will minimize the above expression, which is called the mean squared error (MSE). Note also that, with the choice of z as $E(X)$, MSE will be identical to the definition of the variance of X:

$$MSE = \sigma^2 = E[(X - E(X))^2]$$

A.11 Mean Absolute Error

Similar to the case of the MSE, it can be shown that the value z, that minimizes the mean absolute error:

$$E(\mid X - z \mid)$$

yields the median m for the random variable X. In other words, if m is the median and c is any other number, then

$$E(\mid X - m \mid) \leq E(\mid X - c \mid)$$

Proof: Let us arbitrarily assume that $m < c$. Then,

$$E(\mid X - c \mid) - E(\mid X - m \mid) = \int_{-\infty}^{\infty} (\mid X - c \mid - \mid X - m \mid) f(x)dx$$

$$\begin{aligned}
&= \int_{-\infty}^{m} (c - m) f(x)dx + \int_{m}^{c} (c + m - 2x) f(x)dx + \int_{c}^{\infty} (m - c) f(x)dx \\
&\geq \int_{-\infty}^{m} (c - m) f(x)dx + \int_{m}^{c} (m - c) f(x)dx + \int_{c}^{\infty} (m - c) f(x)dx \\
&= (c - m)[Pr(X \leq m) - Pr(X > m)]
\end{aligned}$$

Since m is the median, Eq.(A.12) must hold true:

$$Pr(X \leq m) \geq \frac{1}{2} \geq Pr(X > m)$$

Hence,

$$E(\mid X - c \mid) - E(\mid X - m \mid) \geq 0$$

Proof under the alternative assumption of $m > c$ is left to the reader.

A.12 Covariance

Covariance between two random variables X and Y is defined as:

$$cov(X, Y) = E[(X - E(X))(Y - E(Y))]$$

Related to the covariance, one can define the correlation coefficient as:

$$\rho(X, Y) = \frac{cov(X, Y)}{\sigma_x \sigma_y}$$

Properties of covariance:

1. $cov(X, Y) = E(X \cdot Y) - E(X)E(Y)$

2. If X, Y are independent random variables, then $cov(X, Y) = \rho(X, Y) = 0$.

3. $Var(X + Y) = Var(X) + Var(Y) + 2cov(X, Y)$

4. $Var(\sum_{i=1}^{n} X_i) = \sum_{i=1}^{n} Var(X_i) + 2\sum_{i=1}^{n}\sum_{j=i+1}^{n} cov(X_i, X_j)$

A.13 Normal Distribution

A random variable X is said to have a Normal (Gaussian) distribution with a mean μ and variance σ^2, if it is distributed according to the following function:

$$f(x \mid \mu, \sigma^2) = \frac{1}{\sqrt{2\pi}\sigma} exp[-\frac{1}{2}(\frac{x-\mu}{\sigma})^2]$$

When plotted, the Normal distribution function looks like a bell as shown in Figure A.1, hence it is frequently referred to as the bell shaped distribution. A random variable X which has a Normal distribution with mean μ and variance σ^2 is commonly denoted by $X \sim N(\mu, \sigma^2)$.

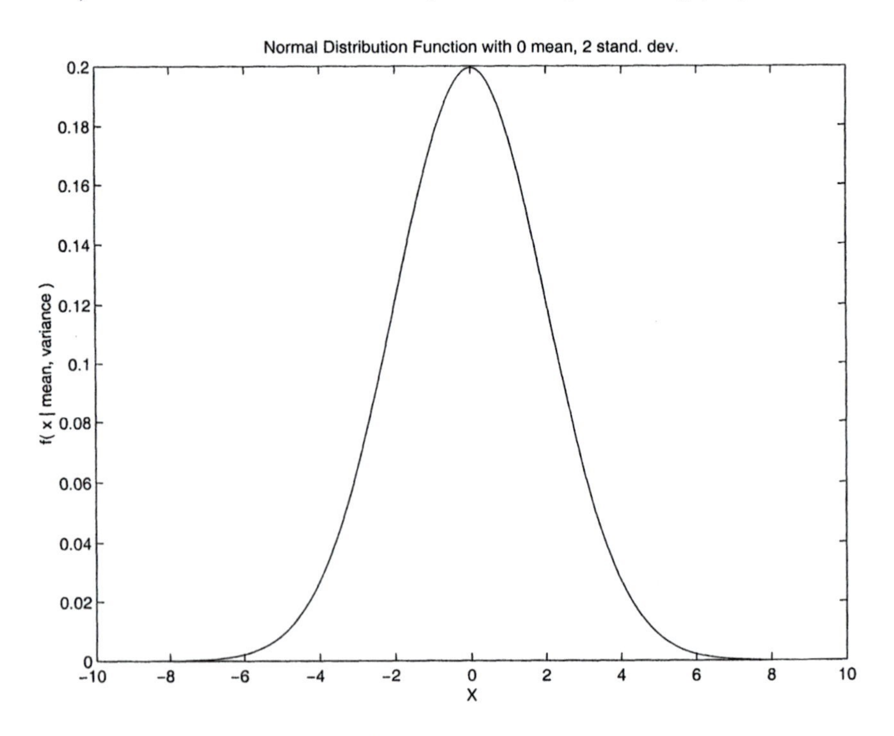

Figure A.1. Normal Distribution function

Theorem

If $X \sim N(\mu, \sigma^2)$ and $Y = aX + b$, then Y will also have a Normal distribution with mean $a\mu + b$ and variance $a^2\sigma^2$.

A.14 Standard Normal Distribution

Normal distribution with 0 mean and 1 variance is called the Standard Normal distribution. The p.d.f. of Standard Normal distribution is given by:

$$\phi(x) = f(x \mid 0, 1) = \frac{1}{\sqrt{2\pi}} exp(-\frac{1}{2}x^2)$$

and the corresponding distribution function d.f. will be given by:

$$\Phi(x) = \int_{-\infty}^{x} \phi(u)du$$

Note that, due to distribution symmetry:

$$\Phi(x) = 1 - \Phi(-x)$$

Example 1.1:

A random variable X is distributed according to a Normal distribution with a mean of 15 and variance of 9.

(a) Find the probability that $X > 16$.

(b) Find the probability that $\mid X - 15 \mid > 4$.

(c) Find x_0 such that $Pr(\mid X - 15 \mid \leq x_0) = 0.90$.

Solution

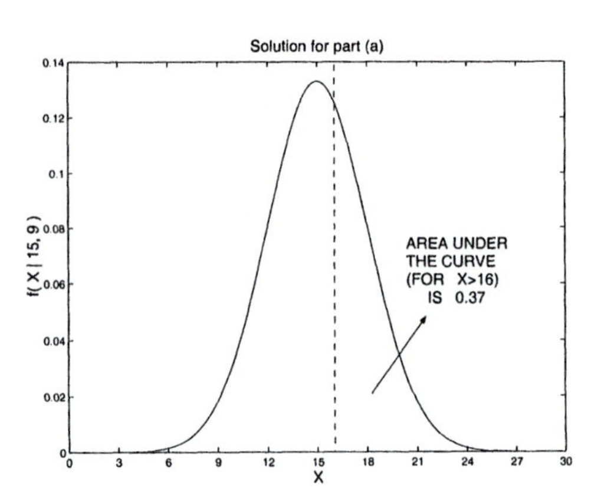

(a)

$$\begin{aligned} Pr(X > 16) &= 1 - Pr(X \leq 16) \\ &= 1 - Pr(\frac{X-15}{3} \leq \frac{16-15}{3}) \\ &= 1 - \Phi(\frac{1}{3}) \\ &= 1 - 0.63 = 0.37 \end{aligned}$$

Note that the value 0.63 is looked up from the Standard Normal distribution table corresponding to the value 1/3.

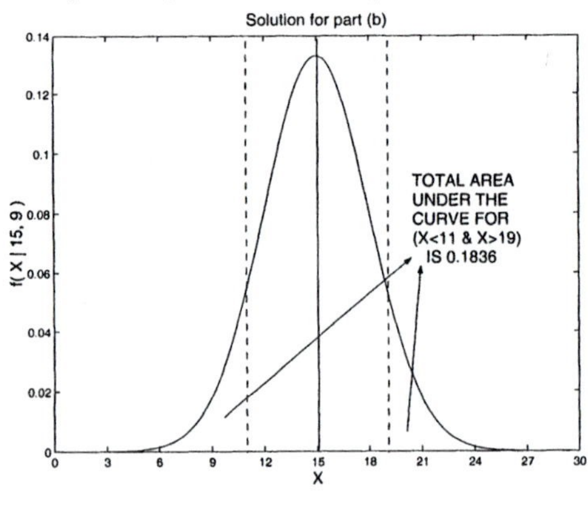

(b)

$$\begin{aligned} Pr(X \leq 11) &+ Pr(X \geq 19) \\ &= 2Pr(X \geq 19) \\ &= 2[1 - Pr(X \leq 19)] \\ &= 2[1 - Pr(\frac{X-15}{3} \leq \frac{19-15}{3})] \\ &= 2[1 - \Phi(4/3)] \\ &= 2[1 - 0.9082] = 0.1836 \end{aligned}$$

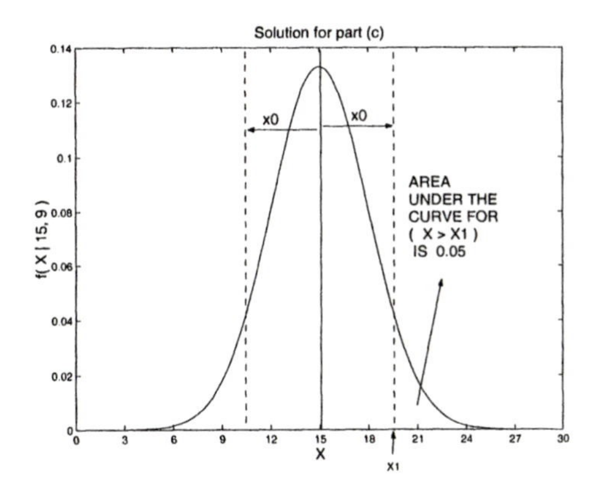

(c)

$$\begin{aligned} Pr(X > x_1) &= 0.05 \\ 1 - Pr(X \leq x_1) &= 0.95 \\ 1 - Pr(\frac{X-15}{3} \leq \frac{x_1 - 15}{3}) &= 0.95 \end{aligned}$$

From the Standard Normal table, look up the value corresponding to 0.95 $\rightarrow$ 1.645.

$$\begin{aligned} \frac{x_1 - 15}{3} &= 1.645 \\ x_1 &= 19.935 \\ x_0 &= x_1 - 15 = 4.935 \end{aligned}$$

A.15 Properties of Normally Distributed Random Variables

1. If $X_1, X_2, \ldots, X_n$ are independent r.v. each with Normal distribution $X_i \sim N(\mu_i, \sigma_i^2)$, then

$$(\sum_{j=1}^{n} X_j) \sim N(\sum_{j=1}^{n} \mu_j, \sum_{j=1}^{n} \sigma_j^2)$$

2. If $Y = (\sum_{i=1}^{n} a_i X_i) + b$, then

$$Y \sim N[(\sum_{i=1}^{n} a_i \mu_i) + b, \sum_{i=1}^{n} a_i^2 \sigma_i^2]$$

A.16 Distribution of Sample Mean

Given a sample $\{X_1, X_2, \ldots, X_n\}$ of random variables, the sample mean is defined as:

$$\bar{X_n} = \sum_{i=1}^{n} \frac{X_i}{n}$$

If the sample is taken from a Normal distribution with mean μ and variance σ^2, then

$$\bar{X_n} \sim N(\mu, \frac{\sigma^2}{n})$$

Example 1.2:

Determine the minimum value of n for which

$$Pr(\mid \bar{X_n} - \mu \mid \leq 1) \geq 0.95$$

if the random sample is taken from a distribution $N(\mu, 9)$.

Solution

Make a change of variable to obtain $Z \sim N(0,1)$, i.e. Standard Normal distribution:

$$Z = \frac{\sqrt{n}}{3}(\bar{X_n} - \mu)$$

Then,

$$\begin{aligned}
Pr(\mid \bar{X_n} - \mu \mid \leq 1) = Pr(\mid Z \mid \leq \frac{\sqrt{n}}{3}) &\geq 0.95 \\
Pr(Z > \frac{\sqrt{n}}{3}) &\leq 0.025 \\
1 - \Phi(\frac{\sqrt{n}}{3}) &\leq 0.025 \\
\Phi(\frac{\sqrt{n}}{3}) &\geq 0.975
\end{aligned}$$

From the Standard Normal table:

$$\frac{\sqrt{n}}{3} \geq 1.96$$

Therefore, n should be at least 35!.

A.17 Likelihood Function and Maximum Likelihood Estimator

Consider the random variables $X_1, X_2, \ldots, X_n$ taken from a distribution with a p.d.f of $f(X \mid \theta)$, where θ is a vector of unknown parameters of this distribution.

Assuming a parameter space Ω which θ belongs, we try to find a region in Ω that θ will most likely to lie. An estimate for θ can be found by observing the random variables X_i's, and choosing the parameter θ that will most likely yield the observed variables. The joint p.d.f. of a set of random observations $x = \{x_1, x_2, \ldots x_n\}$ will be expressed as:

$$f_n(x \mid \theta) = f(x_1 \mid \theta) f(x_2 \mid \theta) \ldots f(x_n \mid \theta)$$

This joint p.d.f is referred to as the *Likelihood Function*, since it yields the distribution of θ for a set of observed variables, x. Variation of the distribution as the parameter θ is changed, will indicate how likely the chosen value of θ is, for a given set of observations. The value of θ, which will maximize the function $f_n(x \mid \theta)$ will be called the *Maximum Likelihood Estimator* **(MLE)** of θ.

A.17.1 Properties of MLE's

1. If $\hat{\theta}$ is MLE of θ, then $g(\hat{\theta})$ will be the MLE of $g(\theta)$.
2. It may not always be possible to express the MLE as an explicit algebraic function.
3. MLE's are consistent estimators, that is the sequence of MLE's will converge to the true unknown value of θ as the sampling size n becomes infinitely large.

In determining the MLE's, it is common to maximize the logarithm of the likelihood function (log likelihood function) instead of the likelihood function itself, in order to simplify the algebra. Since log function is monotonically increasing, the solution of the maximization problem will not be affected by this change of the objective function. The following example shows the procedure of obtaining the MLE for the parameters of a Normal distribution, namely the mean μ and the variance σ^2, based on a finite number of observations.

Example 1.3:

Suppose $\{X_1, X_2, \ldots, X_n\}$ are samples taken from a Normal distribution with unknown μ and σ. Find the MLE of these unknown parameters.

Solution
The likelihood function for n samples from $N(\mu, \sigma^2)$ can be expressed as:

$$f_n(x \mid \mu, \sigma^2) = \frac{1}{\sqrt{2\pi}\sigma^n} exp[-\frac{1}{2\sigma^2}\sum_{i=1}^{n}(x_i - \mu)^2]$$

The log likelihood function will then be given by:

$$\mathcal{L}(\mu, \sigma^2) = -\frac{n}{2} log 2\pi\sigma^2 - \frac{1}{2\sigma^2}\sum_{i=1}^{n}(x_i - \mu)^2$$

Writing the first order optimality conditions for maximizing the log likelihood function $\mathcal{L}$ with respect to the unknown parameters μ and σ^2:

$$\frac{\partial \mathcal{L}}{\partial \mu} = \frac{1}{\sigma^2}\sum_{i=1}^{n}(x_i - \mu) = 0$$

$$\rightarrow \mu = \bar{X}_n \quad \text{sample mean}$$

$$\frac{\partial \mathcal{L}}{\partial \sigma^2} = -\frac{n}{2\sigma^2} + \frac{1}{2\sigma^4} + \sum_{i=1}^{n}(x_i - \mu)^2 = 0$$

$$\rightarrow \sigma^2 = \frac{1}{n}\sum_{i=1}^{n}(x_i - \bar{X}_n)^2 \quad \text{sample variance}$$

It is therefore to be noted that the sample mean and variance constitute MLE's for the unknown parameters of a Normal distribution.

A.18 Central Limit Theorem for the Sample Mean

If the r.v.s. $X_1, X_2, \ldots, X_n$ form a random sample of size n taken from a distribution whose mean is μ and variance σ^2, then for any real number x we can write the following:

$$\lim_{n \to \infty} Pr[\frac{\sqrt{n}(\bar{X}_n - \mu)}{\sigma} \leq x] = \Phi(x)$$

In other words, as the sample size grows, sample mean of any distribution will be distributed more and more like a Normal distribution.

Appendix B

Review of Sparse Linear Equation Solution

Solution of network equations for steady-state, transient or dynamic analysis of power systems operating under normal or emergency conditions all involve solution of large sparse linear equations. In short circuit calculations, each sequence network can be solved separately using the corresponding sequence admittance matrix given below:

$$[Y^s][V^s] = [I^s] \tag{B.1}$$

where superscript s indicates the sequence, i.e. positive, zero or negative. Y^s is the sparse network admittance matrix, I^s is the net bus injection due to the fault and V^s is the bus voltage solution, all defined for the s sequence network.

In power flow solution, nonlinear power balance equations are solved iteratively. At each iteration, a linear equation of the form given below, will be solved in order to obtain the bus voltage corrections, Δx:

$$[J^i][\Delta x^i] = [\Delta S^i] \tag{B.2}$$

where J^i represents the Jacobian of the power balance equations, and ΔS^i is the real and reactive power mismatches at network buses, both evaluated at iteration i.

In solving the state estimation problem using the WLS method, the following set of equations will be solved at each iteration k:

$$H^T R^{-1} H \triangle x^{k+1} = H^T R^{-1}[z - h(x^k)] \tag{B.3}$$

where H is the measurement jacobian evaluated at x^k, R is the measurement error covariance matrix, z is the measurement vector, $h(x^k)$ is the nonlinear measurement function evaluated at x^k, and $\Delta x^{k+1} = x^{k+1} - x^k$ is the incremental change in the state vector at iteration $k+1$. Since, H is a sparse matrix and R is diagonal, the product $H^T R^{-1} H$ will be a sparse and symmetric matrix.

In transient stability studies, the network and machine equations are solved simultaneously at each time step of the simulation period. The machine equations are written as difference equations by discretizing the differential equations of the machines using some numerical integration method, such as trapezoidal rule. The network equations are written as nodal current balance equations in linear form. The resulting system of linear equations will take the following form:

$$\underbrace{\begin{bmatrix} A & B \\ C & Y \end{bmatrix}}_{Jacobian} \begin{bmatrix} \Delta x \\ \Delta V \end{bmatrix} = \underbrace{\begin{bmatrix} R_m \\ \Delta I \end{bmatrix}}_{rhs} \tag{B.4}$$

where the Jacobian matrix is sparse, and the right hand side (rhs) vector is generally full. At each integration step, this linear equation is solved several times iteratively updating the current injections until convergence.

Similarly, in electromagnetic transients simulations, all network branches are modeled by their discrete-time models and the discrete-time network equations given below, are solved at each simulation time step:

$$[G]\,[V] = [I + hist] \tag{B.5}$$

where G is the sparse conductance matrix which depends on network parameters as well as the chosen integration method and step size, V is the node voltage solution, I is the known current injections of the independent current sources, *hist* are the current injections dependent on variables evaluated at previous integration steps.

As evident from the above summary of commonly used analysis tools, solution of sparse linear equations constitutes the essential computational hurdle in almost all power system applications.

While there are numerous methods for the solution of such equations, they can be classified under two broad categories:

1. Direct methods that work on the right hand side vector transforming it into the solution vector after a predetermined and finite number of steps. These steps belong to procedures that are well defined for a given set of equations.

2. Indirect (iterative) methods that start by guessing the solution and repeatedly improving it based on some error criterion. These methods may or may not converge in a reasonable number of iterations

depending on the equations, the iterative scheme employed and machine accuracy.

Almost all power system software is built around direct solvers, and hence only those direct solvers will be reviewed here. Interested readers can look up references [14, 13] for a good review of the iterative solvers.

B.1 Solution by Direct Methods

Consider the linear equation given below:

$$Ax = b \tag{B.6}$$

where A is an nxn matrix, x and b are vectors of order n. Direct solution of x in Eq.(B.6) involves two computational steps:

1. Triangular decomposition of A into its factors:

   $$A = LDU \tag{B.7}$$

   where L and U are lower and upper triangular matrices respectively and their diagonal elements are all equal to 1. D on the other hand, is a strictly diagonal matrix. Also, note that if matrix A is symmetric, then $L = U^T$, hence storage requirements will be reduced accordingly.

   It is also possible to combine LD product into a single lower triangular matrix with non-unity diagonals and express the decomposition using only two matrices as:

   $$A = LU \tag{B.8}$$

   where L in Eq.(B.8) is equal to the product of L and D in Eq.(B.7).

2. Forward and back substitutions:

   Substituting Eq.(B.7) into Eq.(B.6) for A, and defining an intermediate solution vector y as $U\ x$, the solution can be obtained by:

   $$y = D^{-1}L^{-1}b \tag{B.9}$$
   $$x = U^{-1}y \tag{B.10}$$

   Note that, due to the special triangular structure of L and U factors, their inverses need not be explicitly calculated as suggested by the above equations. Instead, a series of substitutions are carried out to transform b into y and subsequently into x, i.e. the final solution. Hence, the above computational steps defined by Eq.(B.9) and Eq.(B.10) are referred to as *Forward Substitutions* and *Back Substitutions* respectively.

Forward and back substitution steps can be studied in terms of *elementary matrices*, possessing some useful properties which will be discussed next.

B.2 Elementary Matrices

There are four types of elementary matrices:

$$L_k^c = \begin{bmatrix} 1 & & & & \\ & \ddots & & & \\ & & 1 & & \\ & & x & 1 & \\ & & \vdots & & \ddots & \\ & & x & & & 1 \end{bmatrix} \qquad L_k^r = \begin{bmatrix} 1 & & & & & \\ & \ddots & & & & \\ & & 1 & & & \\ x & \cdots & x & 1 & & \\ & & & & 1 & \\ & & & & & \ddots & \\ & & & & & & 1 \end{bmatrix}$$

$$U_k^c = \begin{bmatrix} 1 & & & x & & \\ & \ddots & & \vdots & & \\ & & 1 & x & & \\ & & & 1 & & \\ & & & & \ddots & \\ & & & & & 1 \end{bmatrix} \qquad U_k^r = \begin{bmatrix} 1 & & & & & & \\ & \ddots & & & & & \\ & & 1 & & & & \\ & & & 1 & x & \cdots & x \\ & & & & 1 & & \\ & & & & & \ddots & \\ & & & & & & 1 \end{bmatrix}$$

Elementary matrices have the following properties:

1. Their inverses are readily obtained by simply reversing the signs of their off-diagonal entries.

2. Product of elementary lower (upper) triangular matrices yield lower (upper) triangular matrices.

3. Any lower triangular matrix, L with unit diagonals can be written as a product of elementary matrices as:

$$L = L_1^c L_2^c \cdots L_{n-1}^c = L_2^r L_3^r \cdots L_n^r$$

   where L_i^c and L_i^r are formed by using the ith column and row entries of L respectively.

B.3 LU Factorization Using Elementary Matrices

Properly choosing the elements of the lower triangular elementary matrices, a square matrix A can be transformed into an upper triangular form with unit diagonals as shown below:

$$L_n^c L_{n-1}^c \cdots L_2^c L_1^c \cdot A = U$$

Rewriting the above equation:

$$(L_n^c L_{n-1}^c \cdots L_2^c L_1^c)^{-1} \cdot U = A \tag{B.11}$$

$$L \cdot U = A \tag{B.12}$$

Note that, inverse of a lower triangular matrix is also lower triangular, yielding the above defined matrix L.

Decomposition of a non-singular matrix A into its triangular factors may sometimes lead to zero diagonals, causing premature termination of the factorization procedure. Such situations can be resolved by re-ordering the rows and columns of A. This is accomplished by using a permutation matrix P with the following properties:

1. It has a single 1 in each row and column.
2. It is orthogonal, i.e. $P^{-1} = P^T$.
3. Rows (columns) of a matrix A can be reordered by multiplying A by P from the left (right).

In practice, P is never stored as a matrix due to its super sparse structure. Instead, an array of row (column) pointers are used to store the ordering information.

There are more than one ways to decompose a given matrix A into its triangular factors. Here, we will review two of these methods, namely Crout's method and Doolittle's method, both of which can apply to any square matrix. We will then turn our attention to symmetric matrices and review the method of Cholesky for factorization of symmetric matrices.

B.3.1 Crout's Algorithm

This algorithm operates on the columns and rows of matrix A in an alternating manner. At the completion of the procedure, A will be destroyed and replaced by the **T**able **o**f **F**actors, ToF which is a matrix whose lower (upper) triangular part will contain the elements of L (U). If A needs to be kept for other calculations, then a separate matrix ToF can be formed using the same algorithm.

Steps of the algorithm are as follows:

1. Set the first column of ToF equal to the first column of A. This step is redundant if A is being overwritten.

2. Calculate the first row of ToF as:

$$t_{1,k} = a_{1,k}/a_{1,1}$$

where $t_{i,j}$ and $a_{i,j}$ are (i,j)th elements of ToF and A.

3. Set pivot counter $j = 2$.

4. Calculate jth column of ToF:

$$t_{k,j} = a_{k,j} - \sum_{i=1}^{j-1} t_{k,i}t_{i,j} \quad \text{for } k = j, j+1, \ldots, n$$

5. If $j = n$, stop. Else, continue.

6. Calculate jth row of ToF:

$$t_{j,k} = (a_{j,k} - \sum_{i=1}^{j-1} t_{j,i}t_{i,k})/t_{j,j} \quad \text{for } k = j+1, j+2, \ldots, n$$

7. Advance pivot counter, $j = j + 1$. Return to step 4.

Example 2.1:

Given the matrix

$$A = \begin{bmatrix} 4 & 0 & 64 & 0 \\ 1 & -10 & 16 & -20 \\ 2 & 5 & 64 & 10 \\ 0 & -4 & 3 & -16 \end{bmatrix}$$

Find the Table of Factors for A using the Crout's algorithm.

ToF is built following the above described steps of Crout's algorithm. Step by step modification of A into ToF is illustrated below:

$$\begin{bmatrix} 4 & 0 & 64 & 0 \\ 1 & -10 & 16 & -20 \\ 2 & 5 & 64 & 10 \\ 0 & -4 & 3 & -16 \end{bmatrix} \rightarrow \begin{bmatrix} 4 & 0 & 0 & 0 \\ 1 & -10 & 0 & 0 \\ 2 & 5 & 64 & 0 \\ 0 & -4 & 3 & -16 \end{bmatrix} \begin{bmatrix} 1 & 0 & 16 & 0 \\ 0 & 1 & 16 & -20 \\ 0 & 0 & 1 & 10 \\ 0 & 0 & 0 & 1 \end{bmatrix}$$

$$\rightarrow \begin{bmatrix} 4 & 0 & 0 & 0 \\ 1 & -10 & 0 & 0 \\ 2 & 5 & 64 & 0 \\ 0 & -4 & 3 & -16 \end{bmatrix} \begin{bmatrix} 1 & 0 & 16 & 0 \\ 0 & 1 & 0 & 2 \\ 0 & 0 & 1 & 10 \\ 0 & 0 & 0 & 1 \end{bmatrix}$$

$$\rightarrow \begin{bmatrix} 4 & 0 & 0 & 0 \\ 1 & -10 & 0 & 0 \\ 2 & 5 & 32 & 0 \\ 0 & -4 & 3 & -8 \end{bmatrix} \begin{bmatrix} 1 & 0 & 16 & 0 \\ 0 & 1 & 0 & 2 \\ 0 & 0 & 1 & 0 \\ 0 & 0 & 0 & 1 \end{bmatrix}$$

B.3.2 Doolittle's Algorithm

This algorithm is very similar to the Crout's algorithm presented above. Instead of updating alternating rows and columns of the pivot, all elements in the lower right submatrix below the pivot, are updated at each pivoting step. Again, at the completion of the procedure, A will be destroyed and replaced by the **T**able **o**f **F**actors, ToF.

Note that the first two steps of this algorithm are identical to that of Crout's. Steps of the Doolittle's algorithm are given below:

1. Set the pivot counter $k = 1$.

2. Set the pivot column of ToF equal to the pivot column of A. This step is redundant if A is being overwritten.

3. Calculate the pivot row of ToF as:

$$t_{k,j} = a_{k,j}/a_{k,k}$$

   where $t_{k,j}$ and $a_{k,j}$ are (k,j)th elements of ToF and A.

4. Calculate each (i,j)th element of ToF for $i > k$ and $j > k$ according to:

$$t_{i,j} = a_{i,j} - t_{k,j}t_{i,k} \quad \text{forall } i,j > k$$

5. Set $k = k + 1$. If k=n, stop. Else, go to step 2.

Example 2.2:

Given the matrix

$$A = \begin{bmatrix} 2 & 4 & 4 & 2 \\ 3 & 3 & 12 & 6 \\ 2 & 4 & -1 & 2 \\ 4 & 2 & 1 & 1 \end{bmatrix}$$

Find the Table of Factors for A using the Doolittle's algorithm.

ToF is built following the above described steps of Doolittle's algorithm. Step by step modification of A into ToF is illustrated below:

$$\begin{bmatrix} 2 & 4 & 4 & 2 \\ 3 & 3 & 12 & 6 \\ 2 & 4 & -1 & 2 \\ 4 & 2 & 1 & 1 \end{bmatrix} \rightarrow \begin{bmatrix} 2 & 2 & 2 & 1 \\ \hline 3 & 3 & 12 & 6 \\ 2 & 4 & -1 & 2 \\ 4 & 2 & 1 & 1 \end{bmatrix} \rightarrow \left[\begin{array}{c|ccc} 2 & 2 & 2 & 1 \\ \hline 3 & -3 & 6 & 3 \\ 2 & 0 & -5 & 0 \\ 4 & -6 & -7 & -3 \end{array}\right]$$

$$\rightarrow \begin{bmatrix} 2 & 2 & 2 & 1 \\ 3 & -3 & -2 & -1 \\ \hline 2 & 0 & -5 & 0 \\ 4 & -6 & -7 & -3 \end{bmatrix} \rightarrow \left[\begin{array}{cc|cc} 2 & 2 & 2 & 1 \\ 3 & -3 & -2 & -1 \\ \hline 2 & 0 & -5 & 0 \\ 4 & -6 & -19 & -9 \end{array}\right]$$

B.3.3 Factorization of Sparse Symmetric Matrices

Triangular factorization of sparse symmetric matrices can be carried out by Cholesky method. There are two variations to the Cholesky method of factorization:

- Build two identical factors, one being the transpose of the other. This requires a positive definite (p.d.) matrix due to the perfect square diagonal entries.

- Build two identical but transposed factors with unit diagonals and one diagonal factor.

Since the matrices involved in state estimation problem will commonly be positive definite and symmetric, only the first variation will be discussed here.

Consider a sparse p.d. symmetric matrix A. This matrix can be written as a product of three matrices as below:

$$\begin{aligned} A = A_0 &= \begin{bmatrix} d_1 & a_1^T \\ a_1 & B_1' \end{bmatrix} \\ &= \begin{bmatrix} \sqrt{d_1} & 0 \\ \frac{a_1}{\sqrt{d_1}} & I_{n-1} \end{bmatrix} \begin{bmatrix} 1 & 0 \\ 0 & \underbrace{B_1' - \frac{a_1 a_1^T}{d_1}}_{B_1} \end{bmatrix} \begin{bmatrix} \sqrt{d_1} & \frac{a_1^T}{\sqrt{d_1}} \\ 0 & I_{n-1} \end{bmatrix} \\ &= L_1 \cdot A_1 \cdot L_1^T \end{aligned}$$

The above decomposition can now be applied to the submatrix B_1 as below:

$$
\begin{aligned}
A_1 &= \begin{bmatrix} 1 & 0 \\ 0 & B_1 \end{bmatrix} = \begin{bmatrix} 1 & 0 & 0 \\ 0 & d_2 & a_2^T \\ 0 & a_2 & B_2' \end{bmatrix} \\
&= \begin{bmatrix} 1 & & \\ 0 & \sqrt{d_2} & \\ 0 & \frac{a_2}{\sqrt{d_2}} & I_{n-2} \end{bmatrix} \begin{bmatrix} 1 & 0 & 0 \\ 0 & 1 & 0 \\ 0 & 0 & B_2' - \frac{a_2 a_2^T}{d_2} \end{bmatrix} \begin{bmatrix} 1 & 0 & 0 \\ & \sqrt{d_2} & \frac{a_2^T}{\sqrt{d_2}} \\ & & I_{n-2} \end{bmatrix} \\
&= L_2 \cdot A_2 \cdot L_2^T
\end{aligned}
$$

Continuing with this process, A can be written as the following product of elementary matrices:

$$
\begin{aligned}
A &= \underbrace{L_1 \cdot L_2 \cdots L_{n-1} \cdot L_n}_{L} \cdot \underbrace{L_n^T \cdot L_{n-1}^T \cdots L_2^T \cdot L_1^T}_{L^T} \\
&= L \cdot L^T
\end{aligned}
$$

B.3.4 Ordering Sparse Symmetric Matrices

Cholesky factorization scheme proceeds one pivot at a time, modifying the original matrix elements below the pivot row and column according to the following equation:

$$
B' - \frac{aa^T}{d}
$$

where
B' is the lower right corner submatrix below the pivot row and column
a is the column array containing elements of the matrix below the pivot
d is the value of the pivot element

Hence, the matrix aa^T will have as many non-zeros as the square of the number of nonzeros in a. Since, each such nonzero will potentially create a "fill-in" (a nonzero element in a position originally occupied by a zero in A), the choice of pivots during the factorization process will directly affect the sparsity of the resulting factor L.

A commonly used ordering scheme known as the minimum degree or Tinney-2 ordering, yields almost optimum results in terms of maintaining sparsity of L. Tinney-2 scheme of ordering rows/columns can be outlined as follows:

1. Choose the row with the minimum number of nonzeros.

2. Using the chosen row/column as the pivot, carry out one elimination step using Cholesky method, i.e. $B' - aa^T/d$.

3. Go back to step 1 and repeat the procedure until all pivots are processed.

Note that, in finding the Tinney-2 ordering for a given sparse symmetric matrix, Cholesky elimination steps need not be carried out numerically. It is possible to predict the locations of nonzeros in the product aa^T and therefore the "fill-in" elements can be determined symbolically. There is a small chance that a nonzero element may become zero as a result of perfect numerical cancellations during the elimination steps and symbolic processing will not be able to detect it. However, chance of this happening is really small and hence it is not taken into account. Certain applications may call for repeated factorization of a sparse matrix A, whose sparsity pattern remains the same while its elements change numerically. For those cases, the use of symbolic factorization to order the matrix once and subsequently repeating numerical factorization using this ordering, will be computationally more efficient.

Example 2.3:

Let A be a sparse symmetric matrix as given below:

$$A = \begin{bmatrix}
12 & 1 & & & 1 & & & & & & \\
1 & 12 & 1 & & & & & & & & \\
 & 1 & 12 & & & 1 & & & & & \\
 & & & 12 & & & 1 & & 1 & & \\
1 & & & & 12 & 1 & & 1 & & & \\
 & & 1 & & 1 & 12 & & & 1 & & \\
 & & & 1 & & & 12 & & & & 1 \\
 & & & & 1 & & & 12 & & 1 & \\
 & & & 1 & & 1 & & & 12 & 1 & 1 \\
 & & & & & & & 1 & 1 & 12 & \\
 & & & & & & 1 & & 1 & & 12
\end{bmatrix}$$

Applying the Tinney-2 ordering algorithm to matrix A yields the following row/column order:

$$\begin{bmatrix} 8 & 10 & 1 & 3 & 2 & 7 & 11 & 4 & 5 & 9 & 6 \end{bmatrix}$$

B.4 Factorization Path Graph

Factorization of a sparse symmetric matrix can be carried out using the Cholesky method as described above. The procedure involves essentially sequential processing of one pivot row/column at a time until all pivots are processed. Special sparse structure of these matrices makes it possible to further simplify the solution procedures that involve forward and back substitution steps. This can be accomplished by constructing what is referred to as the *factorization path graph* of the matrix for a chosen ordering. Factorization path graph (FPG), also known as factorization tree, is a compact representation of the sparsity structure of the lower triangular (or upper triangular) factor of the sparse matrix A. FPG can be stored in compact form in a single array F whose entries, for $i = 1, \ldots, n-1$, are defined by $F(i) = k$, where k is the row index corresponding to the first nonzero entry below the pivot in column i of the lower triangular factor of A. For convenience, $F(n)$ is set to the dummy value -1, indicating that all paths end at the last row for a connected network. The factorization path for a given row i can be recursively traced as follows:

1. Initialize the list with row i.

2. Substitute i by $F(i)$. If $i < 0$ stop, otherwise continue.

3. Add i to the list and go to step 2.

In the above example, after ordering A according to Tinney-2 algorithm and factorizing it into its lower and upper triangular factors using the Cholesky method, above FPG building procedure can be carried out. The resulting FPG is given in compact form as the following F vector:

$$F = \begin{bmatrix} 2 & 9 & 5 & 5 & 9 & 7 & 8 & 10 & 10 & 11 & -1 \end{bmatrix}$$

Here, $F(i)$ represents the node number that is below the node i in the factorization path graph. Naturally, at each fork, more than one node will be followed by the same node, as is the case for nodes 3 and 4, both of which are followed by 5 in this example. A graphical representation of FPG for this example is shown in Figure B.1. Note that the numbers in the parenthesis in Figure B.1 indicate the row/column indices of the elements in the original (unordered) matrix A. Following the ordering and factorization of A, the FPG can be obtained by tracing the columns of the lower triangular factor L. Figure B.2 shows the FPG tracing procedure on the lower triangular factor L.

B.5 Sparse Forward/Back Substitutions

Following the triangular factorization of a sparse matrix A, solution of Eq.(B.6) will be carried out by the forward and back substitutions as in Eq.(B.9, B.10). Forward substitutions involve sequential updating of the

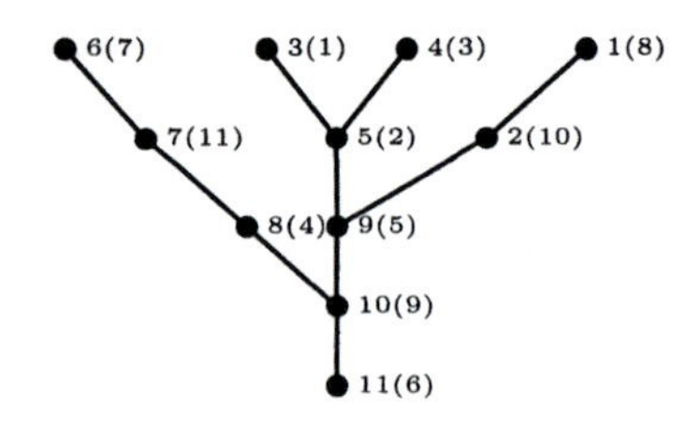

Figure B.1. Factorization Path Graph of matrix A.

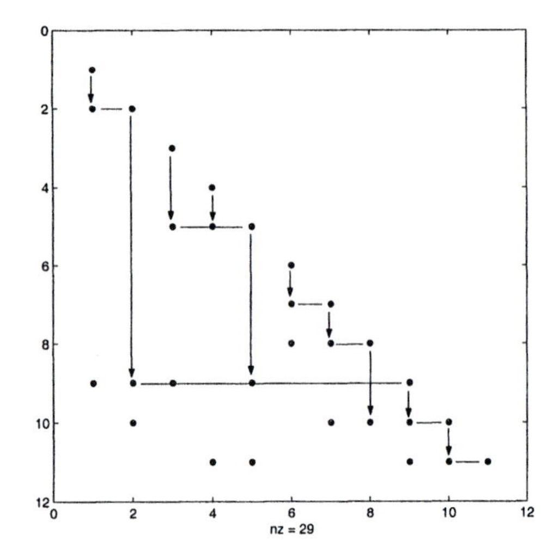

Figure B.2. Obtaining the FPG using the lower factor L.

right hand side vector b, by processing columns of L one at a time as follows:

$$\left.\begin{array}{ccr} b_i & \leftarrow & b_i/L_{i,i} \\ b_{i+1} & \leftarrow & b_{i+1} - L_{i+1,i} \cdot b_i \\ \vdots & \vdots & \vdots \\ b_n & \leftarrow & b_n - L_{n,i} \cdot b_i \end{array}\right\} \quad i = 1, 2, \ldots, n$$

Forward substitution steps can be greatly simplified by noting that the information contained in FPG will indicate which entries of b will have to be updated in processing a given column. In addition, FPG also indicates which columns can be altogether skipped during the forward substitution provided that certain elements of the right hand side vector are zero.

Considering the matrix A of the above example, let us assume that the right hand side vector is a sum of two *singleton* vectors as follows:

$$b = s_1 + s_2 = [1\ 1\ 0 \ldots\ 0]^T$$

where, a *singleton* s_k is defined as a vector whose only nonzero entry is in position k with a value 1.0:

$$s_k = \begin{bmatrix} s_k(1) \\ \vdots \\ s_k(k) \\ \vdots \\ s_k(n) \end{bmatrix} = \begin{bmatrix} 0 \\ \vdots \\ 1.0 \\ \vdots \\ 0 \end{bmatrix}$$

Note that after ordering, the nonzero elements of b will assume positions 3 and 5 in the ordered right hand side vector (see Figure B.1). Looking at the lower triangular factor L in Figure B.2, it can be shown that for the given b, forward substitution steps can be carried out by processing the columns 3, 5, 9, 10 and 11 of L. This is the union of the factorization paths corresponding to the singletons s_3 and s_5 (or s_1 and s_2 in the unordered vector) as can be traced from Figure B.1. Hence, FPG can be used to identify the set of columns to be processed during forward substitution for a given right hand side vector b. Performing the forward elimination process in this way, i.e., skipping columns corresponding to null elements in b, is called *fast forward elimination.*

The same idea can be applied during the backward substitution process when only a few entries of the unknown vector x are wanted. Specifically, it is easy to verify that, in order to obtain x_i, it is sufficient to previously compute the entries x_k for those rows k belonging to the factorization path of i. In this case, the path is traced in the opposite direction. Again, obtaining several elements of x requires that the union of paths corresponding to the needed rows be traced. This procedure is known as *fast back substitution.*

Clearly, the computational cost of sparse vector operations is proportional to the number of rows involved in the process. Hence, bush-shaped trees, rather than cypresses, are preferable from this point of view. Although Tinney-2 ordering scheme usually yields reasonable FPGs, in certain cases it may provide unacceptable results. Several algorithms have been proposed aimed at reducing the average and/or maximum path lengths, while keeping under control the number of fill-ins [7, 3]. Most of them take advantage of the fact that, at each stage of the Tinney-2 method, there are many candidate (minimum degree) nodes to be chosen, and a second tie-break criterion is proposed based on past elimination steps.

B.6 Solution of Modified Equations

In many situations, there is a need to solve an equation system whose coefficient matrix differs from that of a previously solved system by very few elements. When this happens, it is wasteful to solve the modified system from scratch, as significant computational savings can be achieved by taking advantage of the operations already performed to solve the base case.

Let x_0 denote the solution of the original $n \times n$ system

$$Ax_0 = b$$

whose coefficient matrix has been factorized as $A = LU$. Then, the modified system can be in general expressed as follows:

$$(A + M\Delta N^T)x = b \tag{B.13}$$

where Δ is an $m \times m$ matrix and M, N are $n \times m$ sparse matrices, m being directly related to the number of modified branches (or buses). When $m = 1$ the change is referred to as a *rank-1* modification.

For instance, if A represents the bus admittance matrix of a certain system to which a single series branch connecting nodes i and j is added, then Δ reduces to a scalar whose value is the branch admittance, M is a null column vector except for entries $m_i = 1$ and $m_j = -1$, and $N = M$. Removing a branch is equivalent to adding a branch with negative admittance.

Several techniques have been proposed to efficiently deal with this problem, all of which can be grouped within two categories:

- Methods that update the elements of the ToF strictly necessary, i.e., only those which become modified (partial refactorization).

- Methods that modify the independent vector so that it takes care of the changes performed on the system (compensation).

Depending on the application (single or repeated solutions, temporary or permanent changes, etc.) one class of methods will be the most appropriate, but it is not possible in general to provide a rule of thumb in this regard.

Compensation-based methods are well-known in linear circuit theory, where A is the bus admittance matrix and b represents nodal current injections. On the other hand, partial refactorization constitutes a much more recent development based on the factorization path concept. Owing to their apparently different nature and origin, both techniques are customarily presented in separate fashions, usually by means of different notation. In the sequel, however, both categories of methods will be developed from the common framework provided by (B.13), and their similarities will be stressed.

B.6.1 Partial Refactorization

All techniques within this category are based on the important observation that when a certain row (or column) of a structurally symmetric matrix gets modified, only the rows/columns of its ToF belonging to the factorization path of the modified row need to be updated.

The modified matrix can be rearranged as follows

$$A_m = LU + M\Delta N^T = L\underbrace{[I + V\Delta W^T]}_{C}U \tag{B.14}$$

where

$$V = L^{-1}M \quad \text{and} \quad W = U^{-T}N$$

are obtained by means of m fast forward eliminations on the columns of M and N respectively. In (B.14) the matrix within brackets differs from the identity matrix only in the rows/columns pertaining to the path, and can be factorized as,

$$C = L_c U_c$$

from which the ToF of A_m are obtained,

$$A_m = L[L_c U_c]U = (LL_c)(U_c U) = L_m U_m$$

It is worth mentioning that no new fill-ins arise when factorizing matrix C, because all of its non trivial rows/columns share the same non-zero pattern dictated by the respective path. In certain applications (e.g., non permanent changes) it may be advantageous not to explicitly carry out the products LL_c and $U_c U$. In case such products are needed, it should be kept in mind that only the columns (rows) of L_m (U_m) belonging to the involved path differ from those of L (U).

Example 2.4:

Consider the following structurally symmetric matrix,

$$A = \begin{bmatrix} 2 & & & -0.2 & & & -0.1 & \\ & 3 & & & -0.3 & & & \\ & & 4 & & & -0.1 & & -0.1 \\ -0.4 & & & 2.04 & & -0.2 & -0.08 & \\ & -0.9 & & & 1.09 & & -0.1 & \\ & & -0.4 & -0.4 & & 3.05 & -0.08 & -0.19 \\ -0.2 & & & -0.18 & -0.1 & -0.28 & 1.04 & -0.08 \\ & & -0.4 & & & -0.59 & -0.08 & 1.06 \end{bmatrix}$$

whose ToF are,

$$A = LU = \begin{bmatrix} 1 & & & & & & & \\ & 1 & & & & & & \\ & & 1 & & & & & \\ -0.2 & & & 1 & & & & \\ & -0.3 & & & 1 & & & \\ & & -0.1 & -0.2 & & 1 & & \\ -0.1 & & & -0.1 & -0.1 & -0.1 & 1 & \\ & & -0.1 & & & -0.2 & -0.1 & 1 \end{bmatrix} \cdot \begin{bmatrix} 2 & & & -0.2 & & & -0.1 & \\ & 3 & & & -0.3 & & & \\ & & 4 & & & -0.1 & & -0.1 \\ & & & 2 & & -0.2 & -0.1 & \\ & & & & 1 & & -0.1 & \\ & & & & & 3 & -0.1 & -0.2 \\ & & & & & & 1 & -0.1 \\ & & & & & & & 1 \end{bmatrix}$$

Now, the scalar 0.2 is added to a_{33} and a_{66}, and subtracted from a_{63} and a_{36}, yielding a modified matrix which can be expressed as

$$\begin{aligned} &A_m = A + M\Delta N^T = \\ &A + \begin{bmatrix} 0 & 0 & 1 & 0 & 0 & -1 & 0 & 0 \end{bmatrix}^T 0.2 \begin{bmatrix} 0 & 0 & 1 & 0 & 0 & -1 & 0 & 0 \end{bmatrix} \end{aligned}$$

The auxiliary arrays V and W are

$$\begin{aligned} V = L^{-1}M &= \begin{bmatrix} 0 & 0 & 1 & 0 & 0 & -0.9 & -0.09 & -0.089 \end{bmatrix}^T \\ W = U^{-T}N &= \begin{bmatrix} 0 & 0 & 0.25 & 0 & 0 & -0.325 & -0.0325 & -0.0432 \end{bmatrix}^T \end{aligned}$$

Note that the nonzero elements in V and W correspond with the factorization path of row 3 (rows 3, 6, 7 and 8).

The intermediate matrix C is given by

$$C = I + V\Delta W^T = \begin{bmatrix} 1 & & & & & & & \\ & 1 & & & & & & \\ & & 1.0500 & & & -0.0650 & -0.0065 & -0.0086 \\ & & & 1 & & & & \\ & & & & 1 & & & \\ & & -0.0450 & & & 1.0585 & 0.0059 & 0.0078 \\ & & -0.0045 & & & 0.0058 & 1.0006 & 0.0008 \\ & & -0.0045 & & & 0.0058 & 0.0006 & 1.0008 \end{bmatrix}$$

and its ToF,

$$
L_cU_c = \begin{bmatrix}
1 & & 0 & & & & & \\
& 1 & 0 & & & & & \\
& & 1.0000 & & & & & \\
& & 0 & 1 & & & & \\
& & 0 & & 1 & & & \\
& & -0.0429 & & & 1 & & \\
& & -0.0043 & & & 0.0053 & 1 & \\
& & -0.0042 & & & 0.0052 & 0.0005 & 1
\end{bmatrix}.
\begin{bmatrix}
1 & & & & & & & \\
& 1 & & & & & & \\
& & 1.050 & & & -0.0650 & -0.0065 & -0.0086 \\
& & & 1 & & & & \\
& & & & 1 & & & \\
& & & & & 1.0557 & 0.0056 & 0.0074 \\
& & & & & & 1.0005 & 0.0007 \\
& & & & & & & 1.0007
\end{bmatrix}
$$

Consequently, the ToF corresponding to A_m are,

$$
L_mU_m = \begin{bmatrix}
1 & & & & & & & \\
& 1 & & & & & & \\
& & 1 & & & & & \\
-0.2 & & & 1 & & & & \\
& -0.3 & & & 1 & & & \\
& & -0.1429 & -0.2 & & 1 & & \\
-0.1 & & & -0.1 & -0.1 & -0.0947 & 1 & \\
& & -0.0952 & & & -0.1953 & -0.0995 & 1
\end{bmatrix}.
\begin{bmatrix}
2 & & & -0.2 & & & -0.1 & \\
& 3 & & & -0.3 & & & \\
& & 4.2 & & & -0.1 & & -0.1 \\
& & & 2 & & -0.2 & -0.1 & \\
& & & & 1 & & -0.1 & \\
& & & & & 3.1671 & -0.1 & -0.2043 \\
& & & & & & 1.0005 & -0.0993 \\
& & & & & & & 1.0007
\end{bmatrix}
$$

Compared to the original ToF, it is clear that only rows/columns 3, 6, 7 and 8 have changed.

B.6.2 Compensation

The idea is to express the solution of the modified system

$$A_mx = b$$

in terms of the ToF of the original matrix A, rather than computing the ToF of A_m. According to (B.14), the solution of the modified system can be obtained from,

$$x = U^{-1}C^{-1}L^{-1}b \tag{B.15}$$

The inverse of matrix C in the above expression is provided by the so-called *inverse matrix modification lemma*, which states that,

$$C^{-1} = I - V\delta W^T \tag{B.16}$$

where the auxiliary $m \times m$ matrix δ must be previously computed from,

$$\delta = (W^T V + \Delta)^{-1}$$

Whenever Δ is rank deficient the following alternative expression for δ can be used,

$$\delta = \Delta(W^T V \Delta + I)^{-1}$$

Substituting (B.16) into (B.15) yields,

$$x = U^{-1}[I - V\delta W^T]L^{-1}b \tag{B.17}$$

Such a way of arranging the computations is known as *mid-compensation* because the intermediate vector $L^{-1}b$ is "compensated" with the term in brackets, and then back substitution is applied to the resulting vector [2, 1]. It is also possible to compensate the independent vector (*pre-compensation*) or the solution vector (*post-compensation*) by applying

$$x = A^{-1}(I - M\delta A^{-1})b$$

or

$$x = (I - A^{-1}M\delta)x_0$$

respectively. In a majority of applications, mid-compensation is believed to be the best choice.

Example 2.5:

Resorting to the ToF of A in the former example, obtain the solution of

$$A_m x = b$$

with

$$b = \begin{bmatrix} 1 & 2 & 0 & -1 & 2 & 0 & 1 & 0 \end{bmatrix}^T$$

From the matrices obtained above, the scalar δ is first computed

$$\delta = (W^T V + \Delta)^{-1} = 1.3346$$

Next, forward elimination on b yields the intermediate vector

$$y = L^{-1}b = \begin{bmatrix} 1 & 2 & 0 & -0.8 & 2.6 & -0.16 & 1.264 & 0.0944 \end{bmatrix}^T$$

Then, y is compensated as follows

$$\begin{aligned} y_c &= (I - V\delta W^T)y = \\ & \begin{bmatrix} 1 & 2 & -0.0091 & -0.8 & 2.6 & -0.1518 & 1.2648 & 0.0952 \end{bmatrix}^T \end{aligned}$$

Finally, back substitution is applied to y_c,

$$x = U^{-1}y_c =$$
$$\begin{bmatrix} 0.5300 & 0.9394 & 0.0019 & -0.3367 & 2.7274 & -0.0042 & 1.2736 & 0.0945 \end{bmatrix}^T$$

B.7 Sparse Inverse

Although the inverse of a sparse matrix is generally full, there are cases in which only a selected subset of elements of the inverse are needed.

One of such cases arises in sensitivity analysis. Given a linearized model

$$A\Delta x = \Delta u$$

the influence of a certain control action u_j on a dependent variable x_i is determined from:

$$\Delta x_i = s_{ij} u_j$$

where the sensitivity coefficient s_{ij} is the respective element of A^{-1}. The fastest way of computing such a single coefficient is by performing a fast forward elimination on the singleton vector s_j, followed by a fast back substitution ending at the i-th row. A conventional back substitution is justified only when a significant portion of sensitivity coefficients, with respect to the same variable u_j, is required.

In a few but important cases, like short-circuit analysis and bad data identification in WLS state estimation, only the elements of the inverse lying at the same positions as the non-zero elements of the ToF are of interest. Such subset of elements is known as the *sparse inverse*, and the approach based on sparse vector methods is inefficient in this situation, because certain unwanted elements are computed during the fast back substitution processes. The method described below, proposed by Takahashi *et al.* [9], constitutes the cheapest approach to obtain the sparse inverse.

Let us assume for simplicity that A is symmetric. Then, it can be factorized as

$$A = LDL^T$$

and its inverse S can be obtained from,

$$L^T S = D^{-1} L^{-1} \tag{B.18}$$

where the right hand side matrix is lower triangular with diagonal elements d_{ii}^{-1}. The procedure computes columns of the sparse inverse from the right-most column to the left. Within each column j the diagonal element s_{jj} is first computed from,

$$s_{jj} + \sum_{k=j+1}^{n} l_{kj} s_{kj} = d_{jj}^{-1}$$

This is possible because all elements s_{kj} above lie below the diagonal and can be retrieved from the respective transposed elements of columns already computed.

Next, off-diagonal elements s_{ij}, for $i = j-1, j-2, \ldots, 1$, are obtained from,

$$s_{ij} + \sum_{k=i+1}^{n} l_{ki} s_{kj} = 0$$

where again the elements s_{kj} are available from preceding steps (either in the same column or as transposed elements in previously computed columns).

Note that, in both expressions above, off-diagonal elements of S are always multiplied by elements of L, which explains why just the sparse inverse is involved in the computations.

Example 2.6:

Consider the 11×11 matrix of example 2.3, reordered and factorized as shown in figure B.2. Assuming the 11-th column of the sparse inverse is already available, computation of its 10-th column above the diagonal proceeds as follows:

$$\begin{aligned}
s_{10,10} &= d_{10,10}^{-1} - l_{11,10}[s_{11,10}] \\
s_{9,10} &= -l_{10,9} s_{10,10} - l_{11,9}[s_{11,10}] \\
s_{8,10} &= -l_{9,8} s_{9,10} - l_{10,8} s_{10,10} - l_{11,8}[s_{11,10}] \\
s_{7,10} &= -l_{8,7} s_{8,10} - l_{10,7} s_{10,10} - l_{11,7}[s_{11,10}] \\
s_{2,10} &= -l_{9,2} s_{9,10} - l_{10,2} s_{10,10}
\end{aligned}$$

where elements within brackets are taken from their transpose.

B.8 Orthogonal Factorization

Orthogonal decomposition constitutes a numerically more stable alternative to the LU factorization, in particular when the coefficient matrix is very ill-conditioned and computational effort is not the main concern.

Any $m \cdot n$ matrix A of full rank can be factorized into two matrices of the form:

$$A = QR \tag{B.19}$$

where Q is an $m \cdot m$ orthogonal matrix (i.e., $Q^T = Q^{-1}$) and R is an $m \cdot n$ upper trapezoidal matrix (i.e., its first n rows are upper triangular while the remaining $m - n$ rows are null). The alternative expression,

$$Q^T A = R$$

allows practical factorization algorithms to obtain R as a sequence of elementary orthogonal transformations on the columns (rows) of A, as follows:

$$(Q_p^T \ldots (Q_2^T(Q_1^T A))) = R$$

where p is the number of steps required to transform A into R.

Properly partitioning Q and R gives rise to the following reduced form of the QR factorization:

$$A = \begin{bmatrix} Q_n & Q_0 \end{bmatrix} \begin{bmatrix} U \\ 0 \end{bmatrix} = Q_n U \quad \Rightarrow \quad Q_n^T A = U \tag{B.20}$$

It is therefore sufficient to build only the submatrix Q_n rather than the full Q.

Whenever A is square, the unique solution of the linear system $Ax = b$ can be obtained by back substitution on the triangular system,

$$Ux = Q_n^T b$$

It is easy to show that the above system also provides the least-squares solution in the general rectangular case ($m > n$).

Factorization based on the so-called Givens rotations is considered most efficient for large sparse matrices. The most popular version proceeds columnwise from left to right. Each elemental transformation eliminates a single element below the diagonal, in ascending order of rows.

Assume the first $j - 1$ columns, as well as column j up to row $i - 1$, have been processed. Then, elimination of the element a_{ij} is performed by pre-multiplying A times the orthogonal matrix,

$$Q_k = \begin{array}{c} \\ \\ j \\ \\ i \\ \\ \\ \end{array} \begin{bmatrix} 1 & & & & & \\ & \ddots & & & & \\ & & c & & -s & \\ & & & \ddots & & \\ & & s & & c & \\ & & & & & \ddots & \\ & & & & & & 1 \end{bmatrix}$$

(column labels j and i above the columns containing c, s and $-s$, c respectively)

where

$$\begin{aligned} c &= a_{jj}\left(a_{jj}^2 + a_{ij}^2\right)^{-1/2} \\ s &= a_{ij}\left(a_{jj}^2 + a_{ij}^2\right)^{-1/2} \end{aligned}$$

Geometrically, this is equivalent to rotating the columns of A by an angle

$$\theta = \arctan s/c = \arctan a_{ij}/a_{jj}$$

Note that, in practice, the orthogonal matrix

$$Q = Q_1 Q_2 \ldots Q_p$$

is not explicitly built. Instead, the coefficients s and c defining each elemental rotation replace *in situ* the eliminated element a_{ij}.

When applying Givens rotations to large sparse matrices, proper attention must be paid to fill-in elements arising in the process. It is easy to verify that when the element a_{ij} gets eliminated, the resulting non-zero pattern of both rows, j and i, is the union of the original non-zero patterns corresponding to both rows.

Row/column ordering strategies have been developed in an attempt to reduce the number of fill-ins. Such strategies are based on the observation that row permutations on A do not affect $A^T A$, while column permutations on A are equivalent to symmetric permutations on $A^T A$. Therefore, permutation of the columns of A is determined by applying the minimum degree algorithm (Tinney's 2 scheme) to $A^T A$, which reduces the fill-in in U. Intermediate fill-in, i.e., that arising in the factors of Q, is kept under control by ordering the rows of A in ascending order of the index

$$\max\{j : a_{ij} \neq 0\}$$

It is possible to carry out the rotations in such a way that no square roots are necessary (see for instance [12] for implementation details). However, as shown by the results presented in Chapter 3, QU factorization is computationally much more expensive than other techniques based on augmented matrices and LU decomposition.

Example 2.7:

Consider the following matrix

$$A = \begin{bmatrix} 2 & & & & & & & \\ & 1 & & & & & & \\ & & 3 & & 2 & & 1 & \\ & & & 4 & & & & \\ & & & & 3 & & & \\ & & 4 & & & 6 & & 1 \\ & & & 1 & & & 2 & \\ & & & & 2 & & 3 & 1 \end{bmatrix}$$

where the first two columns are already upper triangular. In order to eliminate the element $a_{63} = 4$, the columns of A are rotated by the angle

$$\begin{aligned} \theta &= \arctan a_{63}/a_{33} = 53.13^\circ \\ c &= \cos\theta = 0.6 \\ s &= \sin\theta = 0.8 \end{aligned}$$

This leads to the following elemental orthogonal matrix

$$Q = \begin{bmatrix} 1 & & & & & & & \\ & 1 & & & & & & \\ & & 0.6 & & & -0.8 & & \\ & & & 1 & & & & \\ & & & & 1 & & & \\ & & 0.8 & & & 0.6 & & \\ & & & & & & 1 & \\ & & & & & & & 1 \end{bmatrix}$$

yielding,

$$A_{\text{new}} = Q^T A_{\text{old}} = \begin{bmatrix} 2 & & & & & & & \\ & 1 & & & & & & \\ & & 5 & & 1.2 & 4.8 & 0.6 & 0.8 \\ & & & 4 & & & & \\ & & & & 3 & & & \\ & & & & -1.6 & 3.6 & -0.8 & 0.6 \\ & & & 1 & & & 2 & \\ & & & & 2 & & 3 & 1 \end{bmatrix}$$

Note the extra fill-in elements at rows 3 and 6.

B.9 Storage and Retrieval of Sparse Matrix Elements

While there are numerous schemes developed for sparse matrix storage, some of them apply to matrices with special properties such as symmetric, triangular, banded, etc. The one that will be reviewed here is most general and can be used for any sparse matrix with arbitrary structure. It was suggested by Knuth [8] and is referred to as the *linked list* storage scheme, or simply *Knuth's method* of storage. Knuth's scheme requires 7 arrays for storing an $n \times m$ sparse matrix A. These are described below:

V This is the array containing all the nonzero elements of the sparse matrix in compact form stored in arbitrary order. Dimension of V will be equal to the number of nonzeros in A.

Row This array contains the row index of the corresponding elements stored in V. E.g. the row index of the element stored in $V(i)$ will be $Row(i)$.

Col This array contains the index of the corresponding column elements stored in V. E.g. the column index of the element stored in $V(i)$ will be $Col(i)$.

$NextR$ This array contains pointers to the next nonzero element location in the same row. E.g. if $NextR(i) = k$, then the next nonzero element in the same row as $V(i)$ will be $V(k)$.

$NextC$ This array contains pointers to the next nonzero element location in the same column. E.g. if $NextC(i) = k$, then the next nonzero element in the same column as $V(i)$ will be $V(k)$.

$BeginR$ This array contains pointers to the beginning of each row. E.g. if $BeginR(i) = k$, then the first nonzero element in row i will be $V(k)$.

$BeginC$ This array contains pointers to the beginning of each column. E.g. if $BeginC(i) = k$, then the first nonzero element in column i will be $V(k)$.

As matrix elements are randomly stored in array V, the auxiliary arrays $BeginR$ ($BeginC$) and $NextR$ ($NextC$) are needed to sequentially sweep a given row (column) in ascending order of columns (rows).

Example 2.8:

Consider the (3 × 5) sparse matrix A given below, where the null values are retained for clarity:

$$A = \begin{bmatrix} 0 & 16 & 0 & 340 & 74 \\ 7 & 0 & 0 & 0 & -51 \\ 0 & 44 & 0 & 14 & 23 \end{bmatrix}$$

This matrix can be stored using the following arrays:

$$\begin{aligned}
V &= [\ -51 \quad 44 \quad 340 \quad 23 \quad 14 \quad 7 \quad 74 \quad 16\ ] \\
Row &= [\ 2 \quad 3 \quad 1 \quad 3 \quad 3 \quad 2 \quad 1 \quad 1\ ] \\
Col &= [\ 5 \quad 2 \quad 4 \quad 5 \quad 4 \quad 1 \quad 5 \quad 2\ ] \\
NextR &= [\ -1 \quad 5 \quad 7 \quad -1 \quad 4 \quad 1 \quad -1 \quad 3\ ] \\
NextC &= [\ 4 \quad -1 \quad 5 \quad -1 \quad -1 \quad -1 \quad 1 \quad 2\ ] \\
BeginR &= [\ 8 \quad 6 \quad 2\ ] \\
BeginC &= [\ 6 \quad 8 \quad -1 \quad 3 \quad 7\ ]
\end{aligned}$$

A "−1" is used to indicate the termination of lists in the above pointer arrays. Note that column 3 is empty and the pointer $BeginC(3) = -1$ indicating that there are no nonzero elements to trace in column 3.

In a majority of power system applications, operations with sparse matrices can be arranged in such a way that entries are exclusively accessed by rows (also by columns). Usually, entire rows are retrieved in order to combine them with other rows, perform inner products with column vectors, etc. In such cases, storage requirements can be significantly reduced by getting rid of arrays *Row*, *BeginC* and *NextC*. This will also reduce the overhead caused by the need to keep updated so many auxiliary structures, which is not negligible.

When dealing with square matrices, diagonal elements are customarily stored in a separate array, the array V being reserved just for off-diagonal entries. Also, if the *ToF* are stored on the space previously occupied by A, it may be helpful to create another auxiliary array whose entries are pointers to the first nonzero element above the diagonal of each row.

Of particular interest in power system analysis are square incidence-symmetric matrices, as virtually any problem leads to an equation system which can be rewritten in terms of such matrices. When solving those systems by LU factorization, all stages can be coded in such a way that matrix coefficients above (below) the diagonal are systematically accessed by rows (columns). Consequently, only one half of the matrix non-zero pattern (upper or lower triangle) needs to be stored and handled. Each entry within this structure will contain two transposed elements for numerically unsymmetric matrices, or a single element in the fully symmetric case.

B.10 Inserting and/or Deleting Elements in a Linked List

Above linked list structure allows fast and easy manipulation of the sparse matrices when the sparsity structure of the matrix changes. Change of sparsity structure occurs during the triangular factorization of A, where *fill-ins* are created, and have to be stored as part of the sparse Table of Factors ToF matrix. On the other hand, nonzero elements in an admittance matrix will have to be deleted if a branch is disconnected from the system. That requires efficient elimination of nonzero entries from the link list. In either case, link list can be efficiently updated to reflect the changes in the sparsity structure.

For simplicity, it will be assumed below that matrix elements are linked only by rows (4 arrays out of the 7 possible used).

B.10.1 Adding a nonzero element

Assume that a nonzero value of 23.5 is to be inserted in location (k, j) of a sparse matrix A, whose link list is already formed as described above. Provided the new element will be neither the first nor the last one in row k, it is inserted by performing the following steps:

$$\begin{aligned} V(nzv+1) &\leftarrow 23.5 \\ Col(nzv+1) &\leftarrow j \\ NextR(nzv+1) &\leftarrow NextR(prev) \\ NextR(prev) &\leftarrow nzv+1 \end{aligned}$$

where,
nzv is the current number of matrix entries.
$prev$ is the location of the nonzero entry in row k immediately preceding the added element in the link list.

Note that, before the modification, the following inequality applied:

$$Col(prev) < j < Col(NextR(prev))$$

The reader can easily modify the above procedure to take care of the particular cases in which the new element is the first or last in the considered row.

B.10.2 Deleting a nonzero element

Deletion of a nonzero element from the link list constitutes a simpler task, once its position is identified. Assume that element $A(k,j)$ whose value is stored in $V(i)$ is to be deleted. Then, provided it is not the first element in row k, the following assignment must be carried out:

$$NextR(prev) \leftarrow NextR(i)$$

where,
$prev$ is the location of the nonzero entry in row k which is preceding the deleted element in the link list (i.e., before the change, $NextR(prev) = i$).

Deleting the first element of row k requires simply that

$$BeginR(k) \leftarrow NextR(BeginR(k))$$

Even though it is possible, it is not worth the effort to track and recover the emptied location for future use (garbage collection).

Example 2.9:

Given the sparse matrix A in the previous example:

$$A = \begin{bmatrix} 0 & 16 & 0 & 340 & 74 \\ 7 & 0 & 0 & 0 & -51 \\ 0 & 44 & 0 & 14 & 23 \end{bmatrix}$$

Assuming the reduced row-oriented storage scheme is adopted, the following change (indicated within a box) will have to be made in the pointer array $BeginR$ in order to delete the entry $A(2,1) = 7$, located in $V(6)$. This illustrates the case in which the first entry of a row is deleted.

$$\begin{aligned} V &= \begin{bmatrix} -51 & 44 & 340 & 23 & 14 & 7 & 74 & 16 \end{bmatrix} \\ Col &= \begin{bmatrix} 5 & 2 & 4 & 5 & 4 & 1 & 5 & 2 \end{bmatrix} \\ NextR &= \begin{bmatrix} -1 & 5 & 7 & -1 & 4 & 1 & -1 & 3 \end{bmatrix} \\ BeginR &= \begin{bmatrix} 8 & \boxed{1} & 2 \end{bmatrix} \end{aligned}$$

Now, assume that a new element $A(2,2) = 45.7$ is to be inserted in the matrix A. Then the new pointer arrays will take the following form. Again, the changes are marked within a box:

$$
\begin{aligned}
V &= \begin{bmatrix} -51 & 44 & 340 & 23 & 14 & 7 & 74 & 16 & \boxed{45.7} \end{bmatrix} \\
Col &= \begin{bmatrix} 5 & 2 & 4 & 5 & 4 & 1 & 5 & 2 & \boxed{2} \end{bmatrix} \\
NextR &= \begin{bmatrix} -1 & 5 & 7 & -1 & 4 & 1 & -1 & 3 & \boxed{1} \end{bmatrix} \\
BeginR &= \begin{bmatrix} 8 & \boxed{9} & 2 \end{bmatrix}
\end{aligned}
$$

References

[1] O. Alsaç, B. Stott, W.F. Tinney, "Sparsity-Oriented Compensation Methods for Modified Network Solutions", *IEEE Transactions on Power Apparatus and Systems*, Vol. PAS-102, May 1983, pp. 1050-1060.

[2] F. L. Alvarado, W. F. Tinney, M. K. Enns, "Sparsity in large-scale network computation", *Advances in Electric Power and Energy Conversion System Dynamics and Control* (C. T. Leondes, ed.), Control and Dynamic Systems, vol. 41, Academic Press, 1991, Part 1, pp. 207-272.

[3] R. Betancourt, "An Efficient Heuristic Ordering Algorithm for Partial Matrix Refactorization", *IEEE Transactions on Power Systems*, Vol. 3(3), August 1988, pp. 1181-1187.

[4] S.M. Chan, V. Brandwajn, "Partial Matrix Refactorization", *IEEE Transactions on Power Systems*, Vol. 1(1), Feb. 1986, pp. 193-200.

[5] Duff, Erisman and Reid, "Direct Methods for Sparse Matrices", Oxford Press (1986).

[6] George Liu, "Computer Solution of Large, Sparse, Positive-Definite Systems", Prentice-Hall (1981).

[7] A. Gómez, L.G. Franquelo, "Node Ordering Algorithms for Sparse Vector Method Improvement", *IEEE Transactions on Power Systems*, Vol. 3(1), Feb. 1988, pp. 73-79.

[8] D.E. Knuth, "The Art of Computer Programming: Vol.1, Fundamental Algorithms". Addison-Wesley: Reading, MA (1968).

[9] K. Takahashi, J. Fagan, M.S. Chen, "Formation of a Sparse Bus Impedance Matrix and its Application to Short Circuit Study", PICA Proceedings, May 1973, pp. 63–69.

[10] W. Tinney and R. Walker, "Direct Solutions of Sparse Network Equations by optimally Ordered Triangular Factorization", Proc. of the IEEE, pp.1801-1809, vol. 55, (1967).

[11] W. Tinney, V. Brandwajn, and A. Chan, "Sparse Vector Methods", IEEE Trans. on PAS-104, pp.295-301, (1985).

[12] M. Vempati, I. Slutsker, W. Tinney, "Enhancements to Givens Rotations for Power System State Estimation", *IEEE Transactions on Power Systems*, Vol. 6(2), May 1991, pp. 842-849.

[13] D.M. Young, "Iterative Solution of Large Linear Systems", Academic Press, New York (1971).

[14] R.S. Varga, "Matrix Iterative Analysis", Prentice-Hall, Englewood Cliffs, New Jersey (1962).

Index

RECEIVED
JUN 21 2004